WITHI
CORNELL UI

PLANT BREEDING REVIEWS
Volume 37

Plant Breeding Reviews is sponsored by:

American Society for Horticultural Science
International Society for Horticultural Science
Society of American Foresters
National Council of Commercial Plant Breeders

Editorial Board, Volume 37
I. L. Goldman
C. H. Michler
Rodomiro Ortiz

PLANT BREEDING REVIEWS
Volume 37

edited by
Jules Janick
Purdue University

A JOHN WILEY & SONS, INC., PUBLICATION

Cover design: John Wiley & Sons, Inc.

Cover illustration: Courtesy of the Series Editor

Copyright © 2013 by Wiley-Blackwell. All rights reserved.

Published by John Wiley & Sons, Inc., Hoboken, New Jersey
Published simultaneously in Canada

No part of this publication may be reproduced, stored in a retrieval system, or transmitted in any form or by any means, electronic, mechanical, photocopying, recording, scanning, or otherwise, except as permitted under Section 107 or 108 of the 1976 United States Copyright Act, without either the prior written permission of the Publisher, or authorization through payment of the appropriate per-copy fee to the Copyright Clearance Center, Inc., 222 Rosewood Drive, Danvers, MA 01923, 978-750-8400, fax 978-750-4470, or on the web at www.copyright.com. Requests to the Publisher for permission should be addressed to the Permissions Department, John Wiley & Sons, Inc., 111 River Street, Hoboken, NJ 07030, 201-748-6011, fax 201-748-6008, or online at http://www.wiley.com/go/permission.

Limit of Liability/Disclaimer of Warranty: While the publisher and author have used their best efforts in preparing this book, they make no representations or warranties with respect to the accuracy or completeness of the contents of this book and specifically disclaim any implied warranties of merchantability or fitness for a particular purpose. No warranty may be created or extended by sales representatives or written sales materials. The advice and strategies contained herein may not be suitable for your situation. You should consult with a professional where appropriate. Neither the publisher nor author shall be liable for any loss of profit or any other commercial damages, including but not limited to special, incidental, consequential, or other damages.

For general information on our other products and services or for technical support, please contact our Customer Care Department within the United States at 877-762-2974, outside the United States at 317-572-3993 or fax 317-572-4002.

Wiley also publishes its books in a variety of electronic formats. Some content that appears in print may not be available in electronic formats. For more information about Wiley products, visit our web site at www.wiley.com.

Library of Congress Cataloging-in-Publication Data:

ISBN 978-1-118-49785-2 (cloth)
ISSN 0730-2207

Printed in the United States of America

10 9 8 7 6 5 4 3 2 1

Contents

Contributors	ix
1. Bikram Gill: Cytogeneticist and Wheat Man	1
W. John Raupp and Bernd Friebe	
I. Early Life: Emergence of a Cytogeneticist	2
II. Research	4
III. International Collaborations	23
IV. Educator	24
V. Champion of Wheat Workers	27
VI. The Man	27
VII. Epilogue	29
Acknowledgments	29
Literature Cited	30
2. Synthetic Hexaploids: Harnessing Species of the Primary Gene Pool for Wheat Improvement	35
Francis C. Ogbonnaya, Osman Abdalla, Abdul Mujeeb-Kazi, Alvina G. Kazi, Steven S. Xu, Nick Gosman, Evans S. Lagudah, David Bonnett, Mark E. Sorrells, and Hisashi Tsujimoto	
I. Introduction	39
II. Production and Utilization of Synthetic Hexaploid Wheat	42
III. Impact of Synthetic Hexaploid in Wheat Improvement	58
IV. Conclusions and Future Prospects	100
Acknowledgments	105
Literature Cited	105

3. **Breeding Early and Extra-Early Maize for Resistance to Biotic and Abiotic Stresses in Sub-Saharan Africa**   **123**
   B. Badu-Apraku and M. A. B. Fakorede

   I. Introduction   126
   II. Development of Breeding Populations   131
   III. S_1 Recurrent Selection Program for *Striga* Resistance   146
   IV. Adaptation   164
   V. Development of QPM Populations and Cultivars   169
   VI. Breeding for Combined Tolerance/Resistance to Multiple Stresses in Early and Extra-Early Maize   179
   VII. Inbred-Hybrid Development Program   180
   VIII. Traits for Indirect Selection for Stress Tolerance/Resistance in Contrasting Environments   188
   IX. Future Challenges and Perspectives   193
   Acknowledgments   197
   Literature Cited   197

4. **Almond Breeding**   **207**
   Thomas M. Gradziel and Pedro Martínez-Gómez

   I. Introduction   209
   II. Botany   211
   III. Genetic Diversity   221
   IV. Genetic Improvement   226
   V. Molecular Approaches   238
   VI. Future Progress   248
   Literature Cited   249

5. **Breeding Loquat**   **259**
   Maria L. Badenes, Jules Janick, Shunquan Lin, Zhike Zhang, Guolu L. Liang, and Weixing Wang

   I. Introduction   261
   II. Germplasm   262
   III. Reproductive Physiology   270
   IV. Breeding Objectives   275
   V. Breeding Methods   276
   VI. Future Progress   290
   Literature Cited   291

6. Prognostic Breeding: A New Paradigm for Crop Improvement   **297**
Vasilia A. Fasoula

   I. Introduction   298
  II. Genetic Components of Crop Yield Potential   303
 III. A New General Response Equation   305
 IV. Prognostic Equations for Single Plants and Sibling Lines   307
  V. The Advantages of Prognostic Breeding   335
 VI. The Marriage of Phenotyping with Genotyping   338
VII. Outlook   339
Literature Cited   342

Subject Index   349

Cumulative Subject Index   351

Cumulative Contributor Index   373

Contributors

Osman Abdalla, International Center for Agricultural Research in the Dry Areas (ICARDA), Aleppo, Syria.

Maria L. Badenes, Fruit Breeding Department, Instituto Valenciano de Investigaciones Agrarias, Apartado Oficial 46113, Moncada, Valencia, Spain.

B. Badu-Apraku, International Institute of Tropical Agriculture, P.M.B. 5320, Ibadan, Nigeria.

David Bonnett, International Maize and Wheat Improvement Center (CIMMYT), El Batan, Mexico.

M. A. B. Fakorede, Department of Crop Production & Protection, Obafemi Awolowo University, Ile-Ife, Nigeria.

Vasilia A. Fasoula, Institute of Plant Breeding, The University of Georgia, Athens, GA 30602, USA.

Bernd Friebe, Wheat Genetic and Genomic Resources Center, Department of Plant Pathology, Kansas State University, Manhattan, KS 66506-5502, USA.

Thomas M. Gradziel, Department of Plant Sciences, University of California at Davis, Davis, CA 95616, USA.

Nick Gosman, The National Institute of Agricultural Botany (NIAB), Cambridge CB3 0LE, UK.

Jules Janick, Department of Horticulture and Landscape Architecture, Purdue University, West Lafayette, IN 47907-2010, USA.

Alvina G. Kazi, National Institute of Biotechnology and Genetic Engineering (NIBGE), Faisalabad and National University of Science and Technology (NUST), Islamabad, Pakistan.

Evans S. Lagudah, CSIRO Plant Industry, Canberra, Australian Capital Territory 2601, Australia.

Guolu L. Liang, College of Horticulture and Landscape Architecture, Southwest University, Chongqing 400716, China.

Shunquan Lin, College of Horticulture, South China Agricultural University, Guangzhou 510642, China.

Pedro Martínez-Gómez, Departamento de Mejora Vegetal, CEBAS-CSIC, E-30100 Espinardo (Murcia), Spain.

Abdul Mujeeb-Kazi, National Institute of Biotechnology and Genetic Engineering (NIBGE), Faisalabad and National University of Science and Technology (NUST), Islamabad, Pakistan.

Francis C. Ogbonnaya, International Center for Agricultural Research in the Dry Areas (ICARDA), Aleppo, Syria.

W. John Raupp, Wheat Genetic and Genomic Resources Center, Department of Plant Pathology, Kansas State University, Manhattan, KS 66506-5502, USA.

Mark E. Sorrells, Department of Plant Breeding and Genetics, 240 Emerson Hall, Cornell University, Ithaca, NY 14853, USA.

Hisashi Tsujimoto, Arid Land Research Center, Tottori University, Tottori 680-0001, Japan.

Weixing Wang, College of Horticulture and Landscape Architecture, Southwest University, Chongqing 400716, China.

Steven S. Xu, USDA-ARS, Northern Crop Science Laboratory, Fargo, ND 58102-2765, USA.

Zhike Zhang, College of Horticulture, South China Agricultural University, Guangzhou 510642, China.

Bikram Gill

1

Bikram Gill: Cytogeneticist and Wheat Man

W. John Raupp and Bernd Friebe
Department of Plant Pathology
Wheat Genetic and Genomic Resources Center
Kansas State University
Manhattan, KS 66506-5502, USA

  I. EARLY LIFE: EMERGENCE OF A CYTOGENETICIST
 II. RESEARCH
    A. Tomato Cytogenetics in California
    B. Chromosome Banding Research in Missouri
    C. Wild Wheat Studies at UC Riverside
    D. Sugarcane Breeding in Florida
    E. Germplasm Evaluation and Enhancement in Kansas
    F. Establishing the Wheat Karyotype
    G. Birth of the Chinese Spring Deletion Stocks
    H. FISHing in the Wheat Gene Pool
    I. Gene Rich, High-Recombination Regions
    J. The *Q* Gene Story
    K. Cloning *Lr21*
    L. Hybrid Centromeres
    M. Chromosome Bin Mapping
    N. Comparative Genomics and Wheat Evolution
    O. Sequencing the Wheat Genome
III. INTERNATIONAL COLLABORATIONS
 IV. EDUCATOR
  V. CHAMPION OF WHEAT WORKERS
 VI. THE MAN
VII. EPILOGUE
ACKNOWLEDGMENTS
LITERATURE CITED

Plant Breeding Reviews, Volume 37, First Edition. Edited by Jules Janick.
© 2013 Wiley-Blackwell. Published 2013 by John Wiley & Sons, Inc.

Bikram Gill's brothers proudly refer to him as their "Wheat Man" and, in fact, the Kansas Association of Wheat Growers named him the "Wheat Man of the Year" in 1997. His numerous awards for his wheat research attest to this fame (Table 1.1). Bikram had always wanted to be a scientist of the type who helps people make the world better. Although he thought his botanical training in India was a waste of time, he now knows better. It was his mentor Charlie Rick who taught him that using science to understand nature and serve society is both exciting and rewarding. Nearly five decades later that commitment has helped his dream become reality. Bikram has won more than $20 million in extramural grants to support his research, including significant funding from the Kansas Wheat Commission and the USDA for establishing a gene bank at Kansas State University and wheat genetics research, the McKnight Foundation for Fusarium head blight research, and the National Science Foundation and USDA for wheat genome sequencing. Bikram is the author or coauthor of more than 350 refereed journal publications, 230 abstracts, 17 book chapters, and 54 newsletter items. He has contributed papers to more than 60 conference proceedings and partnered in the release of 54 germplasm lines. He has presented more than 200 lectures both nationally and internationally. He is the coauthor of *Chromosome Biology*. Volume 37 of *Plant Breeding Reviews* is dedicated to Bikram Gill's illustrious and extraordinary career. (A complete list of publications of Bikram Gill is available at http://www.k-state.edu/wgrc/Publications/pubstoc.html)

I. EARLY LIFE: EMERGENCE OF A CYTOGENETICIST

Bikram S. Gill was born on 31 October 1943, in a small village called Dhudike, District Moga, Punjab, India. He was the fifth of 10 children. His parents were farmers; his father also served as a *lambardar* (revenue collector) and a *sarpanch* (mayor) of the village. Bikram was always very interested in education and worked hard on his homework, graduating from high school in 1957 first in his class. He studied at DM College at Moga as a premedical student from 1959 to 1961. Bikram then went on to earn his B.S. degree at Khalsa College, Amritsar, in 1963, followed by B.S. Honors and M.S. Honors degrees in 1966 from Punjab University at Chandigarh where he became really interested in botany. Bikram lectured premedical students at GHG Khalsa College, Gurusar Sudhar, from 1966 to 1968.

In 1968, he was admitted to Brigham Young University where his brother, Gurcharan, was teaching mathematics. His brother had tried to

Table 1.1. Awards, honors, and service of Bikram Gill.

Phi Kappa Phi Award for Academic Excellence, University of California, Davis, 1973
D.F. Jones Postdoctoral Fellowship, University of Missouri, Columbia, 1973–1974
Visiting Professor, CSIRO, Division of Plant Industry, Canberra, Australia, 1986–1987
Visiting Professorship to the German Democratic Republic, U.S. National Academy of Sciences, 1987
International Organizing Committee, Wheat Genetics Symposium, 1988–1998
Chair, International Committee on Wheat Chromosome Banding Nomenclature, 1988–1998
Editorial Board, *Plant Breeding*, 1990
Conoco Distinguished Graduate Faculty Award, Kansas State University, 1990
Elected Fellow, American Society of Agronomy, 1991
Visiting Scholar to India, UNESCO-TOKTEN, 1991
Visiting Professor to Russia and Ukraine, U.S. National Academy of Sciences, 1992
UNDP Visiting Scholar, People's Republic of China, 1993
Board of Directors, Crop Science Society of America, 1994
Editorial board, *Crop Science*, 1994
Elected Fellow, Crop Science Society of America, 1994
Associate editor, *Theoretical and Applied Genetics*, 1995
Visiting Professor, Ludwig Maximillian University, Munich, Germany (DAAD Fellow, Germany–U.S. Exchange Program), 1995–1996
University Distinguished Professor, Kansas State University, 1997
Higuchi Research Achievement Award/Irvin E. Youngberg Award in the Applied Sciences, University of Kansas, 1997
Wheat Man of the Year, Kansas Association of Wheat Growers, 1997
Editorial board, *Genetics*, 1998
Fellow, American Phytopathological Society, 1998
Crop Science Research Award, Crop Science Society of America, 1998
Outstanding Scientist Award, American Association of Agricultural Scientists of Indian Origin, 1999
Listed in the "Century's Top 10 Sikh Scientists," Panj Darya magazine, Punjab, India, 1999
Fellow, American Association for the Advancement of Science, 1999
Fellow, National Academy of Agricultural Sciences of India, 2001
Listed among the world's top highly cited scientists in Animal and Plant Sciences by Thompson Reuters, 2006
Foreign Fellow, National Academy of Sciences, India, 2006
International PI, 111 Project, "Crop Genetics and Germplasm Enhancement," Nanjing Agricultural University, 2008
Friendship Medal, Jiangsu Province, Nanjing, China, 2010
Editorial board, *G3: Genes|Genomes|Genetics*, 2011
Frank N. Meyer Medal for Plant Genetic Resources, Crop Science Society of America, 2011
Editorial Board, Agricultural Research, Official publication of the National Academy of Agricultural Sciences, India, 2012
National Friendship Award, Government of China, Beijing, 2012

talk him into working toward being a medical doctor, but he insisted on botany because of his dream of feeding the world. Working with Howard Stutz at Brigham Young, Bikram developed a chromosome staining technique for cereals that impressed Ralph Anderson very much. Ralph had studied with Charlie Rick at the University of California (UC), Davis, and advised Bikram that was where he needed to be. Bikram began his graduate work with Charlie in 1969. His Ph.D. thesis was on the cytogenetics of tertiary aneuploids with unusual transmission characteristics in tomato. When he came to Kansas State, he was frequently seen wearing a green fishing cap, Charlie's trademark.

After graduating from UC Davis, Bikram moved to the University of Missouri. As a graduate student, he had written a grant proposal for chromosome banding in wheat. Bikram had read about the work that was being done in human cytogenetics with chromosome banding and wanted to achieve the same in wheat. At Missouri, Bikram had the opportunity to work with the late Ernie Sears and Gordon Kimber. During that time, he was introduced to David Apirion from Washington University in St. Louis, and switched to mouse molecular biology. But it was just for a year, because then, following his heart, he was back into wheat research with Giles Waines at the University of California, Riverside. At Riverside, Bikram met Lennart Johnson, an avid collector and researcher of wild wheat species, who introduced him to the world of genetic resources.

On his birthday 31 October 1977, Bikram received a call from Don Meyers, Director of the AREC at the University of Florida, Belle Glade, that he had been hired as an assistant professor to work on sugarcane genetics and breeding. He can still recall that on the first day of work there, he was going out in the field to make selections on a crop he had yet to set eyes on. Thinking he had found his niche with sugarcane, Bikram was surprised when one day he received a call from Gordon Kimber, who told him that Kansas State was looking for a wheat cytogeneticist and that he had already submitted his CV for the position. Barely a year and a half later, he was on his way to Manhattan, Kansas. The world of wheat had called him back, and this time he would not leave.

II. RESEARCH

A. Tomato Cytogenetics in California

Gurdev Khush had left UC Davis to become the rice breeder at the International Rice Research Institute in 1968. Later, he would earn world acclaim as one of the fathers of rice revolution. Bikram filled his slot and learned tomato cytology from him when he made a return visit to Davis in the summer of 1969. Bikram immensely enjoyed his graduate studies at UC Davis and became steeped in tomato cytogenetics research with Charlie Rick (Gill 1983). Charlie mapped the tomato genome, collected wild tomatoes on his frequent explorations to South America, and all his students participated in tomato genetics and wide-hybridization research, in addition to their specific projects. Graduate students shared coauthorship for participating in the on-going research (Rick and Gill 1973) but were solo authors on publications from their thesis research (Gill 1974a,b). Bikram's expertise in tomato cytology won him his first NSF award at K-State in 1980 to study cytogenetic basis of somaclonal variation in potato with similar chromosome morphology. As he was finishing up his graduate studies at UC Davis, Bikram submitted a solo, two-page project to the Research Corporation in New York. Charlie wrote to E.R. Sears at the University of Missouri to sponsor laboratory space.

B. Chromosome Banding Research in Missouri

Chromosomes are characterized by their length, arm ratio, and the presence or absence of secondary constrictions. Conventional staining techniques did not distinguish between morphologically similar chromosomes, so techniques that allowed fast and reliable identification of chromosomes were needed. Bikram's proposal, submitted to Research Corporation in New York, which was founded by D.F. Jones from his patent on hybrid corn, was funded for this research. In the summer of 1973, Bikram began working at the University of Missouri, Columbia, in the laboratory of Gordon Kimber (it turned out that Ernie Sears did all of his work in his office, so all visitors to Missouri worked in Gordon's laboratory. The research was wildly successful (Fig. 1.1); wheat and rye chromosome could be cytogenetically identified for the first time and their chromatin differentiation into euchromatin and heterochromatin states could be rapidly determined (Gill and Kimber 1974a,b).

In the summer of 1974, Bikram attended meetings of the International Congress of Genetics at UC Berkeley followed by a vacation. Upon returning, he was shocked to discover that he could not reproduce the work for which he had published two high-profile papers! Eventually, after recreating his work schedule, Bikram realized that after making chromosome preparations, he used to leave the slides in ethanol and go home for lunch. His new schedule did not include a

Fig. 1.1. Bikram at work in Missouri, 1973, soon after the discovery of the heterochromatic banding patterns of wheat chromosomes. Everyone in the legendary Curtis Hall came to take a peek under the microscope.

lunch break, and so the long, ethanol treatment was skipped. To his surprise, and delight, chromosome spreads dehydrated in ethanol for a long time showed the most beautiful and high-contrast banding patterns he had ever seen. So, just by accident, Bikram had discovered the importance of dehydrating chromosome spreads before C-banding to obtain differentially stained chromosomes (Gill and Kimber 1974a,b).

C. Wild Wheat Studies at UC Riverside

While Bikram was at Missouri, rumors were rife that Lennart Johnson and his student Harcharan Dhaliwal at UC Riverside had discovered the elusive B-genome donor of wheat. Bikram called Harcharan and got a few seeds of this new species with the unfamiliar name *Triticum urartu*. Bikram immediately profiled the *T. urartu* chromosomes with the new staining technique. They were unlike the B-genome and more similar to the A-genome chromosomes of wheat. Bikram let Harcharan know about these results and soon other labs provided evidence of the homology between *T. urartu* and the A genome of polyploid wheat. In fact, Harcharan's own data pointed to the same conclusion, but Johnson would not hear it. Later, molecular work would provide

Fig. 1.2. The WGRC welcomed visits from many prominent wheat scientists throughout the years, such as this group in 1989: (L to R, front row) Bernd Friebe, Ernie Sears, Takashi Endo, and Yashuiko Mukai; (back row) John Raupp, Christine Curtis, Adam Lukaszewski, Bikram Gill, and Bali Ram Tyagi.

conclusive evidence that *T. urartu* was, in fact, the A-genome donor of polyploid wheat. Bikram spent two years at UC Riverside studying wild wheat species and especially endosperm development in wild wheat species hybrids. In due course, Ernie's cytogenetic stocks (Fig. 1.2) and Johnson's wild wheat collection would form the foundation material to launch his career at KSU.

D. Sugarcane Breeding in Florida

Although Bikram stayed in Florida for only a year and a half (January 1978 to May 1979), he immersed himself into sugarcane cytogenetics and breeding research. USDA geneticist C.O. Grassl had assembled a world collection of sugarcane species and had made many intergeneric hybrids at Canal Point, on the banks of Lake Okeechobee. Bikram analyzed the chromosome constitution of these hybrids and published what he considers is an important paper in wide-hybridization research (Gill and Grassl 1986). He traveled to Minnesota to observe *in situ* hybridization research in Ron Phillips' laboratory and to work on a joint manuscript with Charlie Burnham on the development of a tester set of translocations for the tomato genome (Gill et al. 1980). He also submitted a grant proposal on sugarcane cytogenetics to the USDA, which although not funded was well reviewed.

E. Germplasm Evaluation and Enhancement in Kansas

Early on, Bikram realized that the wild relatives and related species are an important reservoir of agronomically interesting genes that can be used in wheat improvement and, in 1979, brought with him a part of the L.B. Johnson collection housed at the University of California, Riverside. At Kansas State, Jimmy Hatchett, a USDA–ARS entomologist, and coworkers already were looking at some synthetic wheats from the Kyoto University gene bank for resistance to a new biotype of Hessian fly, which was the most virulent isolate known at that time. They noticed that one particular line, derived from *Aegilops tauschii*, was highly resistant to Hessian fly (Hatchett et al. 1981). Bikram had 20 additional *Ae. tauschii* lines and, working with Hatchett, found five that were completely resistant to the new biotype (Hatchett and Gill 1981). Genetic analysis revealed that all carried different genes (Hatchett and Gill 1983). This small project was the beginning of a wide-hybridization program that would grow to encompass not only Hessian fly, but many plant diseases and traits as well.

Hari Sharma, a postdoctoral fellow whom Bikram met in Riverside, and John Raupp, the first research assistant, joined Bikram in 1980. Using embryo rescue, Bikram and Hari expanded the range of hybridization of wheat to certain perennial grasses and won Bikram his first USDA competitive grant in 1982. Together with Hatchett, Bikram's team grew to include Lewis Browder, a USDA scientist in Manhattan working with leaf rust; Tom Harvey, a K-State entomologist at Ft. Hays screening for greenbug; and John Moseman, a USDA scientist at Beltsville, Maryland, looking at powdery mildew resistance. The gene bank at Kansas State quickly grew during these initial years and by 1982, the collection contained 1867 lines of wild wheat and in three years had become one of the largest in the United States. From the start, Bikram envisioned the gene bank to be a working collection; extensively evaluated for disease and pest resistance and forming the basis for developing new germplasm for wheat breeders worldwide. The first of several germplasm evaluation papers was published (Gill et al. 1986).

Bikram's vision of a "one-stop shop" for wheat research became reality when he established the Wheat Genetics Resource Center (WGRC) at Kansas State University in 1984. The WGRC has been continuously supported by Kansas wheat growers through Kansas Wheat Commission grants since 1981 and USDA since 1989. Recognized as a center for excellence in wheat research by Kansas Board of Regents in 1984, the WGRC brought together plant pathologists, entomologists, breeders, and USDA personnel with a vision of germplasm conservation and utilization for crop improvement for sustainable production by broadening the crop genetic base; creating and promoting the free exchange of materials, technology, and new knowledge in genetics and biotechnology among the world's public and private organizations; and sponsoring graduate and postgraduate students and visiting scientists for academic training and advanced research in the WGRC laboratories. The WGRC gene bank maintains accessions of all the wild wheat species and, in addition, cytogenetic stocks, the genetic treasures produced by a lifetime of work by wheat scientists. The WGRC established a national and international network to conduct and coordinate genetic studies in wheat. Genes for host–plant resistance to viral, bacterial, fungal, and insect pests and abiotic stresses would be identified, transferred to agronomically useful breeding lines, and deployed. State-of-the-art laboratories, greenhouses, and field plot facilities for teaching and research helped establish the WGRC.

Concurrently with the screening studies, Bikram initiated a direct hybridization program realizing that *Ae. tauschii* was readily amenable to crossing with wheat. Material testing resistant was crossed with wheat, embryo rescued, and segregating progenies produced. Homozygous, resistant germplasm lines could be produced in four generations (Gill and Raupp 1987). *Ae. tauschii* proved to be a goldmine for genes conferring resistance to diseases and pests. In 1984, Stan Cox was hired by the USDA as a plant geneticist and was the needed stimulus for the germplasm enhancement project. Stan helped arrange for the first germplasm releases from the WGRC, which were backcross lines developed by Bikram and Jim Hatchett that carried genes from *Ae. tauschii* for Hessian fly and soil-borne mosaic virus resistance in a Wichita background.

Bikram and John had found that, although the direct cross *Triticum aestivum* (AABBDD)/*Ae. tauschii* (DD) produced completely male-sterile ABDD F_1 plants, only one or two backcrosses to *T. aestivum* as male could produce (along with various aneuploids) some fully fertile euploid AABBDD plants. The A and B genomes of those plants were restored from Wichita wheat while their D genome was a mixture of *T. aestivum*, *Ae. tauschii*, and recombinant chromosomes. In field-testing Bikram's backcross lines, Stan noticed that the Wichita phenotype also was recovered very rapidly with backcrossing, not surprising, since two of the three genomes were immediately restored.

With this in mind, Stan set out to use Bikram's method to cross then-recently released hard winter wheat cultivars, such as 'Karl' and 'TAM 107', as well as new breeding lines, with a set of geographically diverse *Ae. tauschii* accessions and A-genome diploids that John and Bikram had found to have seedling resistance to pests and diseases, most prominently leaf rust. Those crosses led to the development of lines, mostly BC_2-derived, with resistances to leaf rust, soil-borne mosaic virus, spindle-streak mosaic virus, wheat curl mite, Septoria leaf blotch, tan spot, and powdery mildew, the last in cooperation with North Carolina State University and USDA–ARS in Raleigh. Gina Brown-Guedira took over Stan's position in 1997 and continued to work with Bikram on developing germplasm from crosses with both *Ae. tauschii* and *Triticom timopheevii* subsp. *armeniacum*. Forty-nine resistant lines became joint USDA–WGRC germplasm releases between 1985 and 2011, and many more were tested in regional germplasm nurseries and used as parents by breeders.

In the early days of developing germplasm from crosses with progenitor species such as *Ae. tauschii*, Bikram, John, Stan, and others had to make many crosses with aneuploid stocks to locate genes to chromosomes and with other resistant lines for allelism tests to determine whether a novel gene was being transferred. This work became more laborious, time-consuming, and error-prone as the number of D-genome-derived resistance genes grew. So when molecular-marker technology became feasible in wheat, Bikram and Bernd Friebe began using it in the development of new germplasm, streamlining the process considerably. RFLP markers and gliadin proteins also were used to demonstrate the vast reservoir of genetic variation available in *Ae. tauschii* (Lubbers et al. 1991). Compared with the paucity of variation in the D genome of hexaploid wheat (Kam-Morgan et al. 1989), this work provided further rationale for exploiting the wild D genome for wheat improvement.

Bikram realized, however, that some agronomically important traits, such as resistance to wheat streak mosaic virus were not found in species of the primary gene pool of wheat but were present only in the more distantly related species of the tertiary gene pool. Actually, one of Bikram's major responsibilities when he was hired at Kansas State University was to develop wheat germplasm with resistance to wheat streak mosaic virus, which is a serious problem in Western Kansas especially. This work, first initiated by Hari Sharma, produced new hybrids between wheat and several *Agropyron* species (Sharma and Gill 1983). Realizing that gene transfer from these species is more difficult and cannot be achieved by homologous recombination, directed chromosome engineering using the *ph1b* gene was begun.

Bernd Friebe came to the WGRC in early 1989 on a sabbatical to work on standardizing the C-banding nomenclature system for wheat. During this time, C-banding and *in situ* hybridization analyses, both pioneered for wheat in Bikram's laboratory (see below sections), were well established. In combination, these techniques proved to be a very powerful tool for characterizing wheat-alien germplasm. Chromosome banding or fingerprinting identified the chromosomes involved in translocations and genomic *in situ* hybridization or chromosome painting determined the size of the alien segments. One of the first problems they tackled was using such a molecular cytogenetic approach to characterize a set of wheat streak mosaic virus-resistant wheat–*Thinopyrum intermedium* lines that was produced in the 1970s by Wells and coworkers in South Dakota. Screening of this germplasm by chromosome fingerprinting revealed that most had a complete *Th. intermedium* chromosome, either added or replacing wheat chromsomes 4A and 4D in the derived substitution lines, indicating their homoeology to group-4 chromosomes (Friebe et al. 1991). In one of the lines, however, only the short arm of a *Th. intermedium* chromosome was present and translocated to the long arm of wheat chromosome 4D forming a Robertsonian translocation. This line was genetically compensating, and because of the wheat streak mosaic virus resistance gene, it was agronomically useful. The gene was designated as *Wsm1* and conferred immunity to the virus.

Wsm1 was transferred to Kansas winter wheat, but it turned out that a whole *Th. intermedium* chromosome arm had too many deleterious genes and caused a yield penalty. Using the *ph1b* wheat mutant stock to induce homoeologous recombination between wheat and alien chromosomes, they recovered wheat–*Th. intermedium* recombinants with shortened alien segments that still retained the *Wsm1* resistance gene (for review see Qi et al. 2007). The wheat streak mosaic virus-resistant wheat–*Th. intermedium* recombinant chromosomes were transferred to Kansas winter wheat, should have no yield penalty, and can be used in wheat improvement. As a bonus, it turned out that the *Th. intermedium* segment in these recombinant chromosomes also conferred resistance to the *Triticum* mosaic virus, a new virus disease that recently had emerged in the Great Plains (Friebe et al. 2009). Continuing this work, a second source of wheat streak mosaic virus resistance (*Wsm3*) derived from a different *Th. intermedium* chromosome, but only available as a whole-arm wheat–*Th. intermedium* translocation, was released. This line will need further chromosome engineering before the gene can be used in cultivar improvement (Liu et al. 2011).

F. Establishing the Wheat Karyotype

Giemsa C-banding is still widely used to identify plant chromosomes and distinguishes all 21 chromosome pairs of bread wheat.

In Cambridge, England, during the 7th International Wheat Genetic Symposium, Bikram headed a committee to develop a standard karyotype and nomenclature system to describe the chromosome bands of wheat. Together with Bernd Friebe and Takashi Endo, a standard karyotype of wheat was published in 1991 (Gill et al. 1991). For the first time, Giemsa C-banding chromosome fingerprinting proved to be a fast and very reliable technique to identify wheat chromosomes. Bikram and Bernd developed standard karyotypes for most of the related *Aegilops* species, providing insight into the evolutionary relationships among these species (Friebe and Gill 1996). Together with *in situ* hybridization analysis, another technique refined in Bikram's laboratory, the molecular cytogenetic characterization of wheat-alien translocations conferring resistance to diseases and pests was accomplished (for review see Friebe et al. 1996).

Just how powerful were these new techniques? Bob McIntosh visited the WGRC in 1994. He was interested in mapping the rye-derived, leaf rust resistance gene *Lr45* using monosomic analysis. By 1992, he had analyzed 19 of the 21 cross combinations except those involving chromosomes 2A and 1D, suspecting that, most likely, either chromosome was involved in the *Lr45* transfer. Within a few days, Giemsa C-banding showed that the chromosomes involved in the *Lr45* transfer were wheat chromosome 2A and rye chromosome 2R. *In situ* hybridization revealed that the complete long arm and about half of the short arm of rye chromosome 2R was translocated to the distal half of the short arm of wheat chromosome 2A (McIntosh et al. 1995).

G. Birth of the Chinese Spring Deletion Stocks

Forty years ago, Master's degree student Takashi Endo, working under the guidance of Prof. Koichiro Tsunewaki, noticed a strange phenomenon in the fertility of backcross progeny that retained an *Ae. triuncialis* chromosome and that this chromosome was indispensable to fertile gametes, although he was not sure at that time if the alien chromosome itself had a gametocidal effect on gametes lacking the alien chromosome. After visiting S.S. Maan at North Dakota State University in 1981, Endo moved to Bikram's laboratory at Kansas State for 5 months, the beginning of a long-term collaboration. During this time, Endo was introduced to chromosome banding and continued to improve the techniques after his return to Japan. Thanks to chromosome banding, he found that some chromosomes caused sublethal chromosomal breakage in gametes and that resultant chromosomal structural changes could be established in the subsequent generations.

Around 1985, Sir Otto Frankel, a Fellow of the Australian Academy of Science and Honorary Member of the Japan Academy, was visiting Nara, Japan. Over dinner, Endo told Sir Frankel about the chromosomal aberrations, and he mentioned that he would tell the story to Bikram, who was visiting Australia on sabbatical leave. Bikram took an interest and encouraged Endo to produce wheat stocks carrying deletions. Eventually, Endo received a grant sponsored by the Japan Society for the Promotion of Science to visit Bikram's laboratory during his summer vacation for several years. The homozygous or heterozygous deletions were screened and grown at KSU. Root tips of the deletion heterozygotes were sent to Japan and the data sent back to KSU. Later, Bikram obtained funding from the USDA for RFLP mapping of the deletion lines. The deletion line set was released for public use in 1996 (Endo and Gill 1996). Since then, other similar chromosomes have been isolated in common wheat from different *Aegilops* species, and they were named "gametocidal" chromosomes, which were demonstrated to induce chromosomal breakage in gametes lacking them.

H. FISHing in the Wheat Gene Pool

After his pioneering research work of using C-banding to identify individual wheat and rye chromosomes, Bikram continued to pursue the development of new techniques for reliable identification of wheat chromosomes. Lane Rayburn, from the laboratories of J.D. Smith and Jim Price at Texas A&M University, joined Bikram in 1984 and together they published the first results of nonradioactive DNA *in situ* hybridization (ISH) in plants (Rayburn and Gill 1985). Using a biotin-labeling system to map the repetitive DNA probe, pSc119, signals were visualized using a horseradish peroxidase-based detection system. Although unable to identify all 21 wheat chromosomes, the potential of ISH for chromosome identification was established, because different probe or probe combinations can potentially be used to generate different hybridization patterns on individual chromosomes. Thus, ISH-based chromosome identification systems would be more versatile than the traditional chromosome banding systems. The pSc119 probe has since become one of the most frequently used repetitive DNA probes in plant cytogenetics research.

In the 1980s, fluorescence-based detection systems (FISH) became the choice for visualizing ISH signals. Bikram had a clear vision of the potential of FISH and strongly encouraged and supported several students, postdocs, and visiting scientists to use FISH in their research projects. Several influential, FISH-based research papers were published in early 1990s, especially using FISH to detect alien chromosomal segments that were transferred into wheat cultivars or breeding lines (Friebe et al. 1993; Mukai et al. 1993; Jiang et al. 1994). As a graduate student, Jiming Jiang worked with Bikram from 1989 to 1994 developing techniques that combined FISH with chromosome banding techniques by sequentially performing the two techniques on the same chromosome preparations, combining the power of the chromosome identification from both (Jiang and Gill 1993, 1994; Jiang et al. 1994).

FISH mapping on plant chromosomes using DNA probes as small as few kilobases was a highly attractive technique to many plant and animal labs in late 1980s and early 1990s. Such techniques would allow physical mapping of genetic markers, such as the popular RFLP markers, directly on chromosomes and to integrate genetic maps with physical maps. Bikram's lab also attempted to develop FISH techniques for mapping small DNA probes, however, only inconsistent results were obtained in the early experiments.

The Gill lab shifted its attention to using large genomic DNA clones, such as yeast artificial chromosome (YAC) and bacterial artificial chromosome (BAC) clones, as FISH probes. Initial experiments using maize YAC clones as probes did not produce successful results. But BACs from sorghum and rice were instantly successful (Woo et al. 1994; Jiang et al. 1995). Bikram's lab was the first to utilize BAC clones for FISH mapping in plants, and FISH using BAC clones anchored by genetic markers was to become the most popular methodology for integrating genetic linkage maps with chromosomal maps (Jiang and Gill 2006).

I. Gene-Rich, High-Recombination Regions

Initially, C-banding was the primary technique used for isolating and characterizing the deletion stocks. But because a large number of deletions were isolated for each chromosome (an average 20/chromosome), it was difficult to order the deletion breakpoints based on cytology alone. In the late 1980s, Bikram's first graduate student, Lauren Kam-Morgan had carried out a pioneering study that produced the first DNA restriction fragment length polymorphism (RFLP) based linkage map of a wheat chromosome (Kam-Morgan et al. 1989). Another graduate student, Kulvinder Gill, continued this work and made the first RFLP linkage map of *Ae. tauschii* (Gill et al. 1991). While attending a workshop on RFLP mapping in New Delhi, India, in 1991, Bikram remembers teaching during the day, working on a USDA grant proposal, in long hand, at night, and faxing copies during the day, for over three days! The proposal, "Cytogenetically based physical map of wheat genome," was rated as the top proposal by the USDA panel and was funded for the full amount of $300,000.

Kulvinder remained at KSU as a postdoc and later, as a senior scientist, and assumed the responsibility for the group-1, -5, and -6 chromosomes (Gill et al. 1993, 1996a,b). While on a trip to Madison, Wisconsin, Bikram attended the thesis defense of Joanna Werner (a student of the late Stan Peloquin) and recruited her to take charge of group-7 chromosomes (Werner et al. 1992). At Kansas State, a postdoctoral student of Scot Hulbert, Donna Delaney, made physical maps of the group-2 and -3 chromosomes (Delaney et al. 1995a,b). Bikram's graduate student Leigh Mickelson-Young made the group-4 chromosome maps (Mickelson-Young et al. 1995). This large mapping effort led not only to the characterization of the deletion stocks but also revealed gene-rich, high-recombination regions most commonly localized to the distal regions of chromosomes (Werner et al. 1992; Gill et al. 1993). The proximal chromosomal regions showed suppression of recombination and were gene poor. Many wheat-specific genes were associated with distal high-recombination regions (See et al. 2006). C-banding had allowed the partition of wheat chromosomes into the biologically meaningful heterochromatic and euchromatic regions, and deletion stocks permitted the targeted mapping of these regions and opened new possibilities for exploring cereal chromosome biology and the positional cloning of many genes crucial to the biology of the wheat plant.

J. The Q Gene Story

Justin Faris started as a Ph.D. student in Dr. Gill's laboratory in 1995. After much discussion and lab meetings, they concluded that cloning one or more genes from wheat using map-based approaches was imperative. At that time, no wheat genes had been cloned using map-based methods, and Bikram felt it was important to help demonstrate to the community (and funding agencies) that, despite the large genome and polyploid nature of wheat, map-based cloning in wheat was indeed feasible (other wheat scientists, including Beat Keller and Jorge Dubcovsky, also felt this way and initiated the cloning of wheat genes as well). Justin was charged with leading research to clone the Q gene with Li Huang working toward cloning *Lr21*.

The Q gene was targeted for cloning in Bikram's lab for two primary reasons. First, Bikram had a long-standing interest in wheat evolution and domestication. The major domestication gene primarily governing the free-threshing character, Q also pleiotropically influences many other domestication-related characters (for review see Faris et al. 2005).

Second, Bikram had assembled a large number and variety of genetic and cytogenetic stocks involving the critical chromosome 5A to expedite the work. Two deletion lines differing for a submicroscopic segment of 5A containing the Q gene, 5AL-7 and 5AL-23, provided ideal templates for the first step toward cloning, developing markers and saturation mapping of the Q locus. Justin compared the two deletion lines side-by-side using RFLP markers, RNA differential display technology, and AFLPs. Fragments present in 5AL-23 but absent in 5AL-7 were cloned, confirmed, and mapped. From this work, Justin developed 18 markers spanning about 20 cM within the deletion interval defined by the breakpoints the two deletion lines. In 1999, Justin graduated and joined the USDA–ARS at Fargo, North Dakota. This work was published soon thereafter (Faris and Gill 2002).

Although the Q gene was not yet cloned, the foundation was laid and Bikram allowed and encouraged Justin to continue working on Q through close collaboration. A chromosome walk was initiated using the tightly linked markers, a BAC contig spanning the Q gene was assembled, and a candidate gene was identified (Faris et al. 2003). The final phase of cloning the Q gene involved validation of the candidate gene followed by structural, functional, and phylogenetic analysis. Kristin Simons, who received her Ph.D. with Bikram at KSU, split her time between Bikram's lab at KSU and Fargo, ND, in 2004–2006. She showed that Q was an AP2 plant transcription factor, the Q and q alleles differed for a single amino acid, and also that Q alleles are expressed at higher levels than those of q (Simons et al. 2006). Kristin also demonstrated the dosage and pleiotropic effects of Q in transgenic plants and that the mutation that gave rise to the Q allele occurred only once during domestication. These results shed much light on the events that shaped domestication of our modern durum and common wheat cultivars. Collaborative research between the Faris and Gill labs related to Q and domestication continues. The team exploited the cloning of Q to investigate the structure and function of the q homoeoalleles (Zhang et al. 2011).

Bikram's contributions to wheat domestication studies were not limited to the collaboration on Q with Justin. Wanlong Li worked in his laboratory on the genetics of the brittle rachis (*Br*) genes (Li and Gill 2006); and Ph.D. student Shilpa Sood conducted genetic studies on the tenacious glume trait governed by the *Tg* and *Sog* genes in wheat and its relatives (Sood et al. 2009). Bikram's contributions to our current knowledge of the genetics of wheat domestication are quite significant and, at the same time, his guidance served as a launching pad for several scientists' careers.

K. Cloning *Lr21*

During the early 1990s, Stan, John, and Bikram used direct transfer to introgress leaf rust resistance genes into wheat cultivars from five accessions of *Ae. tauschii* and developed several germplasm lines (Cox et al. 1994). Joining Bikram's group in 1996, the project of cloning the *Lr21* resistance gene was given to graduate student Li Huang. Li received her M.S. degree with Moshe Feldman in Israel and had met Bikram on one of his several visits to Nanjing.

The *Lr40* gene was released in 1991 as germplasm WGRC7. The gene mapped to the very distal region of the short arm of wheat chromosome 1D, the same locus as the leaf rust resistance gene *Lr21*. Working with a PCR-based molecular marker, Li noticed identical patterns in WGRC7, an *Lr21* near-isogenic line in Thatcher, and eight other accessions of *Ae. tauschii* shown to have leaf rust resistance gene alleles at the *Lr21* locus. The *Xksud14* marker was a prefect tag for all the alleles at the *Lr21* locus (Huang and Gill 2001). However, one recombinant was detected between the marker and the *Lr21* locus, making the ksud14 sequence not considered as part of the gene.

Logically, Li's effort was devoted to searching for a marker cosegregating with *Lr21* in order to clone the gene. After exhausting all possible approaches, two additional markers were added to the existing map, but none was closer to *Lr21* than *Xksud14-STS*. The mapping progress was running full circle. Discussions with Wanlong Li and Bikram resulted in the hypothesis that ksud14 was part of the gene and that the recombination between *Xksud14-STS* and *Lr21* was intragenic.

In follow-up experiments, working with Harold Trick at Kansas State, Li transformed Fielder, a susceptible wheat cultivar, with a cosmid containing the candidate gene. She also advanced the recombinant plant progeny to select homozygous susceptible and resistant individuals for further fine mapping the regions of recombination. Transgenic Fielder with the *Lr21* candidate gene was resistant. *Lr21* was cloned and shown to encode a 1,080-amino-acid protein with NBS-LRR features (Huang et al. 2003). It was an exciting moment. Cloning *Lr21* demonstrated that map-based cloning was feasible in wheat despite the challenges and complexity of its genome. It turned out that the region where *Lr21* locus resides is a recombination hot spot and intragenic recombinations were possibly a new mechanism to generate new specificities at a simple resistance locus (Huang et al. 2009).

L. Hybrid Centromeres

Peng Zhang came to Bikram's group in 1998 to pursue her Ph.D., because she considered him as one of the world's most respected plant cytogeneticists. At that time, Bikram was interested in how the freshly broken wheat chromosome ends generated in the deletion lines acquired full telomeric repeats and were stabilized (Friebe et al. 2001). Chromosome healing occurred during very early mitotic divisions in the sporophyte by the *de novo* addition of telomeric repeats and was a gradual process. Broken chromosome ends had to pass through several early cell divisions to acquire the full telomeric repeat length.

While making significant progress with the telomere study, Bikram initiated research on centromere structure. Bikram sponsored Jiming Jiang, a Ph.D. student now turned postdoc and funded by a Grant from the USDA, to work in Dr. David Ward's laboratory in Yale University, a pioneer in nonisotopic *in situ* hybridization. Jiming identified a conserved repetitive DNA element located in the centromeres of cereal chromosomes that had high homology with Ty3/*gypsy*-like retrotransposons (Jiang et al. 1996).

Although the DNA element identified by Jiming was conserved in the centromeres of cereal chromosomes, another DNA element, identified by Peter Langridge's group at the University of Adelaide, was specific to rye centromeres (Francki 2001). This element had high homology with that of Ty1-*copia* retrotransposons, and Peter kindly provided the clone to Bikram. Around the same time, another collaborator, Adam Lukaszewski, working at the University of California, Riverside, had produced three generations of Robertsonian wheat–rye translocation chromosomes. With the plant materials and centromeric DNA elements available, Peng was able to analyze the fine structure of the centromeres in these translocation chromosomes and demonstrate their compound structure (Zhang et al. 2001).

Some of the first-generation translocation chromosomes had hybrid centromeres, approximately half rye and half wheat; others had entire rye centromeres. A similar situation was observed in the second generation. All three third-generation translocation chromosomes had hybrid centromeres, indicating that individual subunits capable of interaction with the spindle are distributed throughout the centromere and can break and rejoin at different sites resulting in wheat, rye, or wheat–rye hybrid centromeres. The position of breakage did not affect centromere function and behavior and chromosome segregation during mitosis and meiosis. This study led to a better understanding of the mechanism by which a Robertsonian translocation is formed.

M. Chromosome Bin Mapping

In 1999, the National Science Foundation awarded a 4-year Grant for a wheat genomics project entitled *The Structure and Function of the*

Expressed Portion of the Wheat Genomes, led by Cal Qualset at UC Davis. At that time, expressed sequence tag (EST) analysis was creating exciting prospects for gene discovery in all organisms, irrespective of their genome size. The project was to identify a wheat unigene set by EST analysis and to map them into chromosome regions, or bins, defined by the deletion stocks (Endo and Gill 1996). The fraction length (FL) value of each deletion identifies the position of the breakpoint from the centromere relative to the length of the complete arm. After selecting 101 deletion stocks, the wheat chromosomes were divided into 159 chromosome bins, providing a complete coverage of the wheat genome. Physically mapping nearly 10,000 EST singletons to the wheat deletions presented a big challenge.

Several features of the wheat EST map would distinguish it from other plant EST maps. First, Southern hybridization was used, making it possible to locate diverged duplicated genes sharing 80% or more homology, not possible using a PCR-based approach. Second, the aneuploid stocks used for allocating restriction fragments to individual chromosomes, arms, and bins gave complete genome coverage. For many EST clones, all restriction fragments were mapped to specific chromosome bins. The success of such a large project depended heavily on teamwork and 10 universities were involved. Bikram was selected as the coordinator for EST physical mapping, because he was recognized as an international expert and leader of one of the best plant cytogenetics laboratories in the world.

Bikram worked with Lili Qi, a scientist in his lab with roots going back to Nanjing Agricultural University, and together they successfully organized a workshop to train scientists from the cooperating laboratories. The workshop played an important role in the success of the project.

Of the over 400 deletions lines in Chinese Spring wheat (Endo and Gill 1996), 101 were selected for the bin mapping. Bikram's lab was in charge of characterizing and coordinating the distribution of genetic stocks to the mapping labs. Lili and Bikram characterized 146 genetic stocks with 526 ESTs, which covered entire wheat genome (Qi et al. 2003). Some deletion lines were found to have normal chromosome constitution, however, new deletions and chromosome aberrations were detected in some lines. By the project's end, a total of 7,104 EST unigenes was organized into a chromosome bin map, resulting in approximately 16,000 loci on the different wheat genomes (Qi et al. 2004). This unique resource has provided tools for SNP analysis, comparative mapping, structural and functional analyses, cereal chromosome evolution, and polyploid evolutionary studies, as well as providing a framework for constructing a sequence-ready, BAC-contig map of the wheat genome. The bin-mapped

ESTs are also a good resource for PCR-based marker development for wheat breeding programs.

Later, Bikram and Lili used the bin mapping resources to study centromeric region homologies among the chromosomes of *Brachypodium*, wheat, and rice. These three species, with variable genome complexity and basic chromosome number, are a model system for studying changes in basic chromosome number and the fate of centromeres during evolution. Wheat ESTs and rice centromeric genes were used to identify the centromeres in *Brachypodium distachyon* chromosomes. Six rice centromeres, syntenic to five wheat centromeres, were inactive in *Brachypodium*. The results from BAC FISH using *Brachypodium* BAC clones anchored by wheat pericentromeric ESTs and BAC contig annotation revealed a major impact of centromere inactivation on the loss of centromere retrotransposons and the turnover of centromere-specific satellites during species evolution (Qi et al. 2009, 2010).

N. Comparative Genomics and Wheat Evolution

Wanlong Li worked with Bikram as a Research Assistant Professor from 2000 to 2009. One of his interests focused on comparative genomics and wheat evolution. In the late 1990s, rice had emerged as the model grass species mainly due to its compact genome, conserved gene content, and high colinearity. An important line of evidence of grass gene colineariltry at the sequence level was from the *A1-Sh2* region of the maize, rice, and sorghum genomes. The maize genome is about sixfold that of the rice genome, and the diploid Triticeae genome is nearly twice as large as that of maize. Wanlong looked at this genome inflation, mainly caused by the amplification of transposable elements, using the *A1-Sh2* interval in Triticeae species (Li and Gill 2002). Cloning and mapping these genes in both diploid and hexaploid wheat provided Triticeae researchers a timely caveat for using rice as a model genome for map-based cloning and ushered in more comparative genomic research, which provided a strong argument for sequencing the wheat genome (Gill et al. 2004).

The subject of wheat domestication is close to Bikram's heart and much time was devoted to the study of domestication traits, such as glume toughness, shattering, and seed dormancy. A nonshattering spike, considered to be the first step in domestication, is caused by a mutation at either the spike or spikelet disarticulation locus, and Wanlong mapped this trait in *Aegilops speltoides*, *T. timopheevii*, and *Ae. tauschii* (Li and Gill 2006) and found that it was not colinear in the maize, rice, and sorghum genomes. These results supported nonconserved domestication, where different domestication events targeted different shattering genes, and the existence of multiple genetic pathways underlying seed dispersal. A similar situation was found for a glume toughness genes by Shilpa Sood, another of Bikram's graduate students (Gill et al. 2008; Sood et al. 2009).

Wheat is the classic model for polyploid evolution studies. In the early 1990s, Bikram proposed a nuclear–cytoplasmic interaction (NCI) hypothesis to explain polyploidy speciation (Gill 1991). Later, Jiang and Gill (1994) demonstrated that the emmer and timopheevii lineages of polyploid wheat differ in species-specific chromosome rearrangements and supported the diphyletic origin of polyploid wheats.

In 2001, Bikram initiated and organized the Polyploidy Workshop at the Plant and Animal Conference. Wanlong and Bikram continued to investigate the evolution of polyploid wheat lineages by comparative analysis of the grain *hardness* locus in the short arms of group-5 chromosomes studying the grain softness protein and the puroindolines (Li et al. 2008). In durum wheat, puroindoline genes are absent, causing a super hard grain texture. In hexaploid wheat, the puroindoline genes were introduced from *Ae. tauschii* and restored the soft grain texture. By accident, Wanlong detected puroindoline genes by RFLP analysis in two timopheevii lines and subsequent screening of a large number of accessions indicated that the A-genome copy of the puroindoline gene is present, but the G-genome copy is deleted in all of the timopheevii wheat accessions. Deletion of puroindoline genes also occurred in *Triticum zhukovskyi* and *T. aestivum*. Recurrent deletion of the puroindoline genes suggests a selection against their copy number in the wheat species, most probably because of a negative correlation with seed germination (Li et al. 2008). Now at South Dakota State University, Wanlong and Bikram continue to collaborate toward cloning the brittle rachis gene, *Br2*, and elucidating polyploid wheat phylogeny and evolution.

O. Sequencing the Wheat Genome

Deciphering the genome of wheat presented a complex scientific challenge to the wheat scientific community. Bikram initiated preliminary efforts a decade ago after the rice genome was successfully sequenced in 2002. In 2003, Bikram organized an NSF-funded workshop to discuss strategies to sequence the wheat genome bringing the wheat community together with 63 scientists from 18 countries.

The workshop report summarized the discussions, including the agreement to initiate an international wheat genome project (Gill et al. 2004). Bikram proposed the idea of an international consortium and, in 2005, the International Wheat Genome Sequencing Consortium (IWGSC) was established with the goal of coordinating the effort to sequence the bread wheat genome.

One of six cochairs of the consortium, Bikram proposed developing a physical map of hexaploid wheat to the U.S. National Science Foundation (NSF) but wheat, having a huge, complex, and repeat rich genome did not interest the NSF. In the mid-1990s, Bikram suggested to Jaroslov Dolezel, an expert in flow cytogenetics, that he explore the use of flow cytometry for developing chromosome and chromosome-arm based BAC libraries. Nearly a decade later, the first flow-sorted, BAC library was developed for chromosome 3B, to be followed by libraries for chromosomes 1D, 4D, and 6D, later, arm-specific libraries for the remaining wheat chromosome arms.

Bikram believed that developing a physical map would be a significant step toward sequencing wheat and an excellent genomic resource that would aid in cloning agronomically important genes. Along with Catherine Feuillet, Beat Keller, Jan Dvorak, and Rudi Appels, efforts to pursue funding for sequencing began in earnest. The first project was funded in France and, in 2006, Bikram received a USDA Grant to develop a physical map of the short arm of chromosome 3A. Sunish Sehgal, a graduate of Punjab Agricultural University, joined Bikram for his postdoctorate and was asked to lead the physical mapping project. Later, two other projects were funded by the USDA and the NSF. These grants led to the establishment of a BAC fingerprinting facility at Kansas State University and, over a period of 3 years, nearly 500,000 BAC clones were fingerprinted. Bikram and Sunish assembled the BACs into contigs, which are being anchored to developed physical maps for chromosomes 3A, 1D, 4D, and 6D. Together with Eduard Akhunov at Kansas State, Bikram and Sunish have survey sequenced chromosome 3A using next generation sequencing technology. More than 80% of the genes on chromosome 3A were identified and used to study alternate splicing and rate of evolution in wheat compared to other model grass genomes. To make best use of the wheat genome sequence, Bikram's lab is developing TILLING populations in diploid wheat for gene discovery, which may develop into an efficient trait discovery pipeline in coming years (Rawat et al. 2013).

Apart from the scientific contribution to wheat genome sequencing in his lab, Bikram actively campaigned to persuade many scientists across the globe to contribute to the project. As a result, the Department of

Biotechnology in India has funded the development of a physical map of wheat chromosome 2A.

III. INTERNATIONAL COLLABORATIONS

Bikram's vision of the WGRC was that of an international hub where foreign scientists would come for training and advanced research, more than four-dozen came from 15 countries (Fig. 1.3). The impact, in terms of technology transfer and science education, in those countries has been huge. Especially noteworthy has been collaborative research with scientists in Japan, China, India, and Russia. Kochiro Tsunewaki, who succeeded Hitoshi Kihara (considered along with E.R. Sears as the "father of wheat genetics") at Kyoto University, obtained his Ph.D. at Kansas State University. Ko's student T.R. Endo (who would succeed Tsunewaki at Kyoto University) was the first to visit the WGRC in 1981, laying the foundation of a long-term collaboration that resulted in the production of deletion stocks. Yasuhiko Mukai and Hisashi Tsujimoto were other notable visitors. Dr. Tsunewaki, before he retired, sent his last student, Shuhei Nasuda, for graduate training with Bikram. One of the first visiting scientists from China, as it was opening up to the West, was Peidu Chen from Nanjing Agricultural University. This collaboration expanded with the award of a McKnight Foundation Grant in 1995 for the control of Fusarium head blight

Fig. 1.3. The WGRC in 1992: (L to R, front row) Terri Taylor, Joan Wacker, Donna Delaney, Leigh Young, Gina Brown, and Juan Zhang; (middle row) Bernd Friebe, Bikram Gill, Rama Kota, and Allan Fritz; (back row) Rick Jellen, John Raupp, Patrick Avila, Kulvinder Gill, Trent Schneider, Joe Freiss, Jiming Jiang, Cory Falke, and Duane Wilson.

disease and continues to this day. The first International Fusarium Head Blight Symposium was organized as a part of this program in 2000. More than a dozen scientists came from India and, especially in collaboration with Punjab Agricultural University, led to the transfer and characterization of a cluster of resistance genes from *Ae. geniculata* (Kuraparthy et al. 2007, 2009). From Russia, Nikolai Badaeva came in 1989. Nikolai passed away at a young age, but collaborative research with his spouse, Ekaterina Badaeva, continued for many years, and Bikram is very proud of one of the key papers that came out of this collaboration (Badaeva et al. 1994). The WGRC participated in a USAID project in Morocco for the control of Hessian fly and training of graduate students; Ahmed Amri, the current director of the gene bank at ICARDA, trained at the WGRC. Another notable project was on the production of synthetic wheats at CIMMYT. Bikram enjoyed traveling to these and many other countries for collaborations and scientific presentations and has fond memories of many friends all over the world. Besides the WGRC, which provided a platform for wheat genetics and germplasm research, Bikram, in addition to the Wheat Genome Sequencing workshop, organized first workshops in Plant Cytogenetics and Polyploidy at the Plant and Animal Genome meetings that continue to be held to this day.

IV. EDUCATOR

During his career, Bikram has mentored over 40 M.S. and Ph.D. students, either as major professor or coadvisor (Table 1.2). The academic freedom in the Gill lab has allowed the students and scientists to explore things beyond what may have been done in wheat. His teaching and professional development philosophy is encapsulated into "nuggets of wisdom" that he shares with students. He frequently tells his students that doing science is to be a lifelong student, that there is no substitute for a thorough understanding of the published literature, and that experimenting is 50% of the research, getting it written and published is the other 50%. Bikram made sure his students and scientists actively participated in lab meetings and during their lunch hour in the wheat genetics lobby, gave presentations at professional meetings, wrote grant proposals, and communicated research papers, with no attention as to whether he was listed as the first, the last, or the communicating author in publications from the lab. He has trained new generation of scientists to carry on his mission to make wheat a better crop to feed the world.

1. BIKRAM GILL: CYTOGENETICIST AND WHEAT MAN 25

Table 1.2. Graduate students, postdoctoral fellows, and visiting scientists trained in the WGRC laboratories.

M.S. students			Joey Cainong	Current
Lapitan, Nora L.V.	1984		Bhanu Kalia	Current
Stoddard, Sara L.	1986		Anupama Joshi	Current
Morris, Kay L.	1987		Eric Olson	Current
Henry, Janet K.	1988			
Mickelson-Young, Leigh	1994		*Postdoctoral fellows*	
Miller, Douglas	1994		Sharma, Hari	1980–1983
Zhang, Juan	1995		Rayburn, A. Lane	1983–1985
Wang, Ying Jie	1995		Johnson, Jodie	1985–1986
Huang, Li	1998		Randhawa, Jatinder	1987–1988
Maleki, Lili	2000		Werner, Joanna E.	1990–1992
Khlaasen, Darcey	2002		Kota, Rama	1990–1993
Echalier, Benjamin R.	2003		Lubbers, Edward	1990–1990
Wilson, Jamie J.	2009		Jiang, Jiming	1993–1995
Rothe, Nolan	2010		Gill, Kulvinder	1991–1996
			Jellen, Eric N.	1994–1996
Ph.D. students			Kynast, Ralf G.	1996–1998
Cooper, D. Blake	1986		Boyko, Elena	1996–2001
Lapitan, Nora L.V.	1986		Faris, Justin D.	2001
Kam-Morgan, L.N.W.	1987		Huang, Li	2005–2008
Moffat, John	1988		Quarishi, Umar	2010–2011
Kaleikau, Edward K.	1988		Sehgal, Sunish	2008–
Guenzi, Arron C.	1989		Rawat, Nidhi	2009–
Stein, Ira	1989		Danilova, Tatiana	2010–
Amri, Ahmed	1989			
Gill, Kulvinder S.	1990		*Visiting scientists*	
Rudd, Jackie C.	1992		Endo, Takashi R.	1981–1982,
Jiang, Jiming	1993			1988, 1989,
Fritz, Allan K.	1994			1990, 1992
Brown-Guedira, Gina L.	1995		Chen, Peidu	1982–1983,
Nasuda, Shuhei	1999			1990–1992,
Faris, Justin D.	1999			1995–1996,
Ferrahi, Moha	2001			1997
Owuoche, James O.	2001		Schlegel, Rolf	1984
Huang, Li	2002		Fan, Lu	1984
Brooks, Steven A.	2002		Tsujimoto, Hisashi	1988–1989
Zhang, Peng	2002		Friebe, Bernd Ralf	1989, 1991–2003
Narasiminhamoorthy, Bhrinda	2003		Dhaliwal, Harcharan	1989
Mateos-Hernandez, Maria	2004		Jauhar, Prem P.	1989
Cox, Cindy	2005		Mukai, Yasuhiko	1989–1990, 1992
Simons, Kristin	2005		Tyagi, Bali Ram	1989–1990
Vijayalakshmi, Kolluru	2006		Badaev, Nikolai S.	1989–1990
See, Deven	2007		Badaev, Ekaterina	1990, 1993–1995
Kurapathy, Vasu	2007		Bluethner, W. Dieter	1991
Pumphrey, Michael	2007		Singh, Sukhwinder	1991–1994/03
Sood, Shilpa	2008		Hohmann, Uwe	1992

(*continued*)

Table 1.2. (*Continued*)

Boiko, Elena	1993–1995	Zhou, Bo	1999–2000
Cabrera, Adoracion	1993	Wang, Xiue	1999
Deol, Gurdev S.	1993–1994	Naik, Suresh	2000
Plaha, P.K.	1994	Fang, Shi	2001
McIntosh, Robert A.	1994	Chahal, Gulzar Singh	2002
Cai, Xiwen	1994	Naranjo, Thomas	2002
Wen, Yuxiang	1995	Lamouroux, Didier	2002–2003
Li, Bin	1995	Keller, Beat	2004
Liu, Dajun	1995–1996, 1997	Kumar, Sundip	2004–2005
Köszegi, Bela	1995	Namaganda, Mary	2005
Laddomada, Barbara	1995–1996, 2010–2011	Chang, Zhijian	2007–2008
		Bocchieri, Franchesco	2007–2008
Qi, Lili	1996, 1997–2011	Zhao, Wanchun	2007–2008
Wang, Suling	1996–1997	Bi, Caili	2007–2009
Sutka, Jozsef	1996	Li, Xia	2008
Chen, Qin	1996	Kumar, Sundeep	2008–2009
Smith, C. Michael	1996–1997	Chen, Qian	2009–2011
Chen, Wenpin	1996–1998	Yan, Honfei	2009–2010
Sacco, Francisco	1997	Li, Cheng	2010–2011
Li, Wanlong	1997–2009	Liu, Wenxuan	2011
Linc, Gabriella	1997–1998	Mehdieva, Sabina	2011
Dhar, Manoj Kumar	1998–1999	Abbasov, Mehraj	2011
Ram, Sewa	1998	Li, Chunxin	2012
Nayak, Pritilata	1998–1999		

Jiming Jiang emphasized the impact Bikram had on him as a student, "Every time I look back on my career development, I feel so lucky to have had Bikram as my Ph.D. supervisor. He was always highly positive and excited to listen or see my research progress and problems. His enthusiasm on my work was highly infectious and motivated me to work harder and think deeper. I still remember the moment that I showed Bikram the GISH result of the wheat–*Elymus trachycaulus* z5A translocation line. I was a little puzzled by what I had seen in the microscope. Wow, you are making a big discovery, Jiming!" was Bikram's reaction after he looked through the microscope. He immediately called the z5A chromosome the "zebra" chromosome. This short conversation excited me so much that I felt like a big lottery winner. I had many such short conversations with Bikram, which was the driving force for me to strive to become a good scientist."

Bikram was a recipient of the Conoco Distinguished Graduate Faculty Award at Kansas State University in 1990. He teaches the course "Chromosome and Genome Analysis" for Master's and Ph.D. students.

By allowing students to continue to lead the research after they leave Kansas State, Bikram has been instrumental in launching their research programs by providing a solid line of research to serve as a foundation.

V. CHAMPION OF WHEAT WORKERS

From the beginning, Bikram was determined to ensure that research from his group would yield benefits for agriculture, and that is something that has never changed. Early on, as he studied wheat's many relatives, he appreciated the vast diversity that exists within, not just among species. Recognizing the potential value of that diversity for wheat improvement, Bikram established relationships with wheat breeders, plant pathologists, entomologists, and other wheat workers, not only in Kansas, but also across North America and the world. In his mind's eye, he always had the idea that if he built a germplasm center that it would attract both research funds and top-notch researchers. To accomplish that, it was necessary for him to be the WGRC's high-profile advocate, but in that role, he never sought acclaim for himself, and always made doubly sure that his colleagues received their full share of credit.

Bikram has always enjoyed working with plants, going through field nurseries with breeders, and interacting with them at meetings and breeders' field days. Similar to his breeding colleagues, he is happiest in the greenhouse, watching plants develop and making his own crosses. But he was also one of the first to realize how big a role genomics would come to play in plant breeding, and his desire to learn genomic techniques prompted a sabbatical in Australia early on. That was a big leap of faith into a new area, but as he moved increasingly into genomics, he made sure that the germplasm program served as the anchor, and that the work remained relevant to wheat breeding. The enormous output of useful genetic knowledge and the vast variety of WGRC germplasm releases—and their importance to breeding programs around the world (Gill et al. 2006)—are powerful testimony to Bikram's long relationship with the wheat breeding community.

VI. THE MAN

Upon joining the faculty position at K-State, Bikram visited different research groups in the Great Plains region. In the fall of 1979, while visiting North Dakota State University in Fargo, Dr. Maan introduced him to his niece, Gurinder Kaur (Billo), in a photo album and showed him some of her artwork decorating their house. Bikram and Billo were married on Christmas Eve, 1979, in India. They are the parents of four children. Daugher Ragini Gill (born to Pansy Singh, his wife from a previous marriage) has a B.A. degree in psychology and is the Retirement Counselor and Wellness Program Manager at the University of California, Los Angeles, and a make-up artist in her free time. Of their two sons, Aman is a graduate of the University of California, Berkeley, and is currently working on his Ph.D. in evolutionary biology at New York State University at Stony Brook; and Jason is a graduate of the University of Kansas and is working on his Ph.D./M.D. degrees at the University of Kansas Medical Center and the Stowers Institute in Kansas City. Their daughter Naseeb is a 2011 graduate of the University of Kansas with a B.S. in photomedia and anthropology.

Bikram is a family man through and through. His life's motto has been "raising a family is even more important than doing science, but team work helps." Although he worked hard and nearly always went to the office in the evenings early in his career, he never missed any of his children's extracurricular activities or an opportunity to spend time with family. While writing manuscripts or doing wheat karyotyping, he would pick out a certain chromosome and ask his children to identify its homologue, which they always did, followed by the lesson that one came from dad and one from mom. During high school, both of his sons worked on wheat genetics projects.

He considered his lab members a part of his family as well (Fig. 1.3). He preaches the mottos "for a professor, nothing is more precious than his students" and "for students, make research an ongoing dialogue with your professor. It will be exciting for you and the professor." Toward creating more social and learning opportunities for his lab members, Billo and Bikram frequently host lavish parties in their home on festive occasions, such as Thanksgiving, Christmas, New Years Day, whenever a student graduates, or receptions for visitors. He always says "science is fun" and makes sure that it is!

An avid bicyclist, Bikram rarely drives to work, rain or shine. He enjoys creative writing and reading fiction. His other interests include gardening and cooking with fresh garden produce. For many years, he played volleyball, and WGGRC fielded a softball team for which Bikram was the pitcher. Six of Bikram's brothers are settled in the United States, as are Billo's two brothers, one sister, and their extended families. Bikram and Billo enjoy attending family reunions and social functions. The Gill brothers have renovated their ancestral home in Dhudike, India, which was built by their grandparents, Inder Singh and Bholi Kaur, and parents, Ram Singh and Basant Kaur. In recent years, in preparation for his impending retirement in 2013, Bikram and Billo, with other family members, have traveled during the winter months spending time in Dhudike with his brother's family, who resides in India, as well as lecturing at leading universities. It is a pleasure for Bikram to see his old friends and participate in educational and social projects. Bikram and Billo plan to keep residence in Manhattan, do some community work, and maintain some involvement in the WGGRC and Kansas State. Bikram will continue to do some international consulting and collaborative work but hopes to kick start his new vocation in creative and science fiction writing! Bikram is very passionate about environmental and conservation issues as well as social justice and the public understanding and communicating of science and hopes to write and be active on these issues as well.

VII. EPILOGUE

Bikram Gill's research using related wild species developed wheat many germplasm lines with novel resistance genes to many pests of wheat. His work with plant pathologists, entomologists, breeders, molecular biologists, and cytogeneticists around the world demonstrate the value he placed on establishing collaborations for the betterment of the wheat crop. He is happy that WGRC germplasm and cytogenetic stocks are part of the tool kit and a source of genes for wheat geneticists and breeders worldwide. Recently, he has played an instrumental role in getting the wheat genome sequenced. Today, his students are working with many different crops and in many different venues, public and private. WGRC research formed the basis of a proposal to the Kansas Biosciences Authority that led to the founding of the Heartland Plant Innovations (for-profit) and Earth's Harvest (nonprofit) companies to be housed in the $9M Kansas Wheat Innovation Center on the KSU campus built by Kansas wheat growers. Part of the WGRC research laboratories, including the gene bank, also will be housed in the new facilities.

ACKNOWLEDGMENTS

The material on Bikram Gill's early life comes from several conversations with him during lunch breaks in the spring of 2012. The manuscript was built around letters and conversations with former graduate students, postdocs, and colleagues passing through the WGRC over the years. We greatly thank Stan Cox, Takashi Endo, Justin Faris, Li Huang, Jiming Jiang, Wanlong Li, Lili Qi, Rollie Sears, Sunish Sehgal, and Peng Zhang for their eagerness to help with this project. We could not have produced this manuscript without them. We also thank Bikram's brother, Gurcharan Gill, for his glimpse into Bikram's early life.

LITERATURE CITED

Badaeva, E.D., N.S. Badaev, B.S. Gill, and A.A. Filatenko. 1994. Intraspecific karyotype divergence in *Triticum araraticum* (Poaceae). Plant Syst. Evol. 192:117–145.

Cox, T.S., W.J. Raupp, and B.S. Gill. 1994. Leaf rust-resistance genes *Lr41*, *Lr42*, and *Lr43* transferred from *Triticum tauschii* to common wheat. Crop Sci. 34:339–343.

Delaney, D.E., S. Nasuda, T.R. Endo, B.S. Gill, and S.H. Hulbert. 1995. Cytologically based physical maps of the group-2 chromosomes of wheat. Theor. Appl. Genet. 91:568–573.

Delaney, D.E., S. Nasuda, T.R. Endo, B.S. Gill, and S.H. Hulbert. 1995. Cytologically based physical maps of the group-3 chromosomes of wheat. Theor. Appl. Genet. 91:780–782.

Endo, T.R. and B.S. Gill. 1996. The deletion stocks of common wheat. J. Hered. 87: 295–307.

Faris, J.D., and B.S. Gill. 2002. Genomic targeting and high-resolution mapping of the domestication gene *Q* in wheat. Genome 45:706–718.

Faris, J.D., J.P. Fellers, S.A. Brooks, and B.S. Gill. 2003. A bacterial artificial chromosome contig spanning the major domestication locus *Q* in wheat and identification of a candidate gene. Genetics 164:311–321.

Faris, J.D., K.J. Simons, Z. Zhang, and B.S. Gill. 2005. The wheat super domestication gene *Q*. Wheat Inf. Serv. 100:129–148.

Francki, M.G. 2001. Identification of Bilby, a diverged centromeric Ty1-*copia* retrotransposon family from cereal rye (*Secale cereale* L.). Genome 44:266–274.

Friebe, B. and B.S. Gill. 1996. Chromosome banding and genome analysis in diploid and cultivated polyploid wheats. p. 39–59. In: P.P. Jauhar (ed.), Methods of Genome Analysis in Plants, CRC Press, Boca Raton, FL.

Friebe, B., Y. Mukai, H.S. Dhaliwal, T.J. Martin, and B.S. Gill. 1991. Identification of alien chromatin specifying resistance to wheat streak mosaic virus and greenbug in wheat germplasm by C-banding and in situ hybridization. Theor. Appl. Genet. 81:381–389.

Friebe, B., J.M. Jiang, B.S. Gill, and P.L. Dyck. 1993. Radiation-induced nonhomoeologous wheat-*Agropyron intermedium* chromosomal translocations conferring resistance to leaf rust. Theor. Appl. Genet. 86:141–149.

Friebe, B., J. Jiang, W.J. Raupp, R.A. McIntosh, and B.S. Gill. 1996. Characterization of wheat-alien translocations conferring resistance to diseases and pests: current status. Euphytica 91:59–87.

Friebe, B., R.G. Kynast, P. Zhang, L.L. Qi, M. Dhar, and B.S. Gill. 2001. Chromosome healing by addition of telomeric repeats in wheat occurs during the first mitotic divisions of the sporophyte and is a gradual process. Chromosome Res. 9:137–146.

Friebe, B., L.L. Qi, D.L. Wilson, Z.J. Chang, D.L. Seifers, T.J. Martin, A.K. Fritz, and B.S. Gill. 2009. Wheat–*Thinopyrum intermedium* recombinants resistant to wheat streak mosaic virus and *Triticum* mosaic virus. Crop Sci. 49:1221–1226.

Gill, B.S. 1983. Tomato cytogenetics—a search for new frontiers. p. 457–480. In: M.S. Swaminathan, P.K. Gupta, and U. Sinha (eds.), Cytogenetics of crop plants, MacMillan India Limited.

Gill, B.S. 1991. Nucleo-cytoplasmic interaction (NCI) hypothesis of genome evolution and speciation in polyploid plants. p. 48–53. In: T. Sasakuma and T. Kinoshita (eds.), Nuclear and Organellar Genomes of Wheat Species, Proceedings of the Dr. H. Kihara Memorial International Symposium on Cytoplasmic Engineering in Wheat, Kihara Memorial Yokohama Foundation for the Advancement of Life Science, Japan.

Gill, B.S. 1974a. A novel ditertiary tetrasomic in tomato. Genetics 77:61–70.

Gill, B.S. 1974b. Dosage dependent epistasis in the tomato: its bearing on trisomy abnormalities and evolution. J. Hered. 65:130–132.

Gill, B.S. and C.O. Grassl. 1986. Pathways of genetic transfer in intergeneric hybrids of sugar cane. Sugar Cane 2:2–7.

Gill, B.S., and G. Kimber. 1974a. The Giemsa C-banded karyotype of rye. Proc. Natl. Acad. Sci. (USA) 71:1247–1249.

Gill, B.S., and G. Kimber. 1974b. Giemsa C-banding and the evolution of wheat. Proc. Natl. Acad. Sci. (USA) 71:4086–4090.

Gill, B.S., and W.J. Raupp 1987. Direct genetic transfers from *Aegilops squarrosa* L. to hexaploid wheat. Crop Sci. 27:445–450.

Gill, B.S., R. Appels, A-M. Botha-Oberholster, C.R. Buell, J.L. Bennetzen, B. Chalhoub, F. Chumley, J. Dvorak, M. Iwanaga, B. Keller, W. Li, W.R. McCombie, Y. Ogihara, F. Quetier, and T. Sasaki. 2004. A workshop report on wheat genome sequencing. The International Genome Research on Wheat Consortium Genetics 168:1087–1096.

Gill, B.S., C.R. Burnham, G.R. Stringam, J.T. Stout, and W.H. Weinheimer. 1980. Cytogenetic analysis of chromosomal translocations in the tomato: preferential breakage in heterochromatin. Can. J. Genet. Cytol. 22:333–341.

Gill, B.S., B. Friebe, and T.R. Endo. 1991. Standard karyotype and nomenclature system for description of chromosome bands and structural aberrations in wheat (*Triticum aestivum*). Genome 34:830–839.

Gill, B.S., B. Friebe, W.J. Raupp, D.L. Wilson, T.S. Cox, R.G. Sears, G.L. Brown-Guedira, and A.K. Fritz. 2006. Wheat Genetics Resource Center: The first 25 years. Advan. Agron. 89:74–136.

Gill, B.S., W. Li, S. Sood, V. Kuraparthy, B. Friebe, K.J. Simons, Z. Zhang, and J.D. Faris. 2008. Genetics and genomics of wheat domestication-driven evolution. Isr. J. Bot. 55:223–229.

Gill, B.S., W.J. Raupp, H.C. Sharma, L.E. Browder, J.H. Hatchett, T.L. Harvey, J.G. Moseman, and J.G. Waines. 1986. Resistance in *Aegilops squarrosa* to wheat leaf rust, wheat powdery mildew, greenbug, and Hessian fly. Plant Dis. 70:553–556.

Gill, K.S., B.S. Gill, and T.R. Endo. 1993. A chromosome region-specific mapping strategy reveals gene-rich telomeric ends in wheat. Chromosoma 102:374–381.

Gill, K.S., B.S. Gill, T.R. Endo, and E.V. Boyko. 1996a. Identification and high-density mapping of gene-rich regions in chromosome group 5 of wheat. Genetics 143:1001–1012.

Gill, K.S., B.S. Gill, T.R. Endo, and T. Taylor. 1996b. Identification and high-density mapping of gene-rich regions in chromosome group 1 of wheat. Genetics 144:1883–1891.

Gill, K.S., E.L. Lubbers, B.S. Gill, W.J. Raupp, and T.S. Cox. 1991. A genetic linkage map of *Triticum tauschii* (DD) and its relationship to the D genome of bread wheat (AABBDD). Genome 34:362–374.

Hatchett, J.H. and B.S. Gill. 1981. D-genome sources of resistance in *Triticum tauschii* to Hessian fly. J. Hered. 72:126–127.

Hatchett, J.H., and B.S. Gill. 1983. Expression and genetics of resistance to Hessian fly in *Triticum tauschii* (Coss) Schmal. p. 807–811. In: S. Sakamoto (ed.), Proceedings of the 6th International Wheat Genetics Symposium, Plant Germ-Plasm Institute, Kyoto University, Japan.

Hatchett, J.H., T.J. Martin, and R.W. Livers. 1981. Expression and inheritance of resistance to Hessian fly in synthetic hexaploid wheat derived from *Triticum tauschii* (Coss) Schmal. Crop Sci. 21:731–734.

Huang, L. and B.S. Gill. 2001. An RGA-like marker detects all known *Lr21* leaf rust-resistance gene family members in *Aegilops tauschii* and wheat. Theor. Appl. Genet. 103:1007–1013.

Huang, L., S.A. Brooks, J.P. Fellers, and B.S. Gill. 2003. Map-based cloning of leaf rust resistance gene *Lr21* from the large and polyploid genome of bread wheat. Genetics 164 (2): 655–664.

Huang, L., S. Brooks, W. Li, J. Fellers, J. Nelson, and B.S. Gill. 2009. Evolution of new disease specificity at a simple resistance locus in a weed-crop complex: Reconstitution of the *Lr21* gene in wheat. Genetics 182:595–602.

Jiang, J.M. and B.S. Gill. 1993. Sequential chromosome banding and *in situ* hybridization analysis. Genome 36:792–795.

Jiang, J.M. and B.S. Gill. 1994. Different species-specific chromosome translocations in *Triticum timopheevii* and *T. turgidum* support diphyletic origin of polyploid wheats. Chromosome Res. 2:59–64.

Jiang, J.M. and B.S. Gill. 2006. Current status and the future of fluorescence *in situ* hybridization (FISH) in plant genome research. Genome 49:1057–1068.

Jiang, J.M., B. Friebe, and B.S. Gill. 1994. Chromosome painting of Amigo wheat. Theor. Appl. Genet. 89:811–813.

Jiang, J.M., B.S. Gill, G.L. Wang, P.C. Ronald, and D.C. Ward. 1995. Metaphase and interphase fluorescence *in situ* hybridization mapping of the rice genome with bacterial artificial chromosomes. Proc. Natl. Acad. Sci. (USA) 92:4487–4491.

Jiang, J., S. Nasuda, F. Dong, C.W. Scherrer, S.S. Woo, R.A. Wing, B.S. Gill, and D.C. Ward. 1996. A conserved repetitive DNA element located in the centromeres of cereal chromosomes. Proc. Natl. Acad. Sci. (USA) 93:14210–14213.

Kam-Morgan, L.N.W., B.S. Gill, and S. Muthukrishnan. 1989. DNA restriction fragment length polymorphisms: a strategy for genetic mapping of D genome of wheat. Genome 32:724–732.

Kuraparthy, V., P. Chhuneja, H.S. Dhaliwal, S. Kaur, and B.S. Gill. 2007. Characterization and mapping of cryptic alien introgressions from *Aegilops geniculata* with new leaf rust and stripe rust resistance genes *Lr57* and *Yr40* in wheat. Theor. Appl. Genet. 114: 1379–1389.

Kuraparthy, V., S. Sood, and B.S. Gill. 2009. Molecular genetic description of the cryptic wheat–*Aegilops geniculata* introgression carrying rust resistance genes *Lr57* and *Yr40* using wheat ESTs and synteny with rice. Genome 52:1025–1036.

Li, W. and B.S. Gill. 2002. Colinearity of the *Sh2/A1* region among rice, sorghum and maize is interrupted and accompanied by genome expansion in the Triticeae. Genetics 160:1153–1162.

Li, W.L. and B.S. Gill. 2006. Multiple pathways for seed shattering in the grasses. Funct. Integr. Genomics 6:300–309.

Li, W., L. Huang, and B.S. Gill. 2008. Recurrent deletions of puroindoline genes at the grain hardness locus in four independent lineages of polyploid wheat. Plant Physiol. 146:200–212.

Liu, W., D.L. Seifers, L.L. Qi, M.O. Pumphrey, B. Friebe, and B.S. Gill. 2011. A compensating wheat–*Thinopyrum intermedium* Robertsonian translocation conferring resistance to wheat streak mosaic virus and *Triticum* mosaic virus. Crop Sci. 51:2382–2390.

Lubbers, E.L., K.S. Gill, T.S. Cox, and B.S. Gill. 1991. Variation of molecular markers among geographically diverse accessions of *Triticum tauschii*. Genome 34:354–361.

McIntosh, R.A., B. Friebe, J. Jiang, D. The, and B.S. Gill. 1995. Cytogenetical studies in wheat. XVI. Chromosome location of a new gene for resistance to leaf rust in a Japanese wheat-rye translocation line. Euphytica 82:141–147.

Mickelson-Young, L.A., T.R. Endo, and B.S. Gill. 1995. A cytogenetic ladder-map of the group 4 chromosomes of wheat. Theor. Appl. Genet. 90:1007–1011.

Mukai, Y., B. Friebe, J.H. Hatchett, M. Yamamoto, and B.S. Gill. 1993. Molecular cytogenetic analysis of radiation-induced wheat–rye terminal and intercalary chromosomal translocations and the detection of rye chromatin specifying resistance to Hessian fly. Chromosoma 102:88–95.

Qi, L.L., B. Echalier, B. Friebe, and B.S. Gill. 2003. Molecular characterization of a set of wheat deletion stocks for use in chromosome bin mapping of ESTs. Funct. Integr. Genomics 3:39–55.

Qi, L.L., B. Echalier, S. Chao, G.R. Lazo, G.E. Butler, O.D. Anderson, E.D. Akhunov, J. Dvorák, A.M. Linkiewicz, A. Ratnasiri, J. Dubcovsky, C.E. Bermudez-Kandianis, R.A. Greene, R. Kantety, C.M.LaRota, J.D. Munkvold, S.F. Sorrells, M.E. Sorrells, M. Dilbirligi, D. Sidhu, M. Erayman, H.S. Randhawa, D. Sandhu, S.N. Bondareva, K.S. Gill, A.A. Mahmoud, X-F. Ma, Miftahudin, J.P. Gustafson, E.J. Conley, V. Nduati, J.L. Gonzalez-Hernandez, J.A. Anderson, J.H. Peng, N.L.V. Lapitan, K.G. Hossain, V. Kalavacharla, S.F. Kianian, M.S. Pathan, D.S. Zhang, H.T. Nguyen, D-W. Choi, R.D. Fenton, T.J. Close, P.E. McGuire, C.O. Qualset, and B.S. Gill. 2004. A chromosome bin map of 16,000 expressed sequence tag loci and distribution of genes among the three genomes of polyploid wheat. Genetics 168:701–712.

Qi, L., B. Friebe, P. Zhang, and B.S. Gill. 2007. Homoeologous recombination, chromosome engineering and crop improvement. Chromosome Res. 15:3–19.

Qi, L.L., B. Friebe, P. Zhang, and B.S. Gill. 2009. A molecular-cytogenetic method for locating genes to pericentromeric regions facilitates a genome-wide comparison of synteny between the centromeric regions of wheat and rice. Genetics 183:1235–1247.

Qi, L.L., B. Friebe, J.J. Wu, Y.Q. Gu, C. Qian, and B.S. Gill. 2010. The compact *Brachypodium* genome conserves centromeric regions of a common ancestor with wheat and rice. Funct. Integr. Genomics 10:477–492.

Rawat, N., S.K. Sehgal, A. Joshi, N. Rothe, D.L. Wilson, N. McGraw, P.V. Vadlani, W.L. Li, and B.S. Gill. 2013. A diploid wheat TILLING resource for wheat functional genomics. BMC Pl. Biol. (In press).

Rayburn, A.L., and B.S. Gill. 1985. Use of biotin-labeled probes to map specific DNA sequences on wheat chromosomes. J. Hered. 76:78–81.

Rick, C.M., and B.S. Gill. 1973. Reproductive errors in aneuploids: generation of variant extra chromosomal types by tomato primary trisomics. Can. J. Genet. Cytol. 15:299–308.

Sharma, H.C. and B.S. Gill. 1983. New hybrids between *Agropyron* and wheat. II. Production, morphology and cytogenetic analysis of F1 hybrids and backcross derivatives. Theor. Appl. Genet. 66:111–121.

Simons, K.J., J.P. Fellers, H.N. Trick, Z. Zhang, Y-S. Tai, B.S. Gill, and J.D. Faris. 2006. Molecular characterization of the major wheat domestication gene *Q*. Genetics 172: 547–555.

See, D.R., S.A. Brooks, J.C. Nelson, G.L. Brown-Guedira, B. Friebe, and B.S. Gill. 2006. Gene evolution at the ends of wheat chromosomes. Proc. Natl. Acad. Sci. (USA) 103:4162–4167.

Sood, S., V. Kuraparthy, G.H. Bai, and B.S. Gill. 2009. The major threshability genes soft glume (*sog*) and tenacious glume (*Tg*), of diploid and polyploid wheat, trace their origin to independent mutations at non-orthologous loci. Theor. Appl. Genet. 119:341–351.

Werner, J.E., T.R. Endo, and B.S. Gill. 1992. Toward a cytogenetically based physical map of the wheat genome. Proc. Natl. Acad. Sci. (USA) 89:11307–11311.

Woo, S.S., J.M. Jiang, B.S. Gill, A.H. Paterson, and R.A. Wing. 1994. Construction and characterization of a bacterial artificial chromosome library of *Sorghum bicolor*. Nucleic Acids Res. 22:4922–4931.

Zhang, P., B. Friebe, A.J. Lukaszewski, and B.S. Gill. 2001. The centromere structure in Robertsonian wheat-rye translocation chromosomes indicates that centric breakage-fusion can occur at different positions within the primary constriction. Chromosoma 110:335–344.

2

Synthetic Hexaploids: Harnessing Species of the Primary Gene Pool for Wheat Improvement

Francis C. Ogbonnaya and Osman Abdalla
International Center for Agricultural Research in the
Dry Areas (ICARDA)
P.O. Box 5466
Aleppo, Syria

Abdul Mujeeb-Kazi and Alvina G. Kazi
National Institute of Biotechnology and Genetic Engineering (NIBGE)
Faisalabad
National University of Science and Technology (NUST)
Islamabad, Pakistan

Steven S. Xu
USDA-ARS, Northern Crop Science Laboratory
1605 Albrecht Blvd.
Fargo, ND 58102-2765, USA

Nick Gosman
The National Institute of Agricultural Botany (NIAB)
Huntington Road
Cambridge CB3 0LE, UK

Evans S. Lagudah
CSIRO Plant Industry
GPO Box 1600, Canberra
Australian Capital Territory 2601, Australia

David Bonnett
International Maize and Wheat Improvement Center (CIMMYT)
El Batan, Mexico

Mark E. Sorrells
Department of Plant Breeding and Genetics
240 Emerson Hall
Cornell University
Ithaca, NY 14853, USA

Hisashi Tsujimoto
Arid Land Research Center
Tottori University
Tottori 680-0001, Japan

ABSTRACT

Incorporation of genetic diversity into elite wheat (*Triticum aestivum* L., $2n = 6x = 42$, AABBDD) cultivars has long been recognized as a means of improving wheat productivity and securing global wheat supply. Synthetic hexaploid wheat (SHW) genotypes recreated from its two progenitor species, the tetraploid, *Triticum turgidum* ($2n = 4x = 28$, AABB) and its diploid wild relative, *Aegilops tauschii* ($2n = 2x = 14$, DD) are a useful resource of new genes for hexaploid wheat improvement. These include many productivity traits such as abiotic (drought, heat, salinity/sodicity, and waterlogging) and biotic (rusts, septoria, barley yellow dwarf virus (BYDV), crown rot, tan spot, spot blotch, nematodes, powdery mildew, and fusarium head blight) stress resistance/tolerances as well as novel grain quality traits. Numerous SHWs have been produced globally by various institutions including CIMMYT-Mexico, ICARDA-Syria, Department of Primary Industries (DPI), Victoria-Australia, IPK-Germany, Kyoto University-Japan, and USDA-ARS. This review examines the varied aspects in the utilization of synthetics for wheat improvement including the traits and genes identified, mapped, and transferred to common wheat. It has also been demonstrated that synthetic backcross-derived lines (SBLs, i.e., when SHW is crossed to adapted local bread varieties) show significant yield increases and thus, enhanced yield performance across a diverse range of environments, demonstrating their potential for improving wheat productivity worldwide. This is particularly evident in moisture-limited environments. The use of SBLs, advanced backcross QTL analysis, chromosome introgression lines, and whole genome association mapping is contributing to the elucidation of the genetic architecture of some of the traits. The contribution of transgressive segregation to enhanced phenotypes and the mechanisms including its genetic and physiological basis are yet to be elucidated.

Understanding these would further enhance the utility of SBLs. Considerable progress has been made in the identification of useful quantitative trait loci (QTL) and genes, however the transfer of such rich genetic diversity into elite wheat cultivars is still quite limited. Gaps still exist in data cataloging; and access to such information could serve as an important community resource. Future production of new SHW should extend to under-exploited AB genome tetraploids such as *T. turgidum* ssp. *carthlicum*, *T. turgidum* ssp. *dicoccum*, and *T. turgidum* ssp. *dicoccoides* and identifying gaps in the *Ae. tauschii* germplasm used for existing SHW. Identifying geographical areas where the progenitor species of the existing SHW were collected would assist in guiding future collection missions. The recent advances in molecular technologies with whole genome sequencing becoming affordable will provide researchers with opportunities for more detailed analysis of traits and the deployment of more efficient strategies in the use of the unique exotic alleles derived from SHW for common wheat improvement. Thus, the contribution of SHW and the derived SBLs to wheat cropping systems worldwide is likely to grow in significance. However, these potential benefits are only realizable if phenotyping is equally extensive and effective.

KEYWORDS: *Aegilops tauschii*; direct cross; synthetic hexaploids; *Triticum aestivum*; *Triticum turgidum*; wheat breeding

ABBREVIATIONS
  I. INTRODUCTION
     A. The Importance of Wheat
     B. Wheat Genetic Resources
 II. PRODUCTION AND UTILIZATION OF SYNTHETIC HEXAPLOID WHEAT
     A. Diversity of *Ae. tauschii* and SHW
     B. Interspecific Crosses of *Ae. tauschii* and *T. turgidum*
     C. Direct Cross of Diploid *Ae. tauschii* to Common Wheat
     D. Direct Cross of Tetraploid Wheat to Common Wheat
     E. Factors for Consideration: Direct Versus Synthetic Approaches
        1. Threshability of the Grain
        2. Adaptation of the Durum Parents
        3. Rapid Recovery of Recurrent Parent Genotype
     F. Available Genetic Resources in Synthetic Hexaploid Wheat
     G. Current Strategies for Using Synthetic Hexaploid Wheat in Breeding
        1. Direct Phenotyping
        2. Genetic Analysis via Crosses and Backcrossing
        3. Advanced Backcross-Quantitative Trait Loci (AB-QTL) Analysis
        4. Development and Use of Introgression Lines (ILs) or Substitutions Lines
     H. Expression of *Ae. tauschii* and *T. turgidum* Genes in Synthetic Hexaploid Wheat
III. IMPACT OF SYNTHETIC HEXAPLOID IN WHEAT IMPROVEMENT
     A. Disease and Pests Resistance
     B. Enhancing Yield Productivity
     C. Environmental Stress Tolerance
        1. Improving Salinity Tolerance

        2. Mineral/Nutrient Efficiency
        3. Boron Toxicity
    D. Quality Improvement
        1. Bread Making Quality
        2. Preharvest Sprouting
IV. CONCLUSIONS AND FUTURE PROSPECTS
ACKNOWLEDGMENTS
LITERATURE CITED

ABBREVIATIONS

AFLP	Amplified fragment length polymorphism
AM	Association mapping
BYDV	Barley yellow dwarf virus
CCN	Cereal cyst nematode
cDNA	Complimentary deoxyribonucleic acid
CIMMYT	International Maize and Wheat Improvement Center, Mexico
CSIRO	Commonwealth Scientific and Industrial Organization, Australia
DArT	Diversity arrays technology
DPI	Department of Primary Industries, Australia
DH	Doubled haploid
FAO	Food and Agricultural Organization of the United Nations
ICARDA	International Center for Agricultural Research in the Dry Areas, Syria
IPK-Gatersleben	Institute of Plant Genetics and Crop Plant Research, Germany
NIAB	National Institute of Agricultural Botany, UK
NIBGE	National Institute of Biotechnology and Genetic Engineering, Pakistan
PS	Primary synthetic wheat
QTL	Quantitative trait loci
RT-PCR	Reverse transcription polymerase chain reaction
SBL	Synthetic backcross derived wheat lines
SHW	Synthetic hexaploid wheat
SNP	Single nucleotide polymorphism
SSR	Simple sequence repeats
USDA-ARS	United States Department of Agriculture-Agricultural Research Service

I. INTRODUCTION

A. The Importance of Wheat

Three major crops, wheat (*Triticum aestivum*), rice (*Oryza sativa*), and maize (*Zea mays*) provide about two-thirds of all energy in human diets, and production systems involving these crops provide the mainstay of the global food supply (Cassman 1999) and, as such make an important contribution to food security. Of these top three species, wheat is a staple food for 35% of the world's population and is, therefore, one of the most important. The challenge of feeding a world population of nine billion people, as projected by mid-century, are immense and sustainability of the natural resource base must be preserved (Godfray et al. 2010). To meet the food and fiber needs of this ever-increasing population, far-reaching increases in agricultural output are required. Since existing land has been cultivated for millennia, and only limited land expansion is possible most of the projected increases will have to come from increased productivity. Projected forecasts about climate change point to additional uncertainty and place even more pressure on our natural resources to provide adequate food for everyone.

Wheat provides 21% of the food calories and 20% of the protein to more than 4.5 billion people in 94 developing countries (Braun et al. 2010). It is the world's most widely grown crop with about 218 million hectares harvested and 653 million tons produced in 2010–2011 (Australian Bureau of Agricultural and Resources Economics and Sciences: Agricultural commodity statistics 2011). The crop is grown from Norway and Russia at 65°N to Argentina at 45°S. However, in tropical and subtropical regions, wheat is mostly grown at higher elevations (Dubcovsky and Dvořák 2007). Improving and achieving yield stability across these diverse agroecological zones of production remains a daunting challenge. Improved yield potential, resistance to biotic stresses such as rusts and major abiotic stresses including drought, high temperature (heat), salinity/sodicity, waterlogging, acidity, cold, and soil macro and micronutrient deficiencies or toxicities all have a role in improving overall productivity. However, genetic variation for some of these traits is limited in elite wheat germplasm.

It is projected that the demand for wheat will increase between 30% and 40% by 2030. The current production increases remain at about 1.2% annually, this need to increase to 1.6%–1.8% to be able to meet this increased demand. Of the projected 1.6% increases, 1% would come from genetics and breeding. With limited genetic variation in elite germplasm, it is imperative that a wider spectrum of variability be examined.

B. Wheat Genetic Resources

Harlan and de Wet (1971) proposed the concept of three gene pools, primary, secondary, and tertiary, based on the evolutionary distance between the species and success rate of hybridisation among species. The primary gene pool consists of biological species and crossing within this pool is easy. Common wheat arose recently (6,000–8,000 years ago) from hybridization of tetraploid *T. turgidum* L. ($2n = 4x = 28$, AABB) and diploid, *Ae. tauschii* Coss. ($2n = 2x = 14$, DD) and these two species constitute the primary gene pool (Qi et al. 2007). Gene transfers from these species can be made by homologous recombination either by direct crosses of these species with common wheat or by the production of synthetic wheat (McFadden and Sears 1946; Gill and Raupp 1987). No special cytogenetic manipulation, except embryo rescue, is necessary to produce the F_1 hybrids. On the other hand, gene transfers from *Triticum urartu* or *Triticum monococcum* as well as *Aegilops speltoides* is difficult because of massive hybridization barriers including endosperm abortion, death of hybrids seedlings, female sterility of F_1 hybrids, reduced recombination, and the lack of recombination between 4A of diploid wheat with 4A of polyploid wheat. Thus, these species are described as secondary gene pool by Qi et al. (2007). Extensive reviews of secondary and tertiary gene pools and the transfer of genes from them into common wheat can be found in Jiang et al. (1994), Mujeeb-Kazi and Hettel (1995), Friebe et al. (1996), and Gill et al. (2006, 2008).

Wild relatives of wheat have long been recognized as a source of useful genes for cultivated wheat improvement (Farrer 1904; Kruse 1967, 1969; Mujeeb-Kazi and Kimber 1985; Wang 1989; Mujeeb-Kazi and Asiedu 1990; Valkoun 2001; Gill et al. 2006, 2008). Amongst wild species in the primary pool, considerable genetic variability has been found in both *Ae. tauschii* and *T. turgidum* for potentially adaptive traits, such as resistance to biotic and abiotic stresses (Halloran et al. 2008; Xie and Nevo 2008). One of the strategies in exploiting the wild relatives of wheat has been through synthetic hexaploid wheat (SHW). The SHW, often designated as primary synthetic (PS) wheat, is usually created by artificially hybridizing durum wheat (*T. turgidum* ssp. *durum*; $2n = 4x = 28$, AABB) with accessions of *Ae. tauschii* (syn *Ae. squarrosa*, *T. tauschii*; $2n = 2x = 14$, DD), the donor of the D genome of hexaploid bread wheat (*T. aestivum*; $2n = 6x = 42$, AABBDD), in a process analogous to the natural hybridization event that produced hexaploid wheat. Some SHWs have been produced by crossing different tetraploid wheat ancestors, cultivated, and wild emmer, *Triticum dicoccum* or *T. dicoccoides* or *T. carthlicum;* $2n = 4x = 28$ AABB with *Ae. tauschii* (Lange and Jochemsen 1992; Niwa et al. 2010). The preference for *Ae. tauschii* as the D genome donor in SHW is its close homology with the D genome of hexaploid wheat. It has been demonstrated that wheat and *Ae. tauschii* could be crossed directly and that the D genome chromosomes of the two species pair in meiosis (McFadden and Sears 1946; Alonso and Kimber 1984; Gill and Raupp 1987). Recombination between the D genome of *T. aestivum* and *Ae. tauschii* has been reported to occur at a frequency similar to that of an intraspecific cross (Fritz et al. 1995). The close pairing affinity of *Ae. tauschii* D genome chromosomes with that of *T. aestivum* indicates that only one or two rounds of backcrossing to elite wheat are needed to incorporate both qualitative and quantitative characters from *Ae. tauschii* (Gororo 1999).

There have been two earlier reviews on SHW, van Ginkel and Ogbonnaya (2007) and Yang et al. (2009), the latter with emphasis on the utilization of SHW for wheat improvement in China. Earlier reviews/reports addressing wheat genetic resource exploitation for crop improvement have been spread over the past few decades and go back to Dewey (1984), Sharma and Gill (1983), Sharma (1995), Gill et al. (2006), Mujeeb-Kazi et al. (2008), Trethowan and Mujeeb-Kazi (2008) who documented the contributions of the unique Triticeae gene pool diversity to common wheat improvement. The current review, emphasizes research on the evaluation, identification, transfer, and characterization of traits and genes in SHW and its impact in the context of wheat improvement. The wider adaptation provided by increasing the genomic diversity of bread wheat via SHW provides a means to sustain productivity gains in the face of increasing disease epidemics and worsening climate change scenarios. The current review attempts to address several important questions: (i) can increasing genetic diversity of traits and genes in SHW contribute significantly to overcoming these challenges; (ii) how novel and diverse are genes in SHW; (iii) what is the genetic basis of biotic and abiotic stress tolerance in SHW; (iv) what are the best strategies for exploiting novel diversity in wheat breeding programs; (v) what potential advances in genetic technologies could be utilized to mine new and diverse alleles in SHW for hexaploid wheat improvement; and (vi) where are the current gaps in our knowledge and how can these be filled?

II. PRODUCTION AND UTILIZATION OF SYNTHETIC HEXAPLOID WHEAT

The first published report of the artificial synthesis of hexaploid wheat was by McFadden and Sears (1944). Since then numerous trait transfers aimed at testing expression levels in particular D genome variants at the allohexaploid level have been performed.

A. Diversity of *Ae. tauschii* and SHW

Ae. tauschii Coss. is adapted to a variety of environments such as desert margins, steppe regions, stony hills, wastelands, roadsides, sandy shores, and even humid temperate forests (van Slageren 1994). It is also found in the edges of wheat fields in eastern Turkey, Iraq, Iran, Pakistan, India (Kashmir), China (the Himalaya), Afghanistan, most of central Asia, Transcaucasia, and the Caucasus region (Feldman 2001). The diversity of the D genome of this species is much larger than that of bread wheat. Based on morphology, taxonomists have divided *Ae. tauschii* into two subspecies: ssp. *tauschii* and ssp. *strangulata* (Eig) Tzvel. (Eig 1929; Hammer 1980). Ssp. *tauschii* is further divided into three morphological varieties: *anathera*, *meyeri*, and *typica*, whereas ssp. *strangulata* is monotypic. Ssp. *strangulata* is regarded to be the direct donor of D genome to common wheat (Nishikawa et al. 1980; Dvorak et al. 1998; Pestsova et al. 2000). The diversity of *Ae. tauschii* has been studied using molecular tools such as chloroplast DNA variation (Matsuoka et al. 2008, 2009; Takumi et al. 2009), AFLP (Mizuno et al. 2010), SSR (Naghavi and Mardi 2010), isozymes (Dudnikov and Kawahara 2006), and random amplified polymorphic DNA (RAPD) markers (Okuno et al. 1998). Recently, Sohail et al. (2012) analyzed the diversity using 4,449 polymorphic DArT markers and found the diversity of ssp. *strangulata*, that is the origin of D genome of bread wheat, contains only a limited part of whole diversity of *Ae. tauschii* (Fig. 2.1). Thus, SHWs produced by crosses between tetraploid wheat and any subspecies of *Ae. tauschii* include untapped amount of genetic variation in which useful genes for bread wheat breeding must be present.

B. Interspecific Crosses of *Ae. tauschii* and *T. turgidum*

The crossing of *T. turgidum* (AABB) with *Ae. tauschii* (D^tD^t) followed by the induction of chromosome doubling of the hybrid, via colchicine treatment, produces an amphiploid referred to as a synthetic hexaploid (AABB D^tD^t) (Fig. 2.2) (Kihara et al. 1957; Cox et al. 1990a;

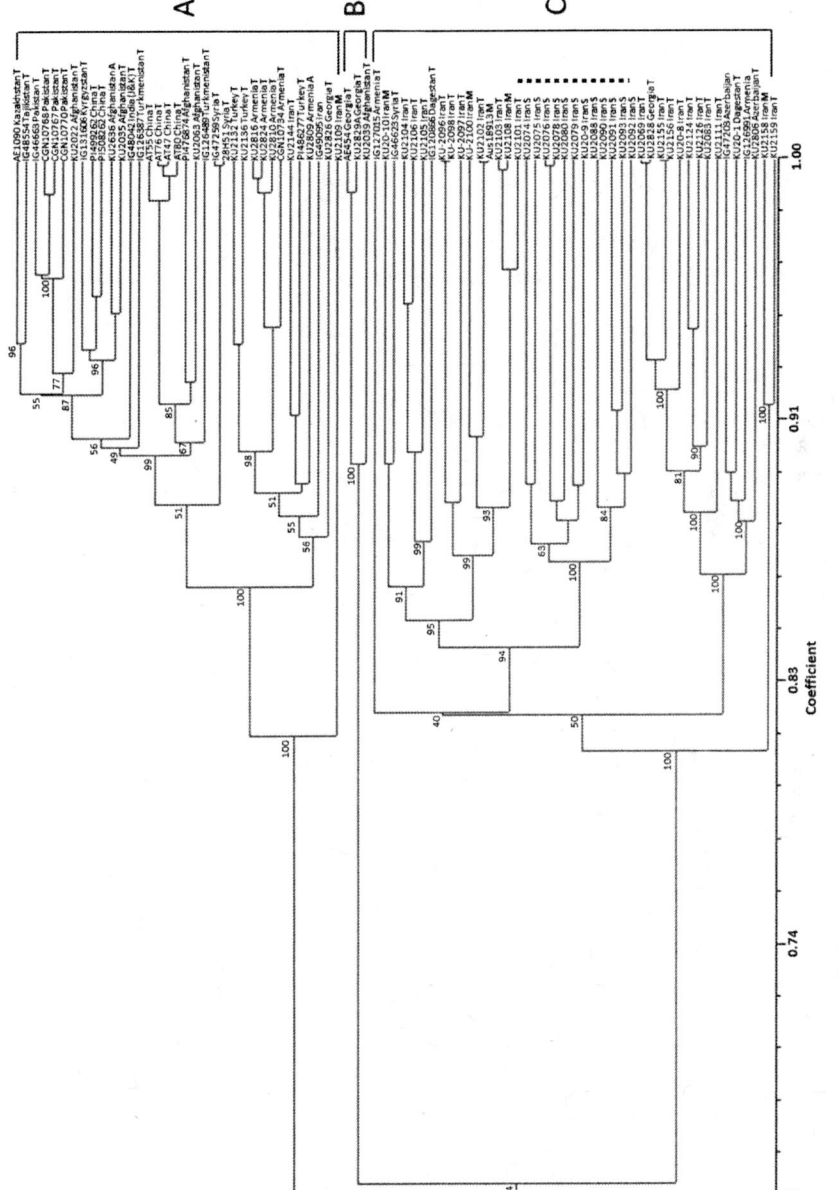

Fig. 2.1. Phylogenetic tree of 81 *Ae. tauschii* accessions constructed using 4449 DArT markers. The accessions were classified largely into three groups, A B and C. Subspecies strangulata that is the D-genome donor of bread wheat is indicated by dotted line. *Source:* Sohail et al. (2012).

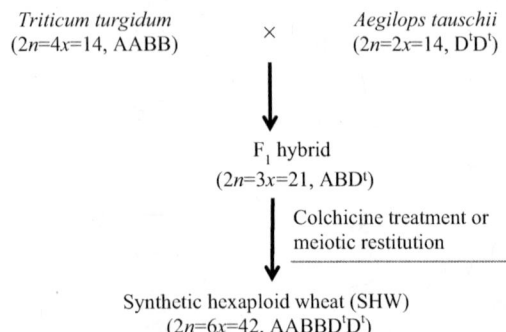

Fig. 2.2. The production of synthetic hexaploid wheat (SHW).

Lange and Jochemsen 1992; Mujeeb-Kazi 1993; He et al., 2003; Matsuoka and Nasuda 2004; Matsuoka et al. 2007). When SHW is crossed with cultivated hexaploid wheat, typically there are 21 chromosome pairs at meiotic metaphase I that are associated primarily as ring bivalents indicating complete chromosome homology and full fertility. In this approach, SHWs are often spontaneously produced from partially fertile hybrid plants with a high frequency of unreduced gametes arising from meiotic restitution (Maan and Sasakuma 1977; Xu and Joppa 1995, 2000).

C. Direct Cross of Diploid *Ae. tauschii* to Common Wheat

The second approach uses direct crosses between *Ae. tauschii* (D^tD^t) and common wheat (AABBDD) to produce a hybrid with the genome constitution $ABDD^t$. The embryo of the hybrid (F_1) seed must be excised within 12–18 days postpollination depending on the prevailing environment and grown on an artificial culture medium to prevent seed abortion. The hybrid is then backcrossed to the common wheat parent and selection made in the progeny for 42-chromosome plants, which have 21 pairs of bivalents at meiosis (Fig. 2.3) (Alonso and Kimber 1984; Gill and Raupp 1987; Mujeeb-Kazi 1993). This eliminates the confounding effects of segregation in the A and B genomes. This method also has the advantage that no colchicine treatment is necessary in the cross between wheat and the *Ae. tauschii* hybrid. However, lines derived from direct-crossed hybrids have the disadvantage of segregating for the D genome, and exhibit instability because of aneuploidy, potentially making genetic analysis more difficult.

2. SYNTHETIC HEXAPLOIDS

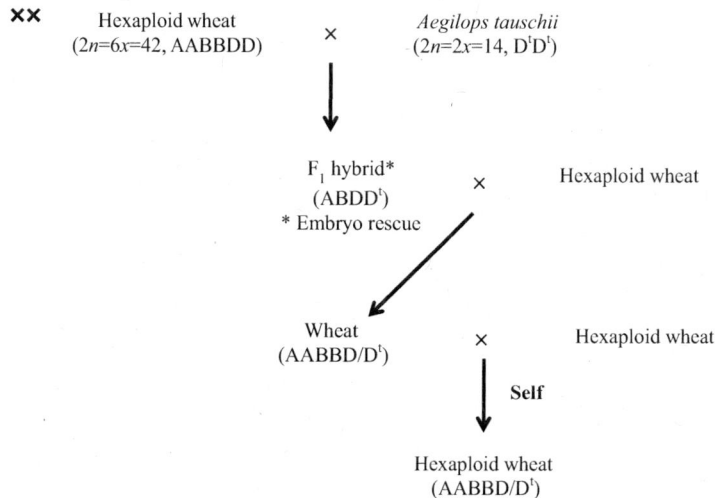

Fig. 2.3. Direct cross for genetic transfer from diploid Ae. tauschii into hexaploid wheat.

D. Direct Cross of Tetraploid Wheat to Common Wheat

Following the determination of the chromosome numbers of bread and durum wheat by Sakamura and Sax in 1918 (Sax 1918; Kihara 1982) the first pentaploid products of crosses between hexaploid and tetraploid wheats were produced and cytologically examined (Sax 1922a; Kihara 1923). Meiotic behaviour of the pentaploid ($2n = 5x = 35$, AABBD) was subsequently described (Kihara 1925) with parental (28 or 42 chromosome) euploid progeny being recovered (Sax 1922b). Similar to the crosses of *Ae. tauschii* to common wheat, direct crosses of tetraploid to hexaploid wheat have been used to transfer useful genes from various tetraploid subspecies into common wheat (Fig. 2.4). The pioneering work of McFadden (1930) in transferring resistance to stem and leaf rust from cultivated emmer wheat (*T. turgidum* ssp. *dicoccum*) into common wheat heralded the use of this approach in interspecific crosses. McFadden (1930) reported that although only one partially fertile F_1 plant from a cross between Yaroslav emmer and common wheat 'Marquis' was initially obtained, the plant produced 100 F_2 shriveled seed. Six true common wheat plants possessing all the desired introgressed characters of the emmer wheat were selected in F_4 generation. Through this procedure, McFadden (1930) successfully introduced *Sr2* (McIntosh et al. 1995) for adult plant stem rust resistance from emmer

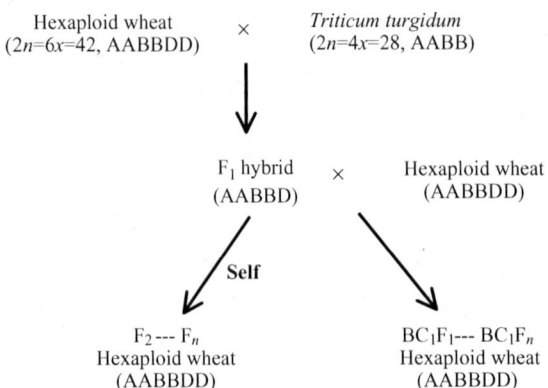

Fig. 2.4. Direct cross for genetic transfer from tetraploid wheat into hexaploid wheat.

wheat into common wheat and named it 'Hope'. $Sr2$ has been deployed in many cultivars developed in the regions where wheat production is vulnerable to stem rust (McIntosh et al. 1995) and continues to confer effective rust resistance more than 80 years later. Using similar breeding scheme, Waterhouse (1933) transferred another stem rust resistance gene $Sr14$ from cultivated emmer cultivar Khapli into hexploid cultivar Steinwedel. Both $Sr2$ and $Sr14$ are presently an important source of resistance to Ug99 lineage races of stem rust (Singh et al. 2011a). Several other genes, including $Lr23$ (Watson and Stewart 1956) for leaf rust resistance from durum wheat, $Yr15$ (Grama and Gerechter-Amitai 1974) for resistance to stripe rust, $Pm30$ (Liu et al. 2002) and $pm42$ (Hua et al. 2009) for resistance to powdery mildew and gene(s) for high grain protein content $Gpc-B1$ and stripe rust resistance $Yr36$ (Kushnir and Halloran 1984; Uauy et al. 2005) from wild emmer, and $Hdic$ (Brown-Guedira et al. 2005b) for resistance to Hessian fly from cultivated emmer, were transferred into common wheat through direct tetraploid–hexaploid crosses. Recently, Xie and Nevo (2008) provided an overview of genes transferred to common wheat from *T. dicoccoides*.

Compared with SHW approach, a common problem in direct cross of tetraploid to hexaploid is the low fertility of the F_1 pentaploid (AABBD) plants. Grama and Gerechter-Amitai (1974) reported that the pentaploid interspecific hybrids between the wild emmer accession G-25 and common wheat cultivars were highly sterile (fertility of 1%–2%). They overcame the sterility by exposing the F_1 progeny to open mass pollination with the common wheat cultivars or using a bridging cross between the progeny of an emmer-durum cross and common wheat cultivars. Presently, the sterility of the pentaploid hybrids is usually overcome by further crossing or backcrossing the hybrids with common wheat cultivars (Liu et al. 2002; Brown-Guedira et al. 2005b; Hua et al. 2009). Direct cross of tetraploid to hexaploid followed by backcross would hasten the recovery of euploid progeny ($2n = 42$) with introgressed genes.

E. Factors for Consideration: Direct Versus Synthetic Approaches

The use of restitution nuclei and the restoration of the parental wheat phenotype by limited backcrossing were reported initially by Alonso and Kimber (1984). Gill and Raupp (1987) subsequently reported that direct hybridization between common wheat and *Ae. tauschii*, followed by two backcrosses to wheat, permitted rapid recovery of the desired genes from *Ae. tauschii* in euploid backgrounds. They postulated that two backcrosses involving a diploid donor and hexaploid recurrent-parent result in approximately 96% restoration of the total recurrent-parent genome, a proportion almost equivalent to that resulting from four backcrosses involving the same ploidy. This suggests that the time scale for germplasm development possessing agronomically acceptable progenies with *Ae. tauschii* as a donor parent is shorter than that usually associated with interspecific crossing. However, aneuploidy is generated in such a progeny and it is necessary to either make further crosses using the direct cross genotype as the male parent and rely on gametic selection or cytologically select euploid plants (Gill and Raupp 1987; Eastwood 1995).

Cox (1997) reported that F_1 plants between wheat and *Ae. tauschii* set approximately one seed per spike, of which about 50% are germinable. Chromosome numbers of BC_1 plants range from 35 to more than 50, of which 25%–50% arise from fertilization of an unreduced female gamete (Cox et al. 1991a,b; Gill and Raupp 1987). Gill and Raupp (1987) and Innes and Kerber (1994) suggested that backcrossing BC_1 plants as males would screen out aneuploid pollen and produce primarily 42-chromosome BC_2 progeny. However, Cox (1997) reported that among all BC_2 plants produced from their studies (Cox et al. 1995), only 60% had 42 chromosomes. Also, Fritz et al. (1995) reported that the rates of transmission of alleles from the diploid parent to BC_2 progenies and recombination among loci are generally at the rates expected under the assumptions of nonpreferential segregation and complete chromosome homology.

Lagudah et al. (1993) argued that in the direct cross approach, an *a priori* assumption is made that the introgressed trait is sufficiently expressed to allow detection and selection in the background of the recurrent wheat parent while in a SHW, the expression of the *Ae. tauschii* trait can be determined. Cox (1997) suggested that several considerations influenced the decision of his group to use the direct backcrossing approach to improve winter wheat in central plains of the United States. Some of those considerations were:

1. Threshability of the Grain. In direct-backcross populations, only one gene conditioning nonthreshability, *Tg* from *Ae. tauschii* segregates whereas the additional gene *q* from the tetraploid parent segregates in crosses between wheat and SHW. Thus, single gene segregation results in a higher frequency of threshable progeny.

2. Adaptation of the Durum Parents. The SHW have been more useful as donor parents in northern Mexico and Australia, where durum and spring wheat are both well adapted. In contrast, durum wheat is not as productive in the central plains of the United States, where winter wheat is produced. Thus, introgression of genes from the nonadapted durum parent of a synthetic into the A and B genomes of hard winter wheat, at the same time genes are being introduced into the D genome, disrupts adaptation to a much greater extent than do direct crosses between hard winter wheat and *Ae. tauschii*.

3. Rapid Recovery of Recurrent Parent Genotype. Genes may be backcrossed directly from *Ae. tauschii* into almost any elite winter wheat cultivar or breeding line, and progeny similar to the winter wheat parent can be selected from BC_2 populations. Recently, the direct hybridization approach has been reexamined where wheat pollen was applied to the female *Ae. tauschii* parent yielding a high 35% embryo formation frequency. However, the F_1 hybrid plant recovery frequency was only 6.8% (Sehgal et al. 2011). During the production of SHW via reciprocal crosses extremely high embryo excision frequencies were observed (between 65% and 80%) against the reciprocal range of 3%–28% (A. Mujeeb-Kazi, unpublished). Different durum wheat parents were used in producing SHW at CIMMYT and a few of these parental lines appear to exhibit very high levels of crossability with *Ae. tauschii*. The durum cvs 'LARU', 'Altar 84', 'Aconchi', 'Croc-1', and a line with the pedigree of 'Cando/R143//Erite/Mexi' were better combiners and later additions included cultivars 'Chen', 'Croc', 'Doy', 'Dvergand', and a line 'D67.2'.

We opine that direct crossing for wheat improvement has not received the attention it deserves. Apart from precision transfers into the D genome only, the speed with which advanced lines of good agronomic phenotype are recovered warrants more attention to this strategy. It is evident from the CIMMYT international nursery that only a few *Ae. tauschii* accessions are represented in their pedigrees. Hence, there is a need to widen the *Ae. tauschii* accessions sources by including collections from areas of particular stresses, for example, drought and thus broaden the variability for such stress tolerance in addition to choice of *Ae. tauschii* on the basis of genotype diversity. These accessions could enter future direct crossing programs that could target breeding goals more effectively. This range in the future could be widened not only for bridge crosses but also for the well-targeted traits.

F. Available Genetic Resources in Synthetic Hexaploid Wheat

Since the first SHW was produced in the 1940s (McFadden and Sears 1944), over 1,500 SHW have been developed globally. The production of SHW on a large scale was conducted at CIMMYT (Mexico) from 1988 to 2010 with an output of approximately 1,300 SHW (Mujeeb-Kazi and Delgado 2001; Mujeeb-Kazi 2003a) using approximately 900 *Ae. tauschii* accessions. The amount of genetic diversity in the *Ae. tauschii* populations were considerably higher than the D genome of bread wheat for many disease and insect resistance, isozymes, and seed storage protein (Valkoun et al. 1990; Dvorak et al. 1998). Additionally, in the 1980s, L.R. Joppa developed a number of spontaneous SHW from partially fertile hybrids between 'Langdon' durum and different *Ae. tauschii* accessions. Among these Langdon-derived SHW, 46 were recently characterized for resistance to various diseases and high molecular weight glutenin subunits (Friesen et al. 2008; Xu et al. 2010).

Amongst the CIMMYT set, two elite sets of 128 SHW were selected based on agronomic performance and disease resistance (Mujeeb-Kazi et al. 2000a,b, 2001a,b,c; Mujeeb-Kazi and Delgado 2001). Both Langdon-derived SHW and elite CIMMYT SHW are widely distributed and are used as base germplasm for genetic study and wheat breeding. A limited number of winter habit SHW were also produced in CIMMYT and are maintained in El Batan, Mexico, and Pakistan. These are not yet in wide distribution but have the potential to aid winter wheat breeding programs. The winter SHW, like the spring habit SHW, also exhibit a wide range in diversity for major phenological traits and allelic diversity profile (Uzma et al. unpublished data). Dreisigacker et al. (2008) reported that on the basis of percentage of polymorphic amplified fragment length polymorphism (AFLP) markers, SHW contained a higher level of molecular diversity than observed in cultivated hexaploid wheat (Lage et al. 2003). This is consistent with the results reported by Zhang et al. (2005), which described significantly higher number of alleles per locus and Nei's gene diversity within SHW than in common bread wheat. In Australia, SHWs were also produced by University of Melbourne, and Department of Primary Industry, Victoria (Eastwood 1995; Gatford 2004), while Kyoto University produced SHW lines in Japan (Matsuoka and Nasuda 2004; Matsuoka et al. 2007). A limited number of SHW lines were also produced at ICARDA (Valkoun 2001).

G. Current Strategies for Using Synthetic Hexaploid Wheat in Breeding

There are a number of approaches to SHW breeding.

1. Direct Phenotyping. In this approach, SHW are evaluated for genetic variation for disease or insect resistance and tolerance to a range of biotic and abiotic stresses. This is the traditional and most widely reported procedure in the literature.

2. Genetic Analysis via Crosses and Backcrossing. Promising SHWs that carry desired traits are crossed with adapted elite wheat cultivars to fast track the development of elite breeder-friendly parental stocks. In some cases, genetic and quantitative trait analyses are undertaken concurrently. This traditional approach involves, first, the screening and identification of SHW with the desired characteristic followed by backcrossing to elite cultivars to generate "synthetic derived backcrossed bread wheat lines '(SBLs)'" (Fig. 2.5), introducing the targeted trait from the SHW donor into adapted germplasm and generating novel recombinant genotypes to widen the existing primary gene pool of common wheat. Such novel genotypes result from either new D^t or $A^d B^d$ (from durum, *T. turgidum* ssp. *dicoccum* or *T. turgidum* ssp. *dicoccoides*) derived alleles or interactions between these new variants and the hexaploid wheat genome. Thus, the impact of SBLs is not exclusive to the introgression of the D genome variants. The additional backcrossing results in considerable time and resources invested in prebreeding efforts before the genetic stock with desirable traits can be available for use in breeding depending on the tetraploid used. For SHW produced using cultivated durum, one BC may be enough. Regardless, population size is probably more important and warrants further investigation to determine the appropriate population size needed to recover desirable traits in an adapted genetic background.

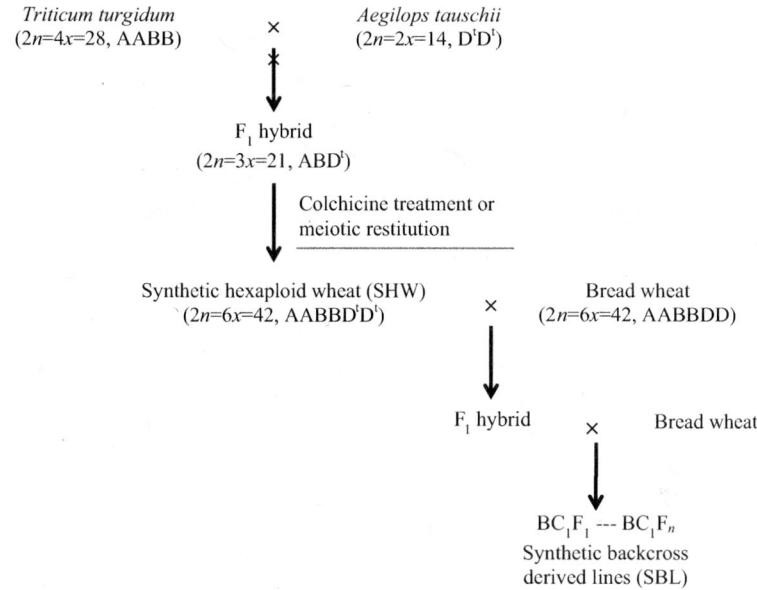

Fig. 2.5. Procedure for the transfer of desirable traits from synthetic hexaploid wheat into elite cultivars (synthetic backcross derived lines, SBLs).

3. Advanced Backcross-Quantitative Trait Loci (AB-QTL) Analysis. Tanksley and Nelson (1996) proposed the advanced backcross AB-QTL approach, which has some advantages over classical QTL analysis. In this method, the detection of QTL and the development of improved breeding lines are conducted simultaneously. Unadapted germplasm from wild species is used to detect exotic alleles superior to the elite alleles, which simultaneously broadens the genetic variation of modern cultivars. Since the AB-QTL strategy is based on BC_2 or BC_3 generations, the portion of the exotic donor genome, and thus linkage drag, is considerably reduced.

A number of AB-QTL populations have been developed using synthetics (Huang et al. 2003, 2004; Narasimhamoorthy et al. 2006; Naz et al. 2008). Disadvantages include smaller classes that reduce power to detect QTL and a large investment in time and labor to develop the populations and considerable investment to phenotype them.

4. Development and Use of Introgression Lines (ILs) or Substitutions Lines. This technique offers the potential for further gains in the use of SHW. Development of specific D genome introgressed segments starting from *Ae. tauschii* single chromosome substitutions or ILs in common wheat (Pestsova et al. 2001) has received limited attention. An IL set represents the complete or nearly complete genome of a wild species in the uniform background of an elite cultivar, where each single line solely contains a single marker-defined segment of the exotic parent (Zamir 2001). Advantages of using ILs include evaluating the effects of introgressed segments in different target recipient backgrounds, the ability to dissect the contribution of the diploid D genome from the other genomes, and the potential to reveal cryptic variation. However, ILs preclude the assessment of epistasis (known to be important for some traits) and cannot dissect the effect of loci on the same segment. One of the impediment in using the IL as a tool in breeding is the time taken to develop them.

There is no documented evidence on the number of backcrosses that are best when making crosses between SHW and adapted common wheat cultivars. Generally, one to two backcrosses appear necessary, and it is advisable to target larger than normal BC_1F_1 and BC_1F_2 populations. To estimate the relative size and genetic structure of BC_1 and BC_2 progeny lines derived from synthetic wheat, the National Institute of Agricultural Botany (NIAB) recently conducted molecular marker analysis of crosses between spring wheat Paragon and four CIMMYT SHW lines (P. Howell, unpublished). Tracing DArT haplotypes of progeny composing individual streams from BC_1 into BC_2 indicated that, on average, over the whole genome, a population size of approximately 300 individuals would capture the whole synthetic genome at BC_1, but an average population of more than 800 individuals would be needed to capture its entirety at BC_2. However, conducting the same analysis on a genome-by-genome basis suggested that the number of lines needed to capture each genome differed markedly with estimated population sizes of 450, 600, and 1500 for the A, B, and D genomes, respectively. Trethowan and van Ginkel (2009) suggested that in a global evaluation of SBL for yield, the SBL possessing a smaller portion of the SHW genome tended to perform better, indicating that at least one backcross is required to improve agronomic type sufficiently to exploit underlying variation for yield and adaptation. The application of appropriate selection pressure against tenacious glumes, hard-threshing and nonadapted plant types at the early segregating generations would result in the identification of good agronomic types with the desired traits. Lage and Trethowan (2008) reported that one or two backcrosses are sufficient to introgress desirable traits from SHW into adapted wheat with minimal linkage drag. They suggested that after one or two backcrosses the coefficient of parental of SHW in derived lines should stabilize between 12.5% and 25%. Zhang et al. (2005) used microsatellite (SSR) markers to quantify introgression patterns of the SHW into locally adapted cultivars and found that only a small segment of the SHW remained after crossing and selection. This indicates that through back- and/or top-crossing and the use of large F_2-populations, it is possible to select progeny with favorable additive elleles from the SHW, thus avoiding undesirable linkage drag (Lage and Trethowan 2008).

One of the impediments in the use of SHW is the relatively high frequency of F_1 hybrid necrosis when crossed with adapted common wheat cultivars. Hybrid necrosis is the gradual premature death of leaves or plants in certain F_1 hybrids of wheat, and it is caused by the interaction of two dominant complementary genes *Ne1* and *Ne2* located on chromosome arms 5BL and 2BS, respectively (Zevan 1981; Nishikawa et al. 1974). When the two dominant alleles for both loci were present either in homozygous or heterozygous condition (i.e., *Ne1Ne1Ne2Ne2*, *Ne1ne1Ne2ne2*, *Ne1Ne1Ne2ne2*, and *Ne1ne1-Ne2Ne2*), necrosis occurs and seedlings plants die early in plant development at the one or two leaf stage. Hexaploid and durum wheat differ in the frequency of alleles at these two loci; hence, crosses between SHW carrying the durum complement of necrosis genes and hexaploid wheat frequently result in hybrid necrosis. The frequency of necrosis is even higher in crosses with *T. turgidum* ssp. *dicoccum* or *T. turgidum* ssp. *dicoccoides*. Consequently, hybrid necrosis quite often limits successful transfer of desirable traits from *Ae. tauschii* into common wheat.

The difficulties in using some of the SHWs in a wheat breeding program was further demonstrated by May and Lagudah (1992) in their attempt to transfer Septoria tritici blotch (STB) resistance identified in *Ae. tauschii* to common wheat. In that study, 14 widely grown and well-adapted Australian cultivars were crossed to the STB-resistant SHW. The F_1 hybrids of six wheat varieties (Egret, Owlet, Vulcan, Suneca, Sunfield, and Hartog) were severely necrotic and hybrids with Millewa segregated for the presence or absence of hybrid necrosis. Seven F_1 hybrids (Grebe, Janz, Kite, M1696, Rosella, Summit, and Teal) were not necrotic. Similarly at NIAB, 50 CIMMYT synthetic wheat lines were crossed as male parents to 11 elite UK winter wheat cultivars (Robigus, Alchemy, Cordiale, Gallant, Humber, Oakley, Q Plus, Scout, Shamrock, Timber, and Viscount). Severe hybrid necrosis occurred in all progeny except those produced from crosses with Robigus (P. Howell, unpublished). Therefore, in attempts to exploit SHW for common wheat improvement it is necessary to use different wheat cultivars to minimize the incidence of necrosis, the failure of the cross, and maximize the chance of finding healthy combinations. It is essential to keep records on necrosis in crosses between common wheat and SHW so that its incidence is minimized by avoiding the use of wheat parents more likely to possess the complementary necrosis genes. This is important to consider for a wider use of diploid and tetraploid species in making new SHW. Chu et al. (2006) mapped *Ne1* and *Ne2* using microsatellite markers and the closely linked markers identified in that study may be used to genotype SHW and wheat parents for *Ne1* and *Ne2*.

H. Expression of *Ae. tauschii* and *T. turgidum* Genes in Synthetic Hexaploid Wheat

Evaluation of SHW provides a means of determining if traits from progenitor species are expressed at a higher ploidy level. Expression of genes for traits of both simple and complex inheritance is reported to change when transferred from the diploid relative to a higher ploidy. Lagudah et al. (1993) reported that the background of the tetraploid genome modulates the expressions of some traits.

Sohail et al. (2011) used 33 *Ae. tauschii* accessions along with their corresponding SHWs produced by crosses with common cultivar of durum wheat, Langdon and compared the expression of drought-tolerance related traits (the net photosynthetic rate, stomatal conductance, leaf water potential, etc.) under well-watered and drought conditions. Although there was large variation in both *Ae. tauschii* and SHWs, the traits of *Ae. tauschii* were not significantly correlated with their corresponding SHW lines, indicating that the these quantitative traits are expressed as a results of complex interaction of many genes in hexaploids. Dreisigacker et al. (2008) earlier reported that tolerance of abiotic stresses such as salinity, heat, and drought cannot be directly defined by the performance of their respective *Ae. tauschii* parents. They proffered the following explanation (i) the homoeologous genes A and B in background of SHW leading to epistatic interactions and differential gene expression; (ii) a modified gene expression caused by genomic changes during artificial hybridization; and (iii) a high genetic diversity within *Ae. tauschii* accessions, which can result in phenotypic variation in the SHW, especially if unintentional selection or drift changes the frequencies of alleles present in the SHW compared with the *Ae. tauschii* parents.

The expression of some traits from *Ae. tauschii* at the allohexaploid level ranged from hypostatic to epistatic effects with the parental tetraploid and the *Ae. tauschii* parents likely to interact as homoeologous, rather than homologous loci. Even in qualitative traits controlled by simpler genetic systems. Hypostatic and recessive gene effects were observed in SHW for a number of characters when comparing the SHW with the respective *T. turgidum* and *Ae. tauschii* parents (for details see Lagudah et al. 1993). Flood et al. (1992) demonstrated that the spring habit of *T. turgidum* was epistatic over the winter habit of *Ae. tauschii* and similarly, the winter habit in common wheat is expressed as a recessive trait. Grain softness is inherited as a dominant trait in common wheat, with the hard grain texture of *T. turgidum* being suppressed by the soft texture of *Ae. tauschii* in SHW.

There have been many studies on the expression of disease resistance genes from *Ae. tauschii* in the hexaploid background. Early researchers suggested that the transfer of resistance to leaf rust (*Puccinia recondita* f. sp. *tritici*) and stem rust (*Puccinia graminis* f. sp. *tritici*) from *Ae. tauschii* into common wheat has been difficult because it is either partially or completely suppressed in SHW. This raises major implications in the use of such sources of resistance for wheat improvement. Kerber and Green (1980) reported suppressor gene(s) on chromosome 7D of wheat that inhibits expression of stem rust resistance in the A and/or B genome. They indicated that the mechanisms underlying this phenomenon are yet not fully understood. The identification of alternative alleles in *Ae. tauschii* that lack the suppressor effect, and crossing them with common wheat (substitution for the corresponding homoallele in common wheat), may allow expression of some rust resistance genes present in the A or B genome, which otherwise would have been suppressed (Lagudah et al. 1993). Alternatively, suppressor alleles may be screened out using molecular markers. Deletion or inactivation through mutation of the suppressor gene might allow expression of genes for resistance already present in hexaploid wheat or in related donor species (Williams et al. 1992).

Work by Kerber and Green (1980) and later, Bai and Knott (1992) indicated that suppressors of disease resistance were in the D genome of common wheat. Additionally, there was evidence that suppressors were disease specific. In the first study, 10 rust resistant wild tetraploid wheats were crossed with both durum and common wheat. Resistance to leaf rust and stem rust was expressed in the hybrids with durum wheats but suppressed in the hybrids with common wheats suggesting that the suppressor(s) may be localized in the D genome of the common wheat. To test this possibility, 12 pentaploid F_1 hybrids between durum and common wheat cultivars were tested with leaf rust race 15. Leaf rust resistance from the durum parents was suppressed in all cases. Further, to localize the genome position of these suppressors, a set of 14 D-genome disomic chromosome substitution lines in 'Langdon' durum were used. The result demonstrated that chromosomes 1B, 2B, and 7B carried genes for resistance to stem rust while chromosomes 2B and 4B possessed genes for resistance to leaf rust and 1D and 3D carried suppressors. Crosses between seven D-genome monosomics of 'Chinese Spring' (CS) with three *T. turgidum* ssp. *dicoccoides* accessions, showed that 'CS' carries suppressors on chromosomes 1D, 2D, and 4D which suppressed the stem rust resistance of all three *T. turgidum* ssp. *dicoccoides* accessions. Therefore, in the absence of any one of the three 'CS' chromosomes carrying suppressors, resistance was expressed indicating that some form of complementary gene interaction is involved. Evidence for specificity was supported by results from a later study (Bai and Knott 1992) in which 'CS' chromosome 3D suppressed resistance to leaf rust race 15 in all three *dicoccoides* accessions, but had no effect on resistance to the other strains. Similarly, a suppressor on CS chromosome 1D suppressed resistance to leaf rust in *dicoccoides* accession PI276000 only. Similar results were obtained in a study by Kerber and Green (1980) who reported that chromosome 7D of cv. Canthatch, carried a dominant effect gene that suppressed resistance to all stem rust strains tested except for race C64. In another study, gene(s) on the A and B genome of cv. Tetra Canthatch suppressed resistance from *Ae. tauschii* accession RL5459 to four out of nine leaf rust races in SHW (Kema et al. 1995).

The expression of seedling resistance to both stripe and leaf rust in SHW was also determined by suppressors present in the A or B and D genomes (Kema et al., 1995). Ma et al. (1995) further demonstrated the presence of suppressors to stripe rust (*Puccinia striiformis*) resistance in A or B and D genomes of *T. turgidum* ssp. *durum* and *Ae. tauschii*. Of 13-hexaploid-wheat lines synthetized by crossing resistant *T. turgidum* ssp. *durum* with susceptible/intermediate *Ae. tauschii*, only six were resistant while others were either intermediate or susceptible. This suggests that resistance was fully expressed, partially expressed, or suppressed depending on the presence of suppressor gene(s) in the *Ae. tauschii* genotypes, again, implicating suppressors within the D genome. Similarly, of 62 SHW derived from crosses of susceptible/intermediate *T. turgidum* ssp. *durum* with resistant *Ae. tauschii*, only 10 showed a level of resistance similar to the *Ae. tauschii* parents while 52 displayed intermediate to susceptible reactions. The high frequency of partial to complete suppression suggests that suppressors of stripe rust resistance are also in A or B genomes. Knott (2000) also found suppressors to stem rust to be present on either A or B genome chromosomes of durum near-isogenic line 'Medea'.

The mechanism of suppression of stem and leaf rust resistance in wheat is genetically complex. Resistance genes located on one genome are inhibited by a suppressor(s) on another. Suppressors of rust resistance are on the D-genome chromosome of wheat and suppressed genes on the A- and B-genome chromosomes of tetraploid wheats and vice versa. The genes also appear to suppress resistance to specific parental genotypes and pathogen isolates (Kema et al. 1995). Badebo et al. (1997) investigated the genetics of suppressors of resistance to stripe rust in SHW derived from *T. turgidum* subsp. *dicoccoides* and *Ae. squarrosa*. They reported that resistance in wild emmer accession, 342-2M was suppressed to race 7E134b in three SHWs due to the D genome in *Ae. squarrosa*. However, they also observed reciprocal suppression consistent with the results of Kema et al. (1995).

In addition to specific suppressor loci, genome-wide processes appear to be active. In this regard, it should be noted that the process of creating new SHW, that is, the process of allopolyploidization, generates two "shocks." One is hybridity, by which two diverged genomes are joined together to form one nucleus. The other is polyploidy, resulting in duplicated genomes. Zhang et al. (2004) indicated that rapid changes associated with the coordination of coexistence of divergent genomes occurred at the molecular level during the formation of polyploidy. Gene expression, for example, gene silencing and gene activation, was changed in the newly synthesized allopolyploid. Gene silencing was a general response in polyploidization and mainly resulted from DNA sequence elimination or methylation.

Analysis of genome-wide gene expression (cDNA-AFLP, RT-PCR, and reverse-Northern analysis) of SHW suggested that gene expression changes are an intrinsic part of polyploidy (He et al. 2003) and that gene loss and, or silencing may be partly or wholly responsible for variation in trait expression (Kashkush et al. 2002). Analysis of SHW developed at CIMMYT, that have undergone numerous rounds of meiosis, suggested that a large proportion (up to 7.7%) of parental cDNA-AFLP fragments were altered in the SHW lines (He et al. 2003). In addition, *Ae. tauschii* genes were reported to suffer a significantly higher level of suppression than *T. turgidum* genes; a similar situation was found in the common wheat Chinese Spring. Moreover, transcript analysis during resynthesis indicates that gene loss and or silencing occurs early in synthetic allohexaploid formation, possibly during the development of F_1 intergeneric hybrids or just after chromosome doubling (Kashkush et al. 2002).

The aforementioned studies suggest that two mechanisms may control gene expression during artificial allohexaploidization in SHW, "suppressor genes" on the alternate parent (Assefa and Fehrmann 2000) or silencing of some genes upon polyploidization (Kashkush et al. 2002). Further research is needed to differentiate between these two mechanisms as the former may be overcome by selecting an alternative parental partner whilst the latter may be more stochastic and overcome merely by repeating the cross. Taken together, one is tempted to suggest that trait expression in *Ae. tauschii* is down regulated to a greater degree than that of tetraploid parents. Therefore a fuller understanding of trait expression in SHW is, therefore, critical to efforts to increase and efficiently utilize the genetic diversity in SHW for common wheat improvement.

III. IMPACT OF SYNTHETIC HEXAPLOID IN WHEAT IMPROVEMENT

Within the last two decades, numerous SHW lines and their derivatives have been produced and evaluated and useful genes/traits have been identified (Mujeeb-Kazi et al. 1996; Ogbonnaya et al. 2005, 2008a; van Ginkel and Ogbonnaya 2007; Trethowan and Mujeeb-Kazi 2008). These include disease resistance genes and traits that contribute to increased yield both under rainfed-drought or moisture limiting conditions and optimal yield expression and the underlying yield components such as large kernel size, high biomass, high above ground biomass, and a large root system that captures water from deep soil zones, tolerance to salinity, waterlogging, and sodicity. Additional traits include improved quality and micronutrient contents such as Fe and Zn.

It was reported that by the year 2003–2004, 26% of all new advanced lines made available through CIMMYT's international screening nurseries to cooperators for either irrigated or semiarid conditions were synthetic derivatives (Gill et al. 2006). However, in the most recent survey carried out as part of this review using data from 7 years, on average 17% of all entries were synthetic derived. However, the percentage varied among nurseries with 9% for Elite Spring Wheat Yield Trial (ESWYT), 11% for International Bread Wheat Screening Nursery (IBWSN), 13% for Semiarid Screening Nursery (SAWSN), and 35% for Semiarid Wheat Yield Trial (SAWYT).

Similarly, using 6 years of data from ICARDA's Spring Bread Wheat Observation Nursery, "SBW-ON," 2007–2012, revealed that an average of 17% of all entries distributed were synthetic derivatives and the percentage between nurseries ranged from 5% in SBW-ON-2007 to 24% in SBW-ON-2012. Lage and Trethowan (2008) in an earlier study using SAWYT 5-15, examined the proportion of SHW derived lines in each SAWYT, as well as their coefficient of parentage (COP). They reported that the first SHW-derived lines appeared in the 5th SAWYT in 1997. Eight percent of the lines were SHW-derived increasing to 46% by the 15th SAWYT. The average of COP of SHW in the SHW-derived lines in the 5th SAWYT was as high as 75% but decreased steadily to an average of 19% by the 15th SAWYT. Several groups around the world have or are currently producing publically available synthetic wheat germplasm; these include CIMMYT-Mexico, ICARDA-Syria, CSIRO-Canberra, University of Sydney-Australia, NIAB-United Kingdom, and USDA-ARS-United States of America. Recently, a population of 215 doubled haploid and recombinant inbred line populations consisting of 1,700 F_6 derived lines from the cross Synthetic W7984 × Opata M85 were produced at Cornell University, University of Missouri and Agriculture and Agri-Food Canada (Sorrells et al. 2011). In recent years, a number of initiatives were established to more systematically explore the utilization of SHW to increase wheat productivity. For example, the Australian Grain Research and Development Corporation (GRDC) funded the Synthetic Evaluation Project from 2001 to 2007—a multiinstitutional project to improve the productivity and sustainability of rainfed wheat production in the Mediterranean environments of Australia. The project aimed to identify sources of disease resistance and abiotic stress tolerance in SHW and to incorporate them into Australian adapted germplasm for yield and drought. This culminated in the organization of the 1st Synthetic Wheat Symposium at the Department of Primary Industries, Horsham, Victoria from September 4–6, 2006. Some 85 delegates, representing 23 research groups from four continents, attended the meeting. The theme of the symposium was "Synthetics for Wheat Improvement," and served as a global forum for current research on the use of synthetics for wheat improvement Imtiaz et al. 2006.

NIAB has been systematically introducing genetic variation into UK wheat from a collection of 440 CIMMYT SHW since 2006 as part of a 4-year program funded by the UK government and private industry. More than 5,000 BC_1 derived F_5 lines in two elite spring (Paragon) and facultative (Xi-19) wheat varieties have been produced. The approach operated on two levels, (1) production of nonselected backcross lines for mapping and gene discovery, (2) development of precommercial germplasm through field selection for introduction into elite UK wheat germplasm by commercial plant breeding companies. In 2011, NIAB received funding from the UK Government for the production and exploitation of new synthetic wheat germplasm as part of a program to "enhance diversity in UK wheat through a public sector prebreeding program." To avoid duplication of effort, at the time of writing, a gap analysis is underway to identify pools of *Ae. tauschii* diversity that are not present or underrepresented in the current CIMMYT and other SHW germplasm.

In 2010, the Mexican government funded the "Seeds of Discovery project" at CIMMYT, part of which includes the exploitation of SHW for improving yield potential, drought and heat tolerance. As a component of the project, CIMMYT and ICARDA are evaluating more than 6,000 SBL for yield potential in the 2012 crop growing season while 1,000 SHW lines are being evaluated for drought and heat tolerance at ICARDA within the same period.

The first documented release of a wheat cultivar derived from SBL by backcrossing SHW obtained from CIMMYT with locally adapted Chinese cultivar was reported in 2006. The cultivar, "Chuanmai 42" which now grows in more than 100,000 ha reportedly out-yielded all commercial varieties over 2 years in Sichuan provincial yield trials, including the commercial check cultivar Chuanmai 107 by 22.7% (Yang et al. 2006, 2009). Similarly, two lines have been released in Uruguay by INIA named 'Genesis 2354' and 'Genesis 2359'. A third variety designated 'NOGAL' was reported to be released in Uruguay and Argentine by a private company. In the United States, one of the most successful uses of SHW in wheat breeding was the deployment of *Gb3* for resistance to greenbug (*Schizaphis graminum* Rondani) in wheat cultivars. *Gb3* is derived from *Ae. tauschii* chromosome 7D and was originally identified in Langdon-derived SHW line Largo (Joppa et al. 1980; Joppa and Williams 1982). It has been deployed in wheat cultivars 'TAM 110' (Lazar et al. 1997) and 'TAM 112' (PI 643143) and provides effective protection from greenbug in the southern Great Plains of the United States (Lu et al. 2010).

Table 2.1 summaries details of some publicly available information on germplasm produced and/or released with useful traits/genes identified and incorporated into common wheat from direct crosses involving *Ae. tauschii*, tetraploids, and SHWs.

Table 2.1. Genes identified and transferred from various sources into bread wheat and germplasm produced including traits whose genes are not yet characterized and localized.

Germplasm	ID	Date of release	Description	Gene(s)	References
			(a) From *Ae. tauschii*		
KS85WGRC01	PI499691 TA5005	10/01/1985	Hessian fly-resistant, hard red winter wheat germplasm, also resistant to soilborne mosaic virus. Pedigree: TA1644 (*Ae. tauschii*)/Newton//Wichita	*H22* 1DL	Gill et al. (1986)
KS89WGRC03	PI535766 TA5013	08/03/1989	Hessian fly-resistant, hard red winter wheat germplasm. Pedigree: TA1642 (*Ae. tauschii*)/2*Wichita	*H23* 6DS	Gill et al. (1991b)
KS89WGRC04	PI535767 TA5014	08/03/1989	Hessian fly, greenbug, and soilborne mosaic virus-resistant, hard red winter wheat germplasm. Pedigree: TA1695 (*Ae. tauschii*)/3*Wichita		Gill et al. (1991c)
KS89WGRC06	PI535796 TA5016	08/03/1989	Hessian fly-resistant, hard red winter wheat germplasm. Pedigree: TA2452 (*Ae. tauschii*)/TA1642 (*Ae. tauschii*)//2*Wichita /3/ Newton	*H24* 3DL	Gill et al. (1991b)
KS89WGRC07	PI535770 TA5017	08/03/1989	Leaf rust-resistant, hard red winter wheat germplasm. Leaf rust resistance governed by a gene on chromosome 1DS. Pedigree: Wichita//TA1649 (*Ae. tauschii*)/ 2*Wichita	*Lr21* 1DS	Gill et al. (1991a)
KS90WGRC10	PI549278 TA5022	08/02/1990	Leaf rust-resistant, hard red winter wheat germplasm. Pedigree: TAM107*3/TA2460 (*Ae. tauschii*)	*Lr39* 2DS	Cox et al. (1992)

(continued)

Table 2.1 (*Continued*)

Germplasm	ID	Date of release	Description	Gene(s)	References
KS91WGRC11	PI566668 TA5024	11/22/91	Leaf rust-resistant, hard red winter wheat germplasm. Pedigree: Century*3/TA2450 (*Ae. tauschii*)	*Lr42* 1DS	Cox et al. (1994b)
KS91WGRC12	TA5025	11/22/91	Adult leaf rust-resistant, hard red winter wheat germplasm. Also segregating for resistance to wheat soilborne mosaic and wheat spindle streak mosaic viruses. Pedigree: Century*3/TA2541 (*Ae. tauschii*)		Cox et al. (1991b)
KS92WGRC15	PI566669 TA5028	08/06/1992	Leaf rust resistant, hard red winter wheat germplasm. Molecular analysis indicates that this line was derived from TA1649, similar to KS89WGRC07. Pedigree: TAM200/KS86WGRC02//Karl	*Lr40* 1DS	Cox et al. (1994b)
KS92WGRC16	PI592728 TA5029	08/06/1992	Leaf rust-resistant, hard red winter wheat germplasm. Pedigree: Triumph 64/3/KS8010-71/TA2470 (*Ae. tauschii*)//TAM200	*Lr21*, *Lr39* 1DS, 2DS	Cox et al. (1997a)
KS92WGRC21	PI566670 TA5034	08/06/1992	Hard red winter wheat germplasm with resistance to powdery mildew and wheat soilborne and spindle streak mosaic viruses. Pedigree: TAM200*3/TA2570 (*Ae. tauschii*)		Cox et al. (1994c)
KS92WGRC22	PI566671 TA5035	08/06/1992	Hard red winter wheat germplasm with resistance to powdery mildew and wheat soilborne and spindle streak mosaic viruses. Pedigree: Century*3/TA2567 (*Ae. tauschii*)		Cox et al. (1994c)

KS93WGRC26	PI572542 TA5039	08/04/1993	Hessian fly-resistant, hard red winter wheat germplasm. Resistance governed by gene $H26$ on chromosome 3D. Pedigree: Karl*3/TA2473 (*Ae. tauschii*)	$H26$ 3DL	Cox et al. (1994a)
KS95WGRC33	PI595379 TA5049	08/03/1995	Septoria leaf blotch-resistant, hard red winter wheat germplasm. See KS96WGRC40, a reselection of this germplasm. Pedigree: KS93U69 (sister line of KS90WGRC10)*3/TA2397 (*Ae. tauschii*)	$Lr41$ 1DS	Cox et al. (1996)
KS96WGRC39	PI604224 TA5055	08/09/1996	Tan spot-resistant, hard red winter wheat germplasm. Pedigree: TAM 107*3/TA2460 (*Ae. tauschii*)		Brown-Guedira et al. (1999)
KS96WGRC40	PI604225	1999	Hard red winter wheat germplasm with resistance to soilborne cereal mosaic virus TAM107*3/TA2460 (*Ae. tauschii*)//TA2397 (*Ae. tauschii*)/3/TAM107*3/TA2460	$SBWMV$ 5DL	Cox et al. (1999), Hall et al. (2009)
KS00WGRC44	TA5068	08/04/2000	Leaf rust-resistant hard red winter wheat germplasm. Resistance due to a single dominant gene. Pedigree: TAM 107*3/TA1715 (*Ae. tauschii*)		Brown-Guedira et al. (2004)
KS04WGRC49	TA5076	08/06/2004	Hard red winter wheat germplasm with unique high-molecular-weight glutenin and gliadin subunits. Pedigree: Karl 92*3/TA2473 (*Ae. tauschii*)		Brown-Guedira et al. (2005a)
NC97BGTD7	PI604033	1999	Wheat germplasm with resistance to powdery mildew. Saluda *3/ *Ae. tauschii* TA2492	$Pm34$ 5DL	Murphy et al. (1999), Miranda et al. (2006)

(*continued*)

Table 2.1 (Continued)

Germplasm	ID	Date of release	Description	Gene(s)	References
NC96BGTD3	PI603250	1998	Wheat germplasm with resistance to powdery mildew. Saluda*3/ Ae. tauschii TA2377	Pm35 5DL	Murphy et al. (1998). Miranda et al. (2007)
RL5713		1987	Wheat germplasm with resistance to leaf rust Thatcher*7//RL5713 (Ae. tauschii)/ MarquisK	Lr32 3DS	Kerber ER (1987)
RL5404		1970	Wheat germplasm with adult plant resistance to leaf rust	Lr22a	Dyck and Kerber et al. (1970)
XX194		1995	Wheat germplasm with resistance to powdery mildew Moroccos183/ AE457-78 (Ae. tausschii)	Pm2	Lutz et al. (1995)
(b) From synthetic hexaploids					
Largo	CI 17895	1982	Hexaploid synthetic wheat with resistance to greenbug. Pedigree: Langdon/Ae. tauschii PI 268210	Gb3 7DL	Joppa et al. (1982)
W7984		2001	Hexaploid synthetic wheat with resistance to greenbug Altar84/TA1651 (Ae. tauschii)	Gb7 7DL	Weng et al. (2005)
L-18913		1987	Hexaploid synthetic wheat with resistance to cereal cyst nematode. Pedigree: Langdon/Ae. tauschii AUS18913	Cre3 2DL	Eastwood et al. (1994), Lagudah et al. (1997)
L-18911		1987	Hexaploid synthetic wheat with high molecular weight glutenin subunits. Pedigree: Langdon/Ae. tauschii AUS18911	GluDx5Dy10 1DL	Lagudah et al. (1987), Mackie et al. (1996)

L-18913		1987	Hexaploid synthetic wheat with high molecular weight glutenin subunits. Pedigree: Langdon/*Ae. tauschii* AUS18913	*GluDx2DyT1*; *Sr46*. 1DL; 2DS	Lagudah et al. (1987), Mackie et al. (1996)
L-18964		1987	Hexaploid synthetic wheat with high molecular weight glutenin subunits. Pedigree: Langdon/*Ae. tauschii* AUS18964	*GluDx5Dy12* 1DL	Lagudah et al. (1987), Mackie et al. (1996)
L-21712		1987	Hexaploid synthetic wheat with high molecular weight glutenin subunits. Pedigree: Langdon/*Ae. tauschii* AUS21712	*GluDx2.1Dy10* 1DL	Lagudah et al. (1987), Mackie et al. (1996)
RL5405		1979	Hexaploid synthetic wheat with resistance to stem rust Tetra Canthatch/RL 5288 (*Ae. tauschii*)	*Sr33* 1DS	Kerber and Dyck (1979)
SW8	PI639730	2004	Hexaploid synthetic wheat with resistance to Hessian fly. Pedigree: Langdon/*Ae. tauschii* Clae 26	*H26* 3DL	Xu et al. (2006)
SW39	PI639732	2004	Hexaploid synthetic wheat with resistance to Hessian fly. Pedigree: Langdon/*Ae. tauschii* RL5561		Xu et al. (2006)
Molly	PI562619	1993	Isogenic line for Hessian fly resistance gene *H13* in winter wheat cultivar Newton. Pedigree: Newton*7/3/KU221-19 (*T. persicum* KU138/*Ae. tauschii* KU2076) Eagle//KS806	*H13* 6DL	Patterson et al. (1994)
CIGM86.940		1998	Hexaploid synthetic wheat with resistance to stripe rust	*Yr28* 4DS	Singh et al. (2000)
CIG87M66-2-1			Hexaploid synthetic wheat with resistance to stem rust RL5289 (*Ae. tauschii*)	*Sr45* 1DS	Marais et al. (1998)

(*continued*)

Table 2.1 (*Continued*)

Germplasm	ID	Date of release	Description	Gene(s)	References
Sear's synthetics					
	G 3489	1989	Hexaploid synthetic wheat with resistance to Septoria tritici blotch *T. dicoccoides*/37-1 (*Ae. tauschii*)	*Stb5* 7DS	Arraiano et al. (2001)
		1989	Hexaploid synthetic wheat with resistance to rootknot nematodes. Pedigree: Produra/*Ae. tauschii* G 3489	*Rkn*	Kaloshian et al. (1989)
		1990	Hexaploid synthetic wheat with resistance to Russian wheat aphid. Pedigree: SQ24 (*Ae. tauschii*)/TD65 (*T. turgidum*)	*Dn3*	Nkongolo et al. (1991)
TAM 110	PI595757	1996	Hard red winter wheat cultivar with resistance to greenbug. Pedigree: TAM 105*4//Amigo*5//Largo (Langdon/*Ae. tauschii* PI 268210)	*Gb3* 7DL	Lazar et al. (1997)
TXGBE272	PI591794	1996	Winter wheat germplasm with resistance to greenbug biotype E. Pedigree: TAM 105*4/Amigo*5//Largo (Langdon/*Ae. tauschii* PI 268210)	*Gb3* 7DL	Lazar et al. (1996)
TXGBE273	PI591795	1996	Winter wheat germplasm with resistance to greenbug biotype E. Pedigree: TAM 105*4/Amigo*5//Largo (Langdon/*Ae. tauschii* PI 268210)	*Gb3* 7DL	Lazar et al. (1996)

TXGBE285	PI591799	1996	Winter wheat germplasm with resistance to greenbug biotype E. Pedigree: TAM 105*4/Amigo*5//Largo (Langdon/Ae. tauschii PI 268210)	Gb3 7DL	Lazar et al. (1996)
TXGBE292	PI591800	1996	Winter wheat germplasm with resistance to greenbug biotype E. Pedigree: TAM 105*4/Amigo*5//Largo (Langdon/Ae. tauschii PI 268210)	Gb3 7DL	Lazar et al. (1996)
TXGH10563B	PI527481	26/02/1989	Winter wheat germplasm with resistance to greenbug biotypes B, C, and E. Pedigree: TAM 105*4/Amigo*5//Largo (Langdon/Ae. tauschii PI 268210)	Gb3 7DL	Porter et al. (1989)
TXGH10989	PI527482	26/02/1989	Winter wheat germplasm with resistance to greenbug biotypes B, C, and E. Pedigree: TAM W-101*4/Amigo*4//Largo (Langdon/Ae. tauschii PI 268210)	Gb3 7DL	Porter et al. (1989)
TXGH13622	PI527483	26/02/1989	Winter wheat germplasm with resistance to greenbug biotypes B, C, and E. Pedigree: TX71A562-6*4/Amigo*4//Largo,(Langdon/Ae. tauschii PI 268210)	Gb3 7DL	Porter et al. (1989)
TAM 112	PI643143	2004	Hard red winter wheat cultivar with resistance to greenbug biotype and leaf rust resistance from Ae. tauschii. Pedigree: TU1254-7-9-2-1 (TAM 200/Ae. tauschii TA2460/TXGH10440 (TAM 105*4/Amigo*5//Largo)		Rudd et al. (2004)

(continued)

Table 2.1 (Continued)

Germplasm	ID	Date of release	Description	Gene(s)	References
XX 186		1995	Hexaploid synthetic wheat with resistance to powdery mildew. Pedigree: Durum 'Santa Marta'/ Ae. tauschii BGRC 1458	Pm19 7D	Lutz et al. (1995)
WX-SYN.B-92-52		1996	Karnal bunt resistant (0% infection VS Check with 65% susceptibility), 6.1 t/ha, 61.3 1,000-grain weight. Pedigree: CIGM87.2768-1B-0Y-0M-0Y (SH12); (Altar 84/T. tauschii (Acc. 198))		Villareal et al. (1996)
WX-SYN.B-92-81		1996	Karnal bunt resistant, 4.2 t/ha, 56.7 1,000-grain weight. Pedigree: CIGM86.953-1B-0Y-0M-0Y (SH46); (Duergand/T. tauschii (Acc. 221))		Villareal et al. (1996)
WX-SYN.B-92-87		1996	Karnal bunt resistant, 5.4 t/ha, 58.6 1,000-grain weight. Pedigree: CIGM87.2762-1B-0Y-0M-0Y (SH10); (Altar 84/T. tauschii (Acc. 10))		Villareal et al. (1996)
WX-SYN.B-92-91		1996	Karnal bunt resistant, 4.5 t/ha, 64.6 1,000-grain weight. Pedigree: CIGM86.949-1B-0Y-0M-0Y (SH 31); (Chen "S"/T. tauschii (Acc. 31))		Villareal et al. (1996)
GP-562	PI610750	2000	Septoria leaf blotch resistant: 2-1 score VS Checks with 4-1, 8-7, & 8-9. Pedigree: CIGM90.248; (Croc/Ae. tauschii (205)//Kauz)		Mujeeb-Kazi et al. (2000b)

GP-563	PI610751	2000	Septoria leaf blotch resistant: 1-1 score. Pedigree: CIGM90.250.1; (Croc/*Ae. tauschii* (205)//Borlaug M95)	Mujeeb-Kazi et al. (2000a,b)
GP-564	PI610752	2000	Septoria leaf blotch resistant: 1-1 score. Pedigree: CIGM90.250.2; (Croc/*Ae. tauschii* (205)//Borlaug M95)	Mujeeb-Kazi et al. (2000a,b)
GP-565	PI610753	2000	Septoria leaf blotch resistant: 2-1 score. Pedigree: CIGM90.358; (Seri M82//Croc/*Ae. tauschii* (224))	Mujeeb-Kazi et al. (2000a,b)
GP-566	PI610754	2000	Septoria leaf blotch resistant: 2-1 score. Pedigree: CIGM90.412; (Croc/*Ae. tauschii* (213)//Papago M86)	Mujeeb-Kazi et al. (2000a,b)
GP-567	PI610755	2000	Septoria leaf blotch resistant: 2-1 score. Pedigree: CIGM90.483; (Altar 84/*Ae. tauschii* (191)//Opata M85)	Mujeeb-Kazi et al. (2000a,b)
GP-568	PI610756	2000	Septoria leaf blotch resistant: 1-1 score. Pedigree: CIGM91.153; (Yaco*2/Croc/*Ae. tauschii* (205)/3/Yaco)	Mujeeb-Kazi et al. (2000a,b)
GP-569	PI610757	2000	Septoria leaf blotch resistant: 2-1 score. Pedigree: CIGM91.191; (Altar/*Ae. tauschii* (224)//2* Yaco)	Mujeeb-Kazi et al. (2000a,b)
GP-570	PI610758	2000	Septoria leaf blotch resistant: 2-1 score. Pedigree: CIGM91.248; (Papago M86//Croc/*Ae. tauschii* (224)/3/2*Borlaug M95)	Mujeeb-Kazi et al. (2000a,b)
GP-571	PI610759	2000	Septoria leaf blotch resistant: 2-1 score. Pedigree: CIGM92.337; (Altar 84/*Ae. tauschii* (191)//Yaco/3/Bagula)	Mujeeb-Kazi et al. (2000a,b)

(*continued*)

Table 2.1 (*Continued*)

Germplasm	ID	Date of release	Description	Gene(s)	References
GP-639	PI 613278	2001	Waterlogging tolerant, 49.3 1,000-grain weight, 6.3% chlorosis VS Check with 76.8% chlorosis. Pedigree: CIGM86.953-SH 19; (Duergand/ *T. tauschii* (Acc. 221))		Villareal et al. (2001)
GP-640	PI613279	2001	Waterlogging tolerant, 44.7 1,000-grain weight, 6.7% chlorosis. Pedigree: CIGM90.863-SH 64; (Botno/ *Ae. tauschii* (Acc. 617)		Villareal et al. (2001)
GP-641	PI613280	2001	Waterlogging tolerant, 37.6 1,000-grain weight, 8.8% chlorosis. Pedigree: CIGM89.567-SH 54; (Ceta/ *Ae. tauschii* (Acc. 895))		Villareal et al. (2001)
GP-642	PI613281	2001	Waterlogging tolerant, 42.4 1,000-grain weight, 9.2% chlorosis. Pedigree: CIGM92.1723-SH 82; (68.111/ Rgb-U//Ward/3/*Ae. tauschii* (Acc. 454))		Villareal et al. (2001)
GP-695	PI613302	2001	Karnal bunt resistant: Immune VS Check with 30% susceptibility, 61.3 1,000-grain weight. Pedigree: CIGM93-183; (Ceta/*Ae. tauschii* (174))		Mujeeb-Kazi et al. (2001a)
GP-696	PI613302	2001	Karnal bunt resistant: Immune, 54.6 1,000-grain weight. Pedigree: CIGM87.2765; (Altar84/*Ae. tauschii* (188))		Mujeeb-Kazi et al. (2001a)
GP-697	PI613302	2001	Karnal bunt resistant: Immune, 61.3 1,000-grain weight. Pedigree: CIGM87.2767; (Altar84/*Ae. tauschii* (192))		Mujeeb-Kazi et al. (2001a)

GP-698	PI613302	2001	Karnal bunt resistant: Immune, 66.3 1,000-grain weight. Pedigree: CIGM90.561; (Yuk/*Ae. tauschii* (217))	Mujeeb-Kazi et al. (2001a)
GP-699	PI613302	2001	Karnal bunt resistant: Immune, 58.3 1,000-grain weight. Pedigree: CIGM88-1239; (Yav2/Tez// *Ae. tauschii* (249))	Mujeeb-Kazi et al. (2001a)
GP-700	PI613302	2001	Karnal bunt resistant: Immune, 64.6 1,000-grain weight. Pedigree: CIGM88.1344; (Doy/*Ae. tauschii* (447))	Mujeeb-Kazi et al. (2001a)
GP-701	PI613302	2001	Karnal bunt resistant: Immune, 65.5 1,000-grain weight. Pedigree: CIGM92-1727; (Doy/*Ae. tauschii* (458))	Mujeeb-Kazi et al. (2001a)
GP-702	PI613302	2001	Karnal bunt resistant: Immune, 64.9 1,000-grain weight. Pedigree: CIGM90.845; (Sca/*Ae. tauschii* (518))	Mujeeb-Kazi et al. (2001a)
GP-703	PI613302	2001	Karnal bunt resistant: Immune, 56.0 1,000-grain weight. Pedigree: CIGM90.846; (Yar/*Ae. tauschii* (518))	Mujeeb-Kazi et al. (2001a)
GP-704	PI613302	2001	Karnal bunt resistant: Immune, 63.7 1,000-grain weight. Pedigree: CIGM90.590; (68.111/Rgb-U// Ward/3/Fgo/4/Rabi/5/ *Ae. tauschii* (629))	Mujeeb-Kazi et al. (2001a)

(continued)

Table 2.1 (Continued)

Germplasm	ID	Date of release	Description	Gene(s)	References
GP-705	PI613302	2001	Karnal bunt resistant: I = 1.59%, 38.8 1,000-grain weight. Pedigree: CIGM90.257-1; (Croc/Ae. tauschii (205)//Flycatcher)		Mujeeb-Kazi et al. (2001a)
GP-706	PI613302	2001	Karnal bunt resistant: I = 0.69, 46.8 1,000-grain weight. Pedigree: CIGM91.61-1; (Croc/Ae. tauschii (224)//Kauz)		Mujeeb-Kazi et al. (2001a)
GP-707	PI613302	2001	Karnal bunt resistant: I = 0.95, 53.8 1,000-grain weight. Pedigree: CIGM90.462; (Altar84/Ae. tauschii (221)//Yaco)		Mujeeb-Kazi et al. (2001a)
GP-708	PI613302	2001	Karnal bunt resistant: I = 0.77, 42.0 1,000-grain weight. Pedigree: CIGM90.248.1; (Croc/Ae. tauschii (205)//Kauz)		Mujeeb-Kazi et al. (2001a)
GP-709	PI613302	2001	Karnal bunt resistant: I = 0.86, 51.8 1,000-grain weight. Pedigree: CIGM250.2; (Croc/Ae. tauschii (205)//Borlaug 95)		Mujeeb-Kazi et al. (2001a)
GP-710	PI613302	2001	Karnal bunt resistant: I = 1.97, 50.0 1,000-grain weight. Pedigree: CIGM990.412; (Croc/Ae. tauschii (213)//Papago M86)		Mujeeb-Kazi et al. (2001a)

GP-711	PI613318	2001	Spot blotch resistant: 2-2 score, 1 grain finish VS Check with 9-9 score & 4 grain finish. Pedigree: CASS97B00040S; (Gan/Ae. tauschii (236)//Doy1/Ae.tauschii (447))	Mujeeb-Kazi et al. (2001b)
GP-712	PI613319		Spot blotch resistant: 2-2 score, 2 grain finish. Pedigree: CASS97B00041S; (Gan/Ae. tauschii (236)//Ceta/Ae. tauschii (895))	Mujeeb-Kazi et al. (2001b)
GP-713	PI613320		Spot blotch resistant: 3-3 score, 1 grain finish. Pedigree: CASS97B00046S; (Scoop1/Ae. tauschii (434)//Ceta/Ae. tauschii (895))	Mujeeb-Kazi et al. (2001b)
GP-714	PI613321		Spot blotch resistant: 3-3 score, 2 grain finish. Pedigree: CASS97B00054S; (Doy1/Ae. tauschii (447)//Ceta/Ae. tauschii (895))	Mujeeb-Kazi et al. (2001b)
GP-715	PI613322		Spot blotch resistant: 3-3 score, 1 grain finish. Pedigree: CASS97B00063S; (68.111/Rgb-U//Ward/3/Fgo/4/Ae. tauschii (629)/5/Ceta/Ae. tauschii (895))	Mujeeb-Kazi et al. (2001b)
GP-716	PI613323		Spot blotch resistant: 3-2 score, 1 grain finish. Pedigree: CIGM990.1291; (Altar/Ae. tauschii 9224)//2*Yaco)	Mujeeb-Kazi et al. (2001b)
GP-717	PI613324		Spot blotch resistant: 2-2 score, 2 grain finish. Pedigree: CASS97B0024S-3DH; (Sabuf//Altar/Ae. tauschii 9224)/3/Yaco/Croc1/Ae. tauschii (205))	Mujeeb-Kazi et al. (2001b)
GP-718	PI613325		Spot blotch resistant: 2-2 score, 2 grain finish. Pedigree: CASS97B00121S; (Bcn//Sora/Ae. tauschii (323))	Mujeeb-Kazi et al. (2001b)

(continued)

Table 2.1 (Continued)

Germplasm	ID	Date of release	Description	Gene(s)	References
GP-719	PI613326		Spot blotch resistant: 3-3 score, 2 grain finish. Pedigree: CASS97B001-1DH; (Opata//Sora//Ae. tauschii (323)		Mujeeb-Kazi et al. (2001b)
GP-720	PI613327		Spot blotch resistant: 2-1 score, 1 grain finish. Pedigree: CASS97B0010-1DHS; Bcn/4/68.111/Rgb-U//Ward/3/Ae. tauschii (325)		Mujeeb-Kazi et al. (2001b)
GP-721	PI613328		Spot blotch resistant: 3-2 score, 1 grain finish. Pedigree: CASS97B00030S-3DH; Bcn//Doy/Ae. tauschii (447)		Mujeeb-Kazi et al. (2001b)
GP-722	PI613329		Spot blotch resistant: 3-2 score, 1 grain finish. Pedigree: CASS94Y00160S; Bcn/4/Rabi//GS/Cra/3/Ae. tauschii (899)		Mujeeb-Kazi et al. (2001b)
			Tan spot, Stagonospora nodorum blotch, Fusarium head blight, leaf rust, plant height, heading time. Pedigree: TA4152-60 (Scoop 1/Ae. tauschii WPI 358)/ND495		Chu et al. (2008a,b; 2009; 2010)
KS99WGRC42	PI635054	2005	Hessian fly resistant hard red winter wheat germplasm. Pedigree: 'Karl 92'/T. turgidum subsp. dicoccum PI 94641//'Jagger'*2/Karl 92.	Hdic 1AS	Brown-Guedira et al. (2005b)
Hope		1930	Stem rust resistant hard red spring wheat germplasm. Pedigree: Yaroslav emmer (T. turgidum subsp. dicoccum)/Marquis	Sr2 3BS	McFadden (1930)

Name	Year	Description	Gene/Locus	Reference
Am3	1992	Hexaploid synthetic wheat with resistance to powdery mildew. Pedigree: *T. turgidum* subsp. *carthlicum* PS5/*Ae. tauschii* Ae38	PmPS5A; PmPS5B 2AL; 2BL	Xu and Dong (1992), Zhou et al. (2005)
K733-18911	1999	Hexaploid synthetic wheat with stripe rust resistance Pedigree: *T. turgidum* subsp. *turgidum*/*Ae. tauschii* AUS18911	*Yr24*; 1BS	McIntosh and Lagudah (2000)
GB4	2007	Winter wheat germplasm with resistance to powdery mildew. Pedigree: 'Laizhou 939'/Am3 (*T. turgidum* subsp. *carthlicum* PS5/*Ae. tauschii* Ae38)	PmPS5A; PmPS5B 2AL; 2BL	Zhou et al. (2007)
Am4	1992	Hexaploid synthetic wheat with resistance to powdery mildew. Pedigree: *T. turgidum* subsp. *carthlicum* PS5/*Ae. tauschii* Ae39	Pm33 2BL	Xu and Dong (1992), Zhu et al. (2005)
PI638740	2005	Wheat germplasm with high grain protein content and resistance to stripe rust. Yecora Rojo (common wheat)*5/*T. turgidum* sp *dicoccoides* FA15-3	*GPC*; *Yr36*. 6BS	Khan et al. (2000), Uauy et al. (2005)
8K118	2009	Wheat germplasm with resistance to powdery mildew. 87-1 (Common wheat)*4//Langdon (durum wheat)/*T. turgidum* subsp. *dicoccoides* IW2	Pm41 3BL	Li et al. (2009)
P63	2009	Wheat germplasm with resistance to powdery mildew. Yanda 1817/*T. turgidum* subsp. *dicoccoides* G303-1M//3'Jing 411	Pm42 2BS	Hua et al. (2009)

A. Disease and Pests Resistance

Significant genetic diversity for a wide range of biotic stresses has been reported in SHW, direct crosses involving *Ae. tauschii* and tetraploid and their derived wheat lines. These include resistance to Hessian fly (Hatchett et al. 1981; Gill et al. 1991b,c; Cox et al. 1994a; Wang et al. 2006; Friesen et al. 2008; Yu et al. 2009, 2010, 2012), greenbug (Joppa et al. 1980; Joppa and Williams 1982; Martin et al. 1982; Gill et al. 1991c; Weng et al. 2005), Karnal bunt (Villareal et al. 1996; Mujeeb-Kazi et al. 2001c, 2008), powdery mildew (Lutz et al. 1994; Hu et al. 2001; Hu and Xin 2001), stem rust (Marais et al. 1994), stripe rust (Ma et al. 1995; Yang et al. 2001; Badebo and Ferhmann, 2005; Ogbonnaya et al. 2008a), leaf rust (Cox et al. 1997a; Assefa and Fehrmann 2000), Septoria nodorum blotch (Loughman et al. 2001), Septoria tritici leaf blotch (STB) (Arraiano et al. 2001), spot blotch (Mujeeb-Kazi et al. 2001a), Fusarium head blight (Mujeeb-Kazi et al. 2001b), tan spot (a.k.a. yellow leaf spot) (Siedler et al. 1994; Xu et al. 2004; Tadesse et al. 2007), cereal cyst nematode (CCN) (*Heterodera avenae* Woll.) (Eastwood et al. 1991) and root lesion nematodes (RLN) (*Pratylenchus thornei* and *Pratylenchus neglectus*) (Thompson et al. 1999; Zwart et al. 2004; Thompson 2008). Thompson (2008) reported the identification of SHWs with varying levels of resistance to *P. thornei*. Of the resistant SHW, with both durum and *Ae. tauschii* parents resistant, six SHW were based on *Ae. tauschii* accessions from subsepecies *typica*, seven from *meyeri*, five from *strangulata* and one from an intermediate *meyeri/typica* accession. There have also been reports of SHW possessing resistance to both *P. neglectus* and *P. thornei* (Zwart et al. 2005, 2006, 2010). Recently, SHWs possessing resistance to several different diseases were identified (see Xu et al. 2004; van Ginkel and Ogbonnaya 2007; Ogbonnaya et al. 2008a; Thompson 2008; Chu et al. 2008a, 2009, 2010; Zwart et al. 2010).

Following the identification of CCN resistance genes in *Ae. tauschii* including the *Cre3* gene, the resistant *Ae. tauschii* accessions were used to produce SHW and subsequently backcrossed with Australian cultivars. The CCN resistant lines carrying the *Cre3* gene from *Ae. tauschii* were developed and advanced to the final stages of evaluation in the Horsham, Victoria based wheat breeding program. The *Cre3* gene is now widely dispersed in wheat germplasm in Australia and CIMMYT, Mexico as well as other countries. A diagnostic PCR based SCAR marker and an SSR marker for selecting *Cre3*, were developed (Ogbonnaya et al. 2001; Martin et al. 2004). Wheat germplasm with an apparent yield advantage over the recurrent parent were developed from the CCN resistant *Ae. tauschii* material. This yield advantage was expressed mostly in environments with high terminal moisture stress and low yield levels. The SBL germplasm continues to be utilized in the Australian wheat breeding programs although no varieties have yet been released from these lines (Eastwood et al. 2006).

Ogbonnaya (2011) summarized the result from the evaluation of 384 SHWs for various diseases and pests including crown rot. The percentage of SHW that displayed resistant reactions varied from 1% for *S. nodorum* glume blotch to 82% for STB. The resistance in SHW for crown rot was significantly better than what exists in current common wheat cultivars.

In the work recently carried out at ICARDA, 914 SHWs were evaluated for resistance to Ug99 stem rust in Ethiopia and to Russian Wheat aphid, Hessian fly and Sunn pest in Syria. A total of 10% of the SHW had adult plant resistance to stem rust indicating the presence of potentially new race-nonspecific genes to be exploited for improving stem rust resistance in wheat. A total of 1%, 2%, and 15% were resistant to Russian wheat aphid, Hessian fly, and Sunn pest, respectively (El Bouhssini et al. 2012). However, the key questions to be answered include, are these genes different from those previously reported in cultivated wheat and what is the genetic basis of their resistance?

Most recently, Yu et al. (2012) evaluated 118 elite CMMYT SHWs and their durum wheat parents for resistance to Hessian fly and identified 52 SHW with high or moderate resistance. Because all the durum parents were susceptible, the resistance genes in these SHW should be contributed by *Ae. tauschii* D genome. They haplotyped the 52 SHW using eight PCR-based markers closely-linked to five resistance genes (*H13*, *H22*, *H23*, *H26*, and *H32*) previously identified in *Ae. tauschii*. The marker data showed that 32 SHW had the same haplotypes as the wheat lines containing the five known resistance genes, but 19 SHW had different haplotypes, suggesting that these lines may contain new genes for resistance to Hessian fly. Thus, haplotype analysis on the uncharacterized SHW using the markers closely linked to the known genes might provide preliminary information on the resistance genes in SHW. As in previous studies, some of the SHW possessed multiple disease and insect pest resistances and thus, provide a valuable resource for identification of resistance genes using SHW-derived mapping populations Table 2.2. The identification of the genetic and molecular basis of major resistance determinants to different pathotypes will aid in the selection of favorable alleles during cultivar development. In parallel with the use of specific pathogen virulence races, SHWs contribute to a diversified wheat gene pool to assist in studies on mechanisms for multiple host pathogen-interactions and ultimately to obtain durable resistance.

Table 2.2. Some mapping populations produced with synthetic hexaploid wheat.

Pedigree	Population type	Number of lines	Number of markers mapped	Traits known to segregate	References
W-7984 (Altar 84/Ae. tauschii WPI 219)/Opata 85	Recombinant inbred	150	>2000 (using 114 RI lines)	Numerous	Börner et al. (2002); Huang et al. (2003); Zwart et al. (2006)
TA4152–19 (Dverd/Ae. tauschii WPI 221)/ND495	Doubled haploid	170		Tan spot, *Stagonospora nodorum* blotch, *Septoria tritici* blotch, Hessian fly	Gu et al. (2010)
TA4152–37 (68.111/RGB-U// WARD/3/Ae. tauschii WPI 629)/ND495	Doubled haploid	120		Tan spot, *Stagonospora nodorum* blotch, *Septoria tritici* blotch, Hessian fly	Gu et al. (2010)
TA4152–60 (Scoop 1/Ae. tauschii WPI 358)/ND495	Doubled haploid	213	410 SSRs + 218 TRAPs + 1 RFLP + 3 phenotypic markers	Tan spot, *Stagonospora nodorum* blotch, Fusarium head blight, leaf rust, plant height, heading time	Chu et al. (2008b)

Cross	Population type	Size	Markers	Trait	Reference
ND495/Largo (Langdon/ Ae. tauschii PI 268210)	Recombinant inbred	236		Tan spot, Stagonospora nodorum blotch, Septoria tritici blotch, greenbug	Gu et al. (2010)
Largo (Langdon/Ae. tauschii PI 268210)/TAM 107	Recombinant inbred	130	20 SSRs linked to Gb3	Gb3 locus for greenbug resistance	Weng et al. (2005)
Kulm/W-7976 (Cando/R143// Mexi'S'/3/Ae. tauschii C122)	Recombinant inbred	103	349 SSRs	Septoria tritici blotch	Ghaffary et al. (2012)
SW2-ICA//Cham-6 (SW2-ICA = Haurani/Ae. tauschii ig 47259)	BCRIL	125	>2000 (DArT + 90 SSRs)	Drought tolerance	Valkoun (2001); Ogbonnaya et al., unpub.
SW2-ICA//Cham-6 (SW2-ICA = Haurani/Ae. tauschii ig 47259) [Syn-10]	BC$_2$F$_6$	136	93 SSRs + 133 AFLPs	Drought, root water-uptake ability	Inagaki et al., unpub.
SW3-ICA//Cham-6 (SW3-ICA = Jennah Khetifa/Ae. tauschii ig 48677)	BC$_2$F$_3$	131		Salinity tolerance, root penetration ability	Inagaki et al., unpub.
SW4-ICA//Cham-6 (SW4-ICA = Jennah Khetifa/Ae. tauschii ig 47259)	BC$_2$F$_3$	136		Drought tolerance, root penetration ability	Inagaki et al., unpub.
Syn10/Syn8 (Syn8, 10, & 15 = SW2-ICA//Cham-6)	BC$_2$F$_6$	218		Drought tolerance, root water-uptake ability	Inagaki et al., unpub.
Syn10/Syn15	BC$_2$F$_6$	138		Drought tolerance, root water-uptake ability	Inagaki et al., unpub.
SYN36//Janz (SYN36 = Altar/ Ae. tauschii Aus18905)	BCRIL	120	150 SSRs	Dormancy and pre-harvest sprouting tolerance	Gatford et al. (2002); Ogbonnaya et al. (2006, 2007a)
SYN37//Janz (SYN37 = Altar/ Ae. tauschii Aus18836)	BCRIL	250	>2000 (DArT + 180 SSRs)	Dormancy and pre-harvest sprouting tolerance	Gatford et al. (2002); Ogbonnaya et al. (2007a); Imtiaz et al. (2008)

(continued)

Table 2.2 (*Continued*)

Pedigree	Population type	Number of lines	Number of markers mapped	Traits known to segregate	References
Aus29639//Yitpi (Aus29639 = Ceta/ Ae. squarrosa (230))	BCRIL	85; 250		Salinity tolerance	Ogbonnaya et al. (2008b)
Aus29639//Annuello (Aus29639 = Ceta/Ae. squarrosa (230))	BCRIL	125	778 DArT + 63 SSRs	Salinity tolerance	Ogbonnaya et al., unpub.
Synthetic W7984/Opata M85	DH	215			Sorrells et al. (2011)
Synthetic W7984/Opata M85	RIL	2000			Sorrells et al. (2011)
Mayoor/TKSN1081//Ae. tauschii (222)/3/FCT	DH	171		Head scab	Trethowan and Mujeeb-Kazi (2008)
Sabuf/3/BCN//Ceta/Ae. tauschii (895)/4/FCT	DH	125		Head scab	Trethowan and Mujeeb-Kazi (2008)
Sabuf/3/BCN//Ceta/Ae. tauschii (895)/4/CNO	DH	102		Head scab	Trethowan and Mujeeb-Kazi (2008)
Sabuf/3/BCN//Ceta/Ae. tauschii (895)/4/OPATA	DH	102		Head scab	Trethowan and Mujeeb-Kazi (2008)
Turaco/5/Chirya3/4/ Siren//Altar/Ae. tauschii (205)/3/ 3* Buc /6/CNO	DH	90		Head scab	Trethowan and Mujeeb-Kazi (2008)
Turaco/5/Chirya3/4/Siren// Altar/Ae. tauschii (205)/3/3* Buc /6/Opata	DH	126		Head scab	Trethowan and Mujeeb-Kazi (2008)

Cross	Generation	Markers	Trait	Reference
CPI/GEDIZ/3/GOO/JO69/CRA/4/Ae. tauschii (208)/5/Opata	DH		188 Drought	Trethowan and Mujeeb-Kazi (2008)
YAV_3/SCO//JO69/CRA/3/YAV79/4/Ae. tauschii (498)/5/Opata	DH		125 Drought	Trethowan and Mujeeb-Kazi (2008)
D67.2/66.270//Ae. tauschii (257) 3/Opata	DH		158 Drought	Trethowan and Mujeeb-Kazi (2008)
GAN/Ae. tauschii (897)//Opata	DH		153 Drought	Trethowan and Mujeeb-Kazi (2008)
DOY 1//Ae. tauschii (458)//Opata	DH		113 Drought	Trethowan and Mujeeb-Kazi (2008)
XX86/Flair (XX86 = T. dicoccum (KU124) * Ae. tauschii 2047)	BC_2F_3	197 SSRs	111	Huang et al. (2004)
TA 4152-4/Karl 92 (PI-564245); (TA4152-4 = Altar 84/Ae. tauschii WX193)	$BC_2F_{2:4}$	151 SSRs	190	Narasimhamoorthy et al. (2006)
Laizhou953/Am3/4*Laizhou953 (Am3 = T. carthlicum PS5/Ae. tauschii Ae38)	BC_4F_3	205 SSRs	97	Liu et al. (2006)
B22 (Batis/Syn022)	BC_2F_3	149 SSRs	250 Milling and baking quality; leaf rust	Kunert et al. (2007); Naz et al. (2008)
Z86 (Zentos/Syn086)	BC_2F_3	149 SSRs	150 Milling and baking quality	Kunert et al. (2007)

(continued)

Table 2.2 (*Continued*)

Pedigree	Population type	Number of lines	Number of markers mapped	Traits known to segregate	References
UC1110/PI610750: UC1110 (pedigree Chukar///Yding// Bluebird/Chanate); PI610750 [CIMMYT numberCYG90.248.1, pedigree Croc1/Ae. tauschii (Synthetic205)//Kauz]	RIL	186	1494 polymorphic probes (SSRs, DArTs, and ESTs)	Yellow rust resistance	Lowe et al. (2011)
Sokoll/Krichauff; Sokoll = Pastor/3/Altar 84/ Ae. squarrosa (Taus)//Opata	DH	150	860 DArT and 111 SSRs	Abiotic stresses, root lesion nematode (*P. thornei*) resistance	Wallwork, Oldach, Linsell et al., unpub.
AUS29529/ HW689DHG48	DH	200	DArT	Crown rot resistance	Mather et al. (2012); Wallwork and Butt (2012)
AUS29529 = SABUF/7/ ALTAR84/AE.SQUARROSA(224)// YACO/6/ CROC_1/AE.SQUARROSA(205)/ 5/BR12*3/4/IAS55*4/CI14123/3/ AS55*4/EG,AUS/IAS55*4/ALD CPI33814/Janz	DH	110		Crown rot resistance	Herde et al. (2012)

CPI133872/Janz	DH	111	DArT and SSRs	Multiple disease resistance: *Septoria tritici* blotch, yellow leaf spot, stripe rust, leaf rust and stem rust resistance to root-lesion nematodes (*P. thornei* and *P. neglectus*)	Zwart et al. (2005, 2010)
BTSchomburgk/AUS33384 AUS33384 = GARZA/BOY// AE.SQUARROSA (427)	DH	194	90K SNP array	*Stagonospora nodorum* blotch and yellow spot	M. Shankar et al., unpub.
Young/AUS33414 AUS33414 = GAN/ AE.SQUARROSA (413)	DH	221	90K SNP array	*Stagonospora nodorum* blotch	M. Shankar et al., unpub.

Recently, association mapping was used to investigate the genetic architecture of resistances to CCN, and root lesion nematodes—*P. neglectus* (PN) and *P. thornei* (PT). Clusters of disease resistance QTL were identified at multiple genomic locations. There were coincident QTL for CCN and PN on chromosome 1A, 1D, 2D, 3D, 5D, 6B, and 7B; CCN and PT on chromosomes 4B and 6A; PN and PT on chromosomes 3B, 6B, and 7A. Of significance is the identification of novel QTL that are involved in the control of root diseases in wheat (Mulki et al. 2012). The loci with combined resistance to multiple diseases such as CCN + PN, CCN + PT, or PN + PT, (Fig. 2.6) constitute valuable resources for the transfer of multiple disease resistance into locally adapted wheat cultivars using marker-assisted selection (MAS).

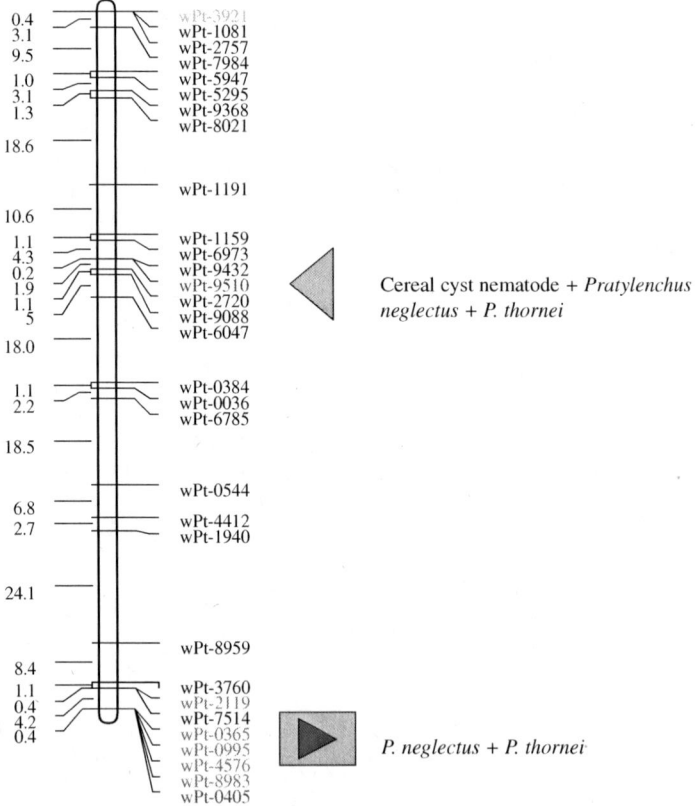

Fig. 2.6. Loci on chromosome 3B that confer multiple root pathogen resistance in synthetic hexaploid.

Due to their wider genetic variation relative to common wheat, SHW are excellent parental materials for developing mapping populations such as recombinant inbred lines and doubled haploids (Xu et al. 2004; Chu et al. 2008b). The "International Triticeae Mapping Initiative" (ITMI) mapping population was developed from the cross between the CIMMYT SHW line W-7984 and the hard red spring wheat (HRSW) 'Opata 85' and it has been extensively used for mapping studies (Nelson et al. 1995; Xu et al. 2004). A consensus genetic map consisting of over 2400 molecular markers has been constructed based on the ITMI population (see Xue et al. 2008). A number of genes or QTLs for resistance to leaf rust, stripe rust, Karnal bunt, Hessian fly, and greenbug have been mapped using the ITMI population (Nelson et al. 1997, 1998; Chhuneja et al. 2006; Sardesai et al. 2005; Weng et al. 2005). Chu et al. (2008b) recently developed a doubled haploid population from the cross between CIMMYT SHW line TA4152-60 (Scoop 1/*Ae. tauschii* WPI 358) with resistance to multiple diseases and a hard red spring wheat line ND495 and used the population to construct a linkage map consisting of 632 markers. Several novel QTLs for resistance to tan spot (Chu et al. 2008a), leaf rust (Chu et al. 2009), and Stagonospora nodorum blotch (Chu et al. 2010) have been identified using this population and the linkage map. The population is currently being used for mapping the genes and QTLs for resistance to Hessian fly, stripe rust, and Fusarium head blight (S.S. Xu, unpublished). Zwart et al. (2010) documented the location of coincident QTLs for both species of RLN on chromosomes 6DS and 2BS in a cross between SHW line CPI133872 and the bread wheat cultivar, 'Janz'.

Naz et al. (2008) developed a 250 BC_2F_3 AB-QTL population derived from a cross between a winter wheat cultivar Batis and SHW designated "Syn 022L" (*T. turgidum* ssp. *dicoccoides* × *Ae. tauschii*) to localize exotic QTL alleles for leaf rust resistance. The QTL analysis revealed six putative QTLs for seedling resistance and seven for adult plant resistance. For field resistance, two loci had stable main effects across environments and five loci exhibited marker by environment interaction effects. The strongest effects were detected at marker locus *Xbarc*149 on chromosome 1D, at which the exotic allele decreased seedling symptoms by 46.3% and field symptoms by 43.6%. Some of the detected QTLs co-localized with known resistance genes, while others appear to be novel resistance loci indicating, that the exotic SHW, Syn 022L may be useful for the improvement of leaf rust resistance in cultivated wheat.

Similarly, a synthetic backcross F_1-derived doubled haploid population between SHW line, CPI133872 and the common wheat cultivar, Janz, was phenotyped for seven disease resistance traits, namely: STB, yellow leaf spot (YLS) also known as tan spot (*Pyrenophora tritici-repentis*), stripe rust, leaf rust, stem rust, and two RLN species in multiple years and/or locations. Zwart et al. (2010) used a multiple-QTL model to identify a tightly linked cluster of foliar disease resistance QTL on chromosome 3DL. Major QTL for resistance to STB and YLS were contributed by the SHW parent CPI133872 and linked in repulsion with the coincident *Lr24/Sr24* locus carried by parent Janz. Additional QTL for YLS were detected in 5AS and 5BL. Consistent QTL for stripe rust resistance were identified on chromosomes 1BL, 4BL and 7DS, with the QTL in 7DS corresponding to the *Yr18/Lr34* region. Three major QTL for *P. thornei* resistance (2BS, 6DS, and 6DL) and two for *P. neglectus* resistance (2BS, 6DS) were detected. They also reported the identification of recombinants combining resistance to STB, YLS, rust diseases and RLN from parents CPI133872 and Janz that constitute valuable germplasm for the transfer of multiple disease resistances into new wheat cultivars.

B. Enhancing Yield Productivity

Wheat is grown under irrigated and rain-fed conditions worldwide and increasing yield is a major thrust of most wheat improvement programs. Arguably, increases in grain yields of common wheat over the last several decades came in part from the use of gibberellic acid-sensitive dwarfing genes (*Rht1* and *Rht2*), which were widely distributed during the "green revolution" (Smale et al. 2002). In addition to reducing plant height, these genes contributed to the repartitioning of photosynthates from vegetative to reproductive tissues. The increased grain yield results from the production of more grains per square meter.

Given the importance of wheat as a food crop, continued genetic gains in yield are necessary, given that there will be a need to provide food for an approximately double the current world population by 2050, with less available resources such as land, oil and fertilizer. To achieve this, the rate of genetic gain in breeding programs can increase by either extending the amount or nature of variation available for selection, or by accelerating the selection process to produce varieties more rapidly (Langridge and Fleury 2011).

Cox et al. (1995) demonstrated the possibility of yield increases in hexaploid wheat from *Ae. tauschii* backcrossed derived populations. The yield increases in some of the progenies were a result of changes in yield components and in particular, increases in 1,000-grain weight relative to the recurrent parent. Similar observations were made in

SBL where higher yielding lines were associated with increased rates of grain filling and higher 1,000-grain weight, especially in lower-yielding environments with drought stress (del Blanco et al. 2000; Gororo et al. 2002).

Similarly, yield advantages of SBLs over elite cultivars of bread wheat have been reported to be as high as 30% in northern Australia and 11% in southern Australia (Dreccer et al. 2007; Ogbonnaya et al. 2007a). In a recent study conducted in northern Australia, the yield advantage of SBLs over Australian bread wheat cultivars was greatest (up to 8%) in low-yielding environments (Rattey and Shorter 2010). Earlier, Gororo et al. (2002) evaluated a set of SBLs in low and high yielding environments in southern Australia. In the high yielding environments, grain yields of the SBLs were similar to those of their Australian recurrent common wheat parent, while in the lower yielding environments, the SBLs produced significantly higher grain yields, up to 149% of their Australian recurrent wheat parent (Gororo et al. 2002). Talbot (2011) investigated the potential of SHWs to improve grain yield in common wheat. Grain yield and its major components were measured in 27 families of SBL grown in five drought-stressed environments in southern Australia. Fourteen of these families, each derived from a different primary synthetic wheat backcrossed to the Australian bread wheat cultivar Yitpi, yielded significantly ($p < 0.05$) more grain on average than their recurrent parent. In higher-yielding environments, grain yield advantages were mostly associated with increases in grain weight, while in lower-yielding environments; grain yield advantages were mostly associated with increases in the number of grains per square meter. For example, in the lowest yielding environments (Minnipa, 276.5 mm; 16 years average rainfall), increases in the number of grains per 320 m^2 (up to 154.3% of that of Yitpi) were attributable to superior grain yields, whereas in the highest yielding environments (Roseworthy, 392.5; 16 years average rainfall), improvements to grain weight (up to 112% of that of Yitpi) were responsible."

Christopher et al. (2006) examined the physiological attributes associated with SBL, that exhibited high yield in CIMMYT trials when grown in a trial conducted at Kingsthorpe, southern Queensland, Australia. They reported that the SBL exhibited a stay-green phenotype and higher yields compared to standard northern cultivars. The yield of SBL examined was 55%, 31%, and 15% higher than the local adapted cultivars—Hartog, Banks, and Baxter, respectively. The yield was also 12% higher than that of the stay-green CIMMYT line Seri-M82. Dreccer et al. (2006) compared the performance of a limited number of SBLs against that of their parents in trials at CIMMYT, Mexico and Horsham, Victoria, Australia. At

CIMMYT, Mexico, increased yield and biomass of SBLs in comparison to recurrent parents was associated with increased water use efficiency and/or increased root length density in the deeper part of the root zone (between 60 and 120 cm) and subsequently higher water use. These differences were not consistent across the Australian trials. In an extensive study of the utility of SBL in northern Australia, Rattey et al. (2008) indicated that SBL appear to offer greatest benefits for improving grain size and the maintenance of grain size across environments, even where grain yield is low. An economic benefit to farmers from larger grains is achieved through lower levels of grain screenings (small grains), especially in environments prone to high screenings levels.

Lage and Trethowan (2008) using international trial data from SAWYT 11 and 12 provided evidence that the synthetic derived wheat germplasm, 'Vorobey' (Croc_1/*Ae. squarrose* (224)//Opata/3/Pastor) was a top performing line in about 48 sites around the world. Vorobey performed well across all environments compared to the best locally adapted check cultivar at each location; trial means ranged from 1 to 8 t/ha. Similarly, in yield trials conducted in Ethiopia, Morocco, and Syria over 3 years and 12 sites comprising of elite ICARDA lines including some SBL, the SBL 'Nejmah-14' (Skauz/Bav92/3/Croc-1/*Ae. squarrosa* (224)//Opata) was the highest yielding and outperformed the best locally adapted check cultivar across sites; trial means ranged from <1 to 6 t/ha under rainfed conditions.

At ICARDA, grain yield of a synthetic backcross population, SW2/2*Cham-6, was investigated over 3 years in a stressed site at Breda-Syria, (BR, 35°16′N, 37°10′E and 354 m above sea level) with a long term average of 266 mm average long term rainfall and nonstressed site, Terbol, Lebanon, (TR, 33°49′N, 35°59′E and, 890 above sea level) 539 mm mean long tern rainfall. Mean grain yield in the stressed site ranged from 0.89 to 1.66 t/ha, with a site mean of 1.23 t/ha compared to Cham-6 with 1.34 t/ha. At Terbol, the nonstress site, mean grain yield ranged from 2.64 to 6 t/ha, with a site mean of 4.22 t/ha compared to Cham-6 with 4.66 t/ha. The performance of the best yielding SBL was 128% in nonstress and 124% in stress sites compared to the recurrent parent Cham-6. About 25% of the SBL higher yielding than Cham-6 across both stress and nonstress sites showed broad adaptation indicating that SBL could be used to provide significant yield advantage and yield stability in diverse environmental conditions.

Yang et al. (2009) reported the release of four cultivars derived from SHW backcrossed to locally adapted cultivars in China. Of the four cultivars, Chuanmai 42 had large kernels and resistance to stripe rust, and produced the highest average yield (>6 t/ha) among all elite cultivars over 2 years in Sichuan provincial yield trials, out-yielding the commercial check cultivar Chuanmai 107 by 22.7%. They reported that Chuanmai 42 increased grain yield by 0.45–0.75 t/ha. Average yields on the fertile Chengdu plain increased to 7.5 t/ha in 2005 and 2006. Based on this level of superiority, Chuanmai 42 also became an important parent in other breeding programs in China.

At NIAB, UK, preliminary results from the program suggest that introduction of genetic variation from SHW should provide similar benefits in terms of improved pathogen resistance and yield potential under UK conditions as it did for CIMMYT, ICARDA and Australian breeding programs. For example, the results from the evaluation of Paragon BC_1F_5 bulks indicated that the best lines yielded between 113% and 119% of the recurrent parent and elite local controls. Similarly, spike productivity, grain size and thousand grain weight (TGW) of selected BC_1F_3 and BC_1F_5 lines were found to be significantly greater than respective recurrent parents (P. Howell, unpublished). Resistance to local stripe rust (30% of SHW tested) pathotypes and head blight caused by *Fusarium graminiarum* (10% of SHW tested) has also been identified both in primary synthetics and derived backcross lines.

The conclusions that can be drawn from these comparisons between the SBLs and elite adapted cultivars under a range of environmental conditions from low rainfall to favorable high rainfall conditions are that: The grain yields of the top SBLs are higher than their recurrent parents and the best locally adapted check cultivars under moisture stress, nonmoisture limiting environments and diverse agroecological zones. These results indicate that SBLs are also responsive to more favorable, higher yield conditions and could be used to improve elite wheat cultivars, for example, for irrigated high-input environments. The significantly enhanced performance of SBLs is not solely associated with mechanisms related to escape from moisture stress. Other characters must play a role including root architecture.

One of the unexplored areas of research is the root systems of SHW and their potential to contribute to improved productivity. This is partly because of lack of efficient phenotyping for root traits. Palta et al. (2011) investigated the root systems of three SBLs, AUS33687, AUS33435, and AUS33684 measured at ear emergence (Z59 in Zadoks' growth scale for cereals; Zadoks et al. 1974) on a deep sandy soil at Wongan Hills, Western Australia (WA). The root system of AUS33684 neither had seminal roots as deep as those of AUS33435 nor as dense a root at depth as AUS33687, but its root system had a profuse branching in the top 0.35 m of the soil profile (Fig. 2.7). The root proliferation in AUS33684 they posited led to a higher total root length density (in the 0.0–1.2 m

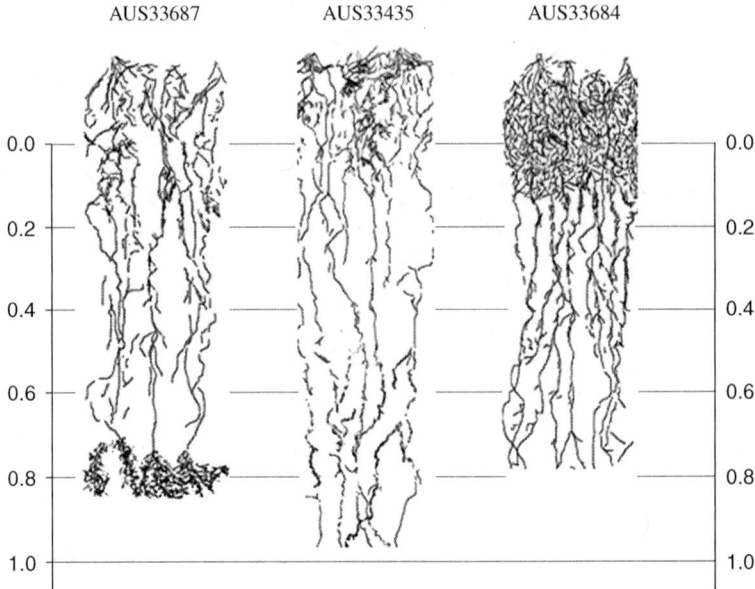

Fig. 2.7. Rooting patherns of three synthetic backcross derived lines illustrating desirable traits for root distribution at depth (AUS33687), depth of rooting (AUS33435), and larger root system (AUS33684). *Source*: Palta et al. (2011).

soil layers) than in AUS33687 and AUS33435 (Fig. 2.6). They argued that total root length density is considered to be directly related to the amount of water uptake and to indicate the size of the root system. However, there was no indication of the yield performance of the three SBL genotypes and therefore the potential benefit of the diversity in root systems to yield improvement in their studies. Nevertheless, they argued that vigorous or large root system contributes to adaptation in dry environments and dry seasons where crop growth depends on seasonal rainfall. However, a large root system may be of less value in environments where crop growth is dependent on stored soil water where access to more soil water runs the risk of exhausting soil water before completing grain filling (Palta et al. 2011). However, the exemption would be when water can be accessed from deeper in the profile. Reynolds et al. (2007) reported that improved performance of SBL compared to recurrent parents was in part due to its increased root biomass at depth and better water extraction capacity—an important attribute on soils where wheat is grown on stored moisture. Lopez and Reynolds (2011) reported that SBL showed on average a 26% yield increase when compared with the parental hexaploid wheat under terminal drought. Among the traits associated with improved performance was earliness to flowering, greater root mass at depth, greater water extraction capacity and increased water use efficiency at anthesis. They reported some degree of independence between these traits indicating the potential for the pyramiding of favorable alleles. In a hydroponic based study carried out under controlled glasshouse, 20 out of 100 SHW had higher initial aerial and root biomass than current Australian check cultivars used. The SHW exhibited significantly thicker root systems than elite common wheat cultivars with some possessing longer root systems (Dreccer and Ogbonnaya, unpublished). Deeper rooting depth impacts water use and could be beneficial in exploiting water at depth under drought conditions.

C. Environmental Stress Tolerance

1. Improving Salinity Tolerance. Wheat production areas globally are divided into irrigated and rainfed sectors. The former is beset with a variable salinity level constraint that penalizes yield outputs. It is estimated that 20% of the irrigated land in the world is presently affected by salinity excluding the regions classified as arid and desert lands (Yamaguchi and Blumwald 2005). In rainfed production systems where transient salinity occurs, yields can be well below theoretical for the rainfall received, when subsoil salinity is present, and unused water at harvest is one of its symptoms (Sadras et al. 2002). In Australia for example, the cost of transient salinity and associated constraints in sodic soils for the farming economy has been estimated at ca. AU$1.5 billion per year (Rengasamy 2002).

Conventional diversity for salinity tolerance has been limiting with just one land race Kharcia 65 playing a major role in salt tolerant varietal development in India where the cultivars KRL1-4 and later KRL 19 emerged. An encouraging transfer strategy emerged when the D genome of wheat was shown to be of significance for salinity tolerance (Gorham et al. 1987; Shah et al. 1987) paving the way for follow-up experiments and exploitation of the D genome synthetics.

Evaluation of an elite set of SHW from CIMMYT resulted in several SHW being categorized as tolerant (Pritchard et al. 2002; Mujeeb-Kazi 2003b). Independent studies indicated and confirmed that SHW possessed considerable variation for salinity tolerance based on Na^+ exclusion (Dreccer et al. 2004). In their study, some SHW exhibited superior differential response to salt compared to the common wheat check in a hydroponic system at a concentration of 100 mM NaCl

(Dreccer et al. 2004). In a further confirmation of these results, more than 300 SHW were evaluated over 4 years. The SHW displayed significant genetic variation for sodium tolerance, often exhibiting much lower Na^+ concentrations and higher K^+/Na^+ ratios in leaves than most Australian commercial common wheat cultivars. However, these studies did not evaluate the two mechanisms that could contribute to it: (i) the ability of transporters in the leaf sheath to retain Na^+ there and (ii) the size of the sheath versus the blade results in the exclusion of Na^+ from blade. In addition to Na^+ exclusion, SHW with significantly superior tissue tolerance than elite wheat cultivars after exposure to 250 mM NaCl were identified (Ogbonnaya et al. 2009).

Earlier studies established that relative to durum wheat, salinity tolerance of common wheat is associated with a relatively high ability to exclude Na^+ from the leaf blades resulting in an overall increase in the K^+/Na^+ ratio, in some cases associated with an increase in K^+ uptake (Gorham et al. 1987; Dvořák et al. 1994). Figure 2.8 shows the successful transfer of salinity tolerance measured as Na^+ exclusion in SHW into an elite Australian common wheat cultivar, Yitpi with some of the SBLs showing significantly enhanced Na^+ exclusion compared to either the SHW or the recurrent common wheat cultivar.

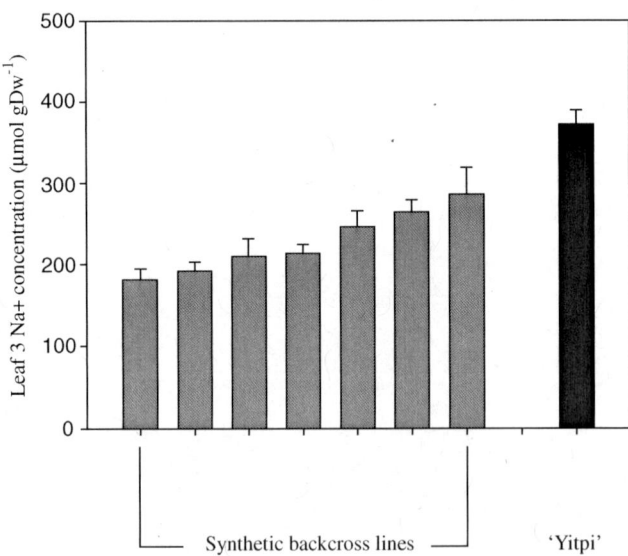

Fig. 2.8. Synthetic backcross derived lines with lower Na+ exclusion than the recurrent elite Australian wheat cultivar, 'Yitpi'.

A major locus, *Kna1*, controlling K^+/Na^+ uptake has been shown to be present on chromosome 4DL (Dubcovsky et al. 1996). QTL analysis of SBL using two of the SHW indicates that different mechanisms contribute to enhanced salinity tolerance of SBL. The results suggest that in addition to the locus on 4D, other loci on chromosomes 2B, 3B, 4B, 6D, 7A, and 7D influence salinity tolerance, either as Na^+, K^+/Na^+ ratio or as tissue tolerance. There were coincident QTL for traits such as predicted mean tissue necrosis (tissue tolerance) and, K^+/Na^+ on chromosomes 3B and 7A. Ogbonnaya et al. (2008b) reported the identification of minor gene for Na^+ exclusion in synthetic derived population 'AUS29639// Yitpi' on chromosome 7A. For predicted mean necrosis, a group of markers on 4D consistently explained most of the variation at 21%. A similar trend was observed for the predicted mean Na^+ with 8%–18% phenotypic variance and 6%–14% of phenotypic variance for K^+/Na^+ ratio on chromosome 4B. The identification of such loci independent of *Kna1* suggests that the salt tolerance in SHW could be deployed to mitigate the impact of salinity and help stabilize and improve wheat production in saline soils.

2. Mineral/Nutrient Efficiency. The search for genetic resources that accumulate high levels of iron (Fe) and zinc (Zn) led to in-depth evaluation of wheat landraces (Monasterio and Graham 2000), tetraploid and diploid progenitors of hexaploid wheat for their enhanced micronutrient status (Cakmak et al. 1999; Genc and McDonald 2004). *Ae. tauschii, T. turgidum* ssp. *dicoccoides, T. monococcum*, and *T. boeoticum* were among the most promising sources of high Fe and Zn levels in the grain. Some of these genotypes showed values as high as 142 mg/g of Zn; however, in some locations manure had been applied in the past cropping cycles that probably contributed to the high Zn values. Cakmak et al. (1999) concluded that SHWs represent invaluable sources of useful genes for improvement of mineral efficiency, and these genes can be readily transferred to wheat. A recent study evaluated a set of high yielding lines under field conditions (Oury et al. 2006). The Zn values generally ranged between 15 and 35 mg/g but increased to 43 mg/g in some genotypes, while the Fe concentrations ranged from 20 to 60 mg/g. Taking into account the bioavailability, the daily intake and the estimated average requirements, some general estimations were made by "Harvest Plus" to set a breeding target for wheat. For example, in Pakistan and northern India the target is to increase Fe and Zn levels by 25 and 10 mg/g, respectively, above the mean of genotypes grown in the region. This translates into total Zn and Fe levels in the grain of 45 and 60 mg/g, respectively. From these results, whilst there is sufficient unexplored genetic variability in SHW to develop wheat varieties with increased mineral levels in the grain, it is worthy of note that Zn and Fe are concentrated in the aleurone layer of the grain, it is therefore likely that low yield and high mineral content may be coupled together. This is because the aleurone/endosperm ratio changes with increasing yield with implication in for breeding.

However, it should also be kept in mind that mineral bioavailability is limited by the presence of phytic acid (PA) in the aleurone layer, which forms insoluble complexes with dietary cations, thus hindering their intestinal absorption (Cheryan and Rackis 1980). PA breakdown strongly depends on the phytase activity in the flour. Consequently both mineral and phytase concentrations should be taken into account in breeding programs. Recently, Ram et al. (2010) reported higher genetic variability of phytase in SHW compared to Indian cultivars. Since there was a larger genotypic effect on phytase levels and also larger variation in activity as compared to phytate levels, there is greater scope for manipulating phytase levels than phytate content through breeding in wheat. The release of cultivars with high mineral concentrations complemented with high intrinsic phytase activity could greatly improve the nutritional value of common wheat, provided that whole-wheat flour is utilized to preserve the source of the minerals. CIMMYT nearly a decade ago had screened some wheat progenitor genetic resources and identified accessions of *T. turgidum* ssp. *dicoccum* with elevated levels of iron and zinc. These tetraploids were used to develop SHW in the CIMMYT wide crossing unit and those stocks (*T. turgidum* ssp. *dicoccum/Ae. tauschii*) entered the wheat-breeding program. A nursery set has been deployed in India and Pakistan that appears very promising though results are being awaited.

3. Boron Toxicity. Boron toxicity limits wheat yield in some environments by limiting root growth and affecting water uptake. Dreccer et al. (2003) evaluated the variability for boron tolerance in 49 entries of the primary SHW set and identified boron tolerance in 26 of the entries that was similar to the tolerant conventional check 'Frame' (Fig. 2.9). Further, 21 out of 87 SHW performed as well as Halberd, the most boron tolerant cultivar in Australia, and were significantly different from elite check cultivars. It is not known, however, if the genes controlling boron tolerance are different from those identified earlier in the primary wheat gene pool.

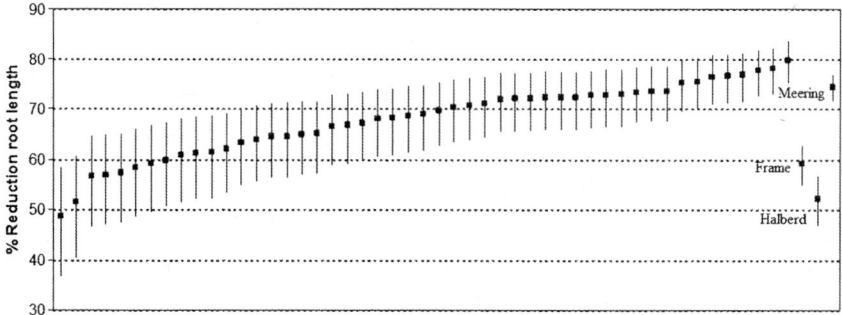

Fig. 2.9. Genetic variation for boron tolerance in synthetic hexaploid wheat (SHW). *Source*: Dreccer et al. (2003).

D. Quality Improvement

The SHWs possessed significantly more genetic variation for quality traits including those associated with colour and colour stability of Asian noodles such as near zero extremes for polyphenol oxidase and lipoxygenase, than currently available in common wheat (Mares and Mrva 2008).

1. Bread Making Quality. The major determinants of end use quality in wheat include grain hardness or endosperm texture, protein content and protein quality. Functional differences in grain and flour quality between hexaploid and tetraploid (*T. turgidum* ssp. *durum*) wheats have been attributed to the influence of the D genome (Ogbonnaya et al. 2005). With the growing interest during the last decade in exploitation of the D genome SHWs and expected varietal outputs it is crucial that quality is considered in the breeding scheme particularly when an alien resource is utilized.

Reports of William et al. (1993), Peña et al. (1995), Hsam et al. (2001), and Tang et al. (2008) have shown that unique and beneficial quality glutenin subunits exist. Desirable sub-units at the $Glu\text{-}D^t1$ locus are $5+10$, $1.5+10$, $2.1+10$, $1.5+12$, and $3+10$. In the sets of SHW evaluated, 50%–60% of these subunits were observed in Elite-1 and Elite-II subsets of SHW (Bibi et al. 2011). The superiority of two *Ae. tauschii* encoded subunits 1Dx1.5 and 1Dx2.1 for grain quality improvement prompted the development of an allele-specific molecular marker (Lu et al. 2005a,b) to facilitate their early selection in segregating generations when SHW were targeted for breeding.

In recent studies at CIMMYT and in Australia (van Ginkel and Ogbonnaya 2007), it has become clear that synthetic derivatives carrying excellent bread-making quality can be bred if the common wheat parent(s) in the cross have good baking quality. Variation at the *Ha* locus on chromosome 5D, a major determinant of milling yield, has revealed complete deletions in the *Pina* and *Pinb* genes present in the A and B genomes of common wheat. Phenotypic differences in grain hardness/softness are largely due to the locus on the D genome resulting from deletions in puroindoline a (*Pina*) and/or mutations in puroindoline b (*Pinb*) (Ogbonnaya et al. 2005 and references cited therein). *Ae. tauschii* is generally soft-grained and so are the SHWs. Characterization of the *Ae. tauschii* and SHWs identified eight different *Pina* alleles and six unique *Pinb* alleles that are all associated with a soft endosperm (Gedye et al. 2004; Massa et al. 2004; Lillemo et al. 2006).

To further understand the mechanism of grain hardness formation in *Ae. tauschii* and SHW, a total of 57 new SHWs derived from crosses of *T. turgidum* ssp. *dicoccoides* and *Ae. tauschii* were analyzed to compare puroindoline alleles to grain hardness using gene-specific PCR primers and the single kernel characterization system (SKCS) of Perten 4100 (Chen et al. 2006; Li et al. 2007). Four allelic variations of *Pina* and *Pinb* were identified, that is, *Pina-D1a*, *Pina-D1c*, *Pina-D1h*, and *Pina-D1j*. Eight (14%) of the lines contained *Pina-D1a/Pina-D1j* while the remaining 86% of the lines, 49, carried *Pina-D1c/Pina-D1h*. Kernel texture differed significantly among the SHWs; SKCS hardness ranged from 10.5 to 42.6, with significant differences between the parental types, *T. turgidum* ssp. *dicoccoides* or *Ae. tauschii* (Chen et al. 2006; Li et al. 2007). However, variation in kernel texture did not strictly correspond to puroindoline genotypes *Pina-D1a/Pina-D1j* and *Pina-D1c/Pina-D1h*, indicating genes besides those found at the *Ha* locus affected kernel texture (Chen et al. 2006; Li et al. 2007).

Narasimhamoorthy et al. (2006) using an AB-QTL population of hardwinter wheat by synthetic cross identified a major hardness QTL which coincided with the well-known *Ha* and *Gsp* puroindoline genes on 5DS, in addition to a novel QTL on chromosome arm 3BL that have not been previously reported for grain hardness.

The overall visco-elastic properties that confer bread making characteristics of hexaploid wheat in contrast to the pasta making characteristics of durum wheat reflects the impact of the D genome. While various aspects of grain, flour, and dough quality have been studied to further partition components of the D genome that exert major influence, most of the studies have revealed the significant role of gluten and puroindoline proteins in flour functionality. Although variation at the *Glu-D^{t}1* locus that controls HMW glutenins in *Ae. tauschii* is much greater than the corresponding locus in *T. aestivum*, none of the novel variants transferred into common wheat from *Ae. tauschii* has so far been found to improve bread-making characteristics compared to what already exists in common wheat varieties (Bennett 1994).

The predominant influence of the D-genome in determining the hardness/or softness of common wheat is of some concern to breeders who appear hesitant in using SHW for improvement of common wheat due to fear of compromising the quality of their elite germplasm. In response to these concerns, Garg et al. (2010) investigated the end use quality, in particular bread-making quality by dough strength and storage protein profile, of SBL derived from backcrossing SHW with elite commercial Australian varieties. The SBLs investigated were predominantly from the following lines: Altar 84/*Ae. squarrosa* (211)//Janz/3/Westonia, Croc-1/*Ae. squarrosa* (224)//Kulin/3/Westonia and Cpi/Gediz/3/Goo//Jo69/Cra/4/*Ae. squarrosa* (208)/5/2*Westonia. The results from the study indicated that all the SBLs were prime-hard like parental cultivars or even harder. This suggests that backcrossing using appropriate parents obviates the concerns of breeders in the use of SHW because the results demonstrate the full recovery of hard-grain textured SBLs due largely to the use of hard grain parents and a favorable selection of hard genotypes.

Mares and Mrva (2008) evaluated SHW, SBLs with bread wheat and durum wheat cultivar for traits that determine colour and colour stability in Asian noodles and the frequency of a genetic defect know as late maturity α-amylase (LMA). They reported that the SHW harboured substantial genetic variation for quality traits associated with colour and colour stability of Asian noodles including near zero extremes for polyphenol oxidase and lipoxygenase. They indicated that these extremes represent a significant advantage compared with current bread wheat cultivars and are similar to the best durum wheat. While alternative strategies for reducing polyphenol oxidase and lipoxygenase are available, the SHW provide a useful resource for wheat breeders attempting to develop improved wheat cultivars for the Asian noodle market. However, they found that most SHW were prone to late maturity α-amylase and mature grain contained unacceptably high levels of α-amylase. Elimination of this genetic defect, or selection within breeding populations for low or non-LMA, is both time consuming and labour intensive and presents a significant obstacle to exploitation of variation for other traits. Nonetheless, as proof of concept of the potential of exploiting SHW, they recovered near-zero polyphenol oxidase (PPO) SBL, free from LMA, from backcross populations involving a high LMA SHW. In perhaps the most comprehensive evaluation of the quality of SBL, Nelson et al. (2006) used QTL analysis to study the milling and baking quality of 114 recombinant inbred lines of the ITMI SHW (W7985) × common wheat (Opata 85) crosses grown in US, France, and Mexico. The traits investigated included kernel-texture, protein concentration and quality, dough strength, and mixing traits. They reported that some of the RILs had consistently superior quality traits than their parents, indicating the potential of a new combination of alleles from the diploid and tetraploid parents, especially alleles of known storage proteins for improvement of quality traits in wheat cultivars. For example, mixogram traits were influenced most by chromosomal regions containing gliadin or low-molecular-weight glutenin loci on chromosome arms 1AS, 1BS and 6DS, with the SHW contributing the favorable alleles. Similarly, Kunert et al. (2007) used the advanced backcross QTL (AB-QTL) strategy to locate QTLs for baking quality traits in two BC_2F_3 populations of winter wheat. The BC_2F_3 were from SHW produced from *T. turgidum* spp. *dicoccoides* and *Ae. tauschii*. The exotic allele improved trait performance at 14 QTLs (36.8%), while the elite genotype contributed a favorable effect at 24 QTLs (63.2%). The favorable exotic alleles were mainly associated with grain protein content, though the greatest improvement of trait performance due to the exotic alleles was achieved for the traits falling number and sedimentation volume. At the QTL on chromosome 4B, the exotic allele increased the falling number by 19.6%; and, at the QTL on chromosome 6D, the exotic allele increased sedimentation volume by 21.7%, indicating that the SBL derived from wild emmer × *Ae. tauschii* carried favorable QTL alleles for baking quality traits.

2. Preharvest Sprouting. Most cultivated hexaploid wheat varieties are characterized by low embryo dormancy and are sensitive to preharvest sprouting (PHS). Thus, PHS is one of the most important quality constraints in wheat production in environments characterized by summer rainfall and high humidity. PHS resistance is inherited as a quantitative trait controlled by a large number of genes, and is significantly influenced by genetic background, environmental conditions and their interactions. Ogbonnaya et al. (2007b) characterized the extent of genetic diversity of 28 wheat lines, including durum and SHW, reputed to possess PHS resistance. Across most current varieties, the alleles at some of the QTL regions associated with PHS tolerance were largely the same, depending on the gene pool used, while the natural variation present in *Ae. tauschii* provided a novel source of allelic diversity. This knowledge of the different haplotypes can be used to design optimum strategies for the incorporation of PHS resistance into cultivated varieties. Consistent with previous studies, their results indicated clear differences in the level of dormancy between red-grained and white-grained genotypes. However, the white-grained genotypes were more variable, with and a few genotypes displayed levels of dormancy comparable to red-grained genotypes.

Gatford et al. (2002) reported that *Ae. tauschii* accessions and the SHW derived from them expressed high levels of seed dormancy. In an attempt to incorporate PHS tolerance into wheat, a SBL population was generated from a cross between an Australian white common wheat cultivar, 'Janz,' and a SHW, 'Syn36'. The resultant SBLs were grown in field trials at Horsham in 2004/2005 and evaluated for grain dormancy (GI), sprouting index (SI), and visibly sprouted seeds (VI), the latter two following artificial weathering. There were significant ($P < 0.01$) genotypic differences among the SBLs across the three indices studied using a weighted germination index of 0–1, where 1 represents extreme susceptibility. The range in value in the BC_1F_6 population for GI value after 14 days (GI-14) were 0.13–0.85 compared to 'Syn36' the resistant parent with a GI-14 value of 0.30, and 'Janz', the susceptible parent with GI-14 value of 0.86, indicative of transgressive segregants towards high dormancy. A similar trend was observed for all measures of PHS tolerance. Some of the SBLs were identified with significantly superior preharvest sprouting tolerance across all three measures (GI, SI, and VI), compared to the current Australian PHS tolerant standard SUN325 with a GI-14 of 0.46. A similar result was obtained in a SBL involving SHW, 'Syn37' and the cultivar, 'Janz' (Imtiaz et al. 2008). In a further elucidation of genetic architecture of preharvest sprouting resistance (PHSR), Imtiaz et al. (2008) identified seven QTL controlling resistance to PHS on chromosomes 1D, 2D, 3D, 6D, and 4A. The SHW parent, 'Syn37' contributed resistance on 1D, 2D, 3D, and 4A, while the susceptible cultivar, 'Janz' contributed alleles on chromosome 6D, which indicated the presence of complementary alleles in the two parents. A large component of the phenotypic variation was explained by red grain color (RGC) depending on the indices used. They mapped *RGC* or *R-D1b* locus at a distance of 5 cM from the marker *wms1200* located on chromosome 3D, with the highest level of variation explained by GI (43%), VI (17%), and SI 8%. On aggregate, the red-seeded genotypes overall displayed higher levels of dormancy than the white seeded genotypes. Similarly, in a cross between a PHS tolerant SHW and common wheat cultivar, '88-1643', Ren et al. (2008) reported the identification of a major QTL associated with PHS tolerance designated *Qphs.sau-2D* within the marker intervals between *Xgwm261* and *Xgwm484* in two consecutive years which accounted for 26%–28% of the phenotypic variation for PHS tolerance. Some white-grain synthetic-derived wheat lines that are highly resistant to preharvest sprouting were developed in Sichuan and Heilongjiang, China and Horsham, Australia using SHW produced with highly dormant *Ae. tauschii* accessions (Lan et al. 2005; Li et al. 2005; Imtiaz et al. 2008).

IV. CONCLUSIONS AND FUTURE PROSPECTS

Wheat breeding targets are linked with the increasing population trends and with 2030 being very close, about 730 million tons of wheat will be required to feed a population of about nine billion people. With genetic diversity being paramount for improving productivity through breeding, looking at resources beyond bread wheat is imperative and SHWs and the SBLs provide a rich source of genetic diversity for improving wheat production. The potential benefit of SHW gained prominence in the mid-1980s with the simultaneous use of direct crossing of *Ae. tauschii* with wheat at Kansas State University (Gill et al. 2006) and the random production of SHW at CIMMYT aimed at capturing the value of the three different genomes for the improvement of common wheat. Two decades later, the exploitation of genetic variability in SHW is being translated into the successful identification, mapping, characterization, and transfer of several agronomically important traits including disease and pest resistance into common wheat. Recently, new sources of resistance were identified in *Ae. tauschii* against the stem rust race, Ug99, (Rouse et al. 2011). This was also evident in the reports of the successful development of high yielding Ug99 resistant SBLs in CIMMYT (Singh et al. 2011a,b) and ICARDA (Abdalla et al. 2011). Similarly, Zegeye et al. (2012) reported the identification of both major and minor gene resistance in SHWs against the stripe rust *Yr27* virulent race that devastated wheat crops in Ethiopia and Syria in 2010 thus providing genetic resource that could be useful in broadening the genetic bases of stripe resistance in wheat.

In the short-term this trend is expected to continue by deploying populations derived from SHW using SSR, DArT, and SNP markers for genetic mapping, and trait identification. However, while the approach that first involves screening for targeted biotic and abiotic stresses would continue, emphasis must now include the use of genome-wide approaches that involves low-cost sequencing technologies.

With about 1,500 SHWs catalogued globally, the validation of existing genes and the discovery of new genes for resistance and tolerance to abiotic and biotic stresses in SHW can be enhanced by employing genotyping by sequencing of the SHW and undertaking of genome-wide association studies. The release of the 5× sequence coverage of the wheat genome (Science Daily, August 27, 2010) will facilitate efforts towards full sequencing of the bread wheat genome which, if accomplished, will accelerate gene discovery, allow even more targeted mining of novel genes/alleles in SHW and will enable their subsequent introgression into bread wheat with impact on breeding efficiency. However, these potential outcomes are equally depended on precision phenotyping.

One of the key features of SHW is the availability of lines with multiple disease resistance (MDR). However, this potential has yet to be exploited. It would be necessary to assemble a set with MDR and conduct allelism tests with previously described genes. The understanding of the mechanisms involved in conferring MDR is likely to have implications for resistance spectra and durability associated with specific MDR loci in wheat improvement. As pathogens mutate or new stresses become threats this unique diversity in SHW will gain further significance. Where major genes are involved, it is quite likely that a pathogen shift will defeat single gene resistance particularly for rust diseases. The availability of adult plant resistance QTL in the SHW, if proven to be race nonspecific, provides the opportunity to use gene pyramiding as a means of improving durability. Ultimately, the cloning of the genes underlying MDR will lead to more precise and targeted incorporation of MDR genes into wheat varieties with negligible linkage drag.

It has also been demonstrated that SHW and their derived SBLs (i.e., when crossed to adapted local varieties) show significant yield increases and thus enhanced yield performance across a diverse range of environments, demonstrating their potential for use in improving wheat productivity worldwide. This is particularly evident in moisture-limited environments. It is crucial that this potential be realized at the farm level. These should be used to accelerate the development of improved varieties that can provide high yields and improved livelihoods, and at the same time meet the challenges of marginal environments and the threat of climate change in rainfed wheat production systems. However, the genetic and physiological bases of enhanced performance are little understood, thus key questions associated with understanding and unraveling the basis of enhanced yield performance remain to be answered. The precise contribution of genes or genomic segments derived from either the $A^d B^d$, D^t or specific interactions with the ABD genomes of hexaploid wheat remain largely undefined. In the absence of any *a priori* basis for selecting SHW that are predictive of the outcomes of significant yield boost in SBLs there is the need to analyze and define genomic regions with consistent contributions to yield increases/yield potential in introgression lines from different recipient wheat backgrounds across different environments. While there are now many reports of potentially beneficial QTLs identified in SBLs, most are yet to be validated in diverse genetic background, followed by fine mapping and ultimately cloning.

Future production of new SHWs should extend to evaluating under-exploited tetraploids A and B genome such as *T. turgidum* ssp. *carthlicum*, *T. turgidum* ssp. *dicoccum*, and *T. turgidum* ssp. *dicoccoides* and identifying gaps in the existing *Ae. tauschii* germplasm used for existing SHWs. Gosman et al. (2011) used 15 SSRs to explore the relationship between the D-genomes of *Ae. tauschii* (255 accessions), CIMMYT SHWs (50 genotypes) and a 190 elite diverse cultivars of *T. aestivum* from UK, France, Germany, Serbia Croatia and the Russia. Results using STRUCTURE (V2.1) at both $K=3$ and $K=7$ placed bread wheat (coloured purple) into a single exclusive sub-population (Fig. 2.10). The D-genome donors of the CIMMYT SHWs and *Ae. tauschii* accessions were partitioned between the remaining groups. Clade A is a mixture of ssp. *tauschii* and ssp. strangulata encompassing *Ae. tauschii* accessions from the Caucuses and the region south of the Caspian Sea with no apparent geographical split observed for sub-clades 1 and 2. Bread wheat is mostly associated with clade A supporting the hypothesis that it evolved around the shores of the Caspian Sea (Fig. 2.10). From this analysis, the available CIMMYT SHWs mostly exploit clade A and, therefore, by inference mostly capture D-genome diversity from *Ae. tauschii* accessions more closely related to bread wheat. Clade B encompasses accessions to the north (sub-clade B1)and east (sub-clade B2) of the Caspian Sea which is apparently under-exploited within current CIMMYT SHWs germplasm. Identifying geographical areas where the progenitor species of the existing SHW where collected would assist in guiding future collection missions. Coupled with ecoenvironmental and disease prevalence data, such efforts would guide future SHW and wide crossing programs toward a more targeted approach to exploiting the rich vein of genetic diversity provided by the primary, secondary and tertiary gene pools. Such a targeted approach is encapsulated in the focused identification of germplasm strategy (FIGS).

FIGS develop environmental profiles, using statistical approaches and GIS technologies, to predict where selection pressures are likely

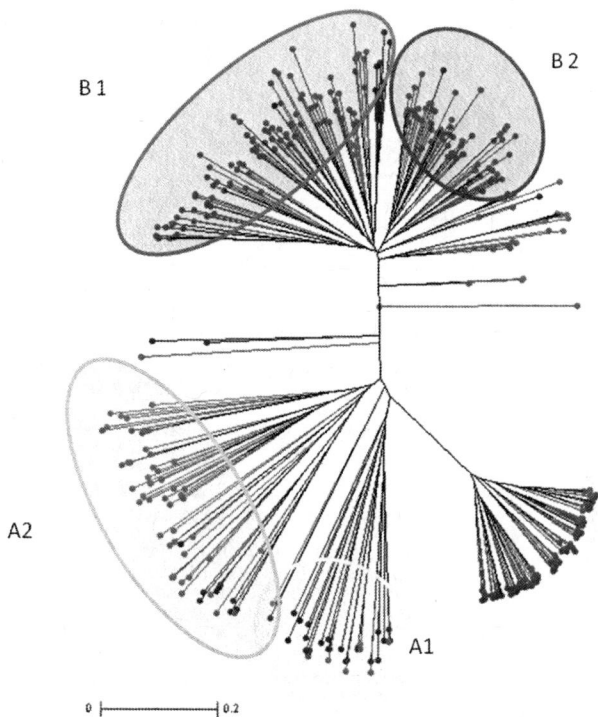

Fig. 2.10. Dendrogram of 255 *Ae. tauschii* accessions, 50 CIMMYT synthetic hexaploid wheat and 190 elite T. aestivum cultivars based on the genetic distance (percentage difference) calculated from data of 15 microsatellites, using UPGMA as the clustering method. *Source*: Courtesy Gosman et al. (2011).

to occur for specific traits. Germplasm collection sites that fall within the appropriate environmental profile are identified, and the resulting high probability trait-specific subset of germplasm is advanced for screening (Mackay and Street 2004). Thus, implementation of FIGS could result in the production of new SHWs focused on maximizing allelic diversity for key abiotic and biotic stress tolerances that continue to impede wheat productivity. Targeted traits include drought, heat, salinity, and multiple disease resistance, all of which can contribute to sustainable wheat production in vulnerable dry-land communities that in turn can contribute to poverty reduction and increased food insecurity.

On the other hand, many studies have indicated inconsistency of the traits in diploid (*Ae. tauschii*) and allohexaploid (SHW) levels, suggesting difficulty of selection of useful traits in diploid level.

Furthermore, the expected traits of SHWs may not always appear in the SBLs produced by backcrosses with elite wheat cultivars. This is because the genetic background of SHWs are largely different from that of elite cultivars. For the enhancement of the genetic diversity of bread wheat through SHWs, the genes of *Ae. tauschii* should be evaluated in the genetic background of elite wheat cultivars. For this purpose populations including large diversity of *Ae. tauschii* in genetic background of elite cultivars are necessary. Tsujimoto (2010) proposed to produce multiple synthetic derivative (MSD) populations by backcrossing elite cultivar and SHW lines produced using variety of *Ae. tauschii* accessions (Fig. 2.11). Once MSD population is produced, it can be the source of genes for breeding with any breeding objectives.

Arguably, the potential of SHW and SBLs described here demonstrate that such germplasm can provide an unquestionable benefit to wheat

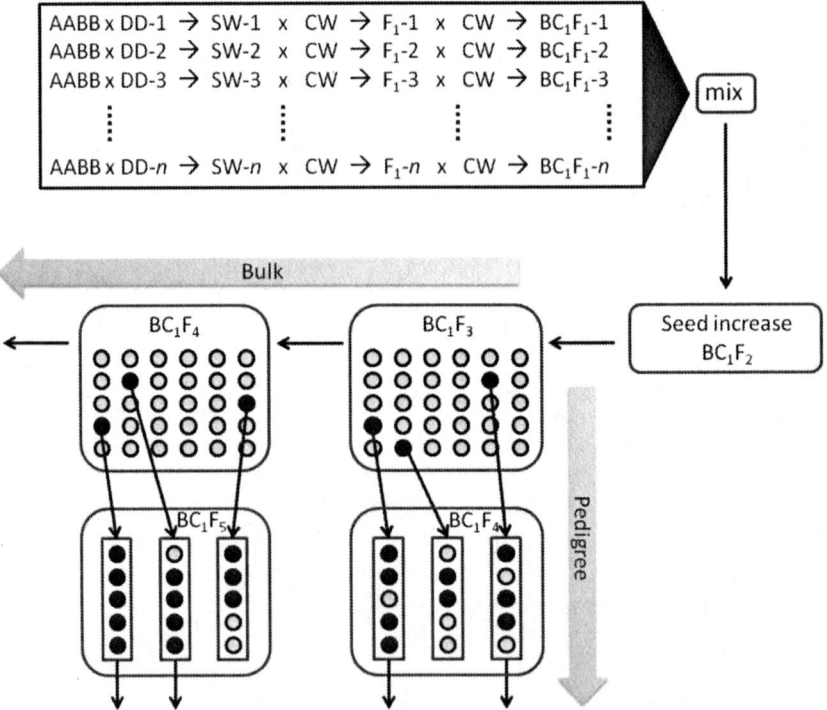

Fig. 2.11. Multiple Synthetic Derivatives (MSD) wheat population with variation from many accession of *Aegilops tauschii*. AABB, durum wheat cultivar; DD-1 to DD-n, *Ae. tauschii* accessions; CW, a leading cultivar of common wheat from a target region; SW-1 to SW-n, synthetic wheat lines. *Source*: Tsujimoto (2010).

improvement. Nonetheless, currently only a limited number of varieties have been released from SBLs. Major limitations include some of the perceived undesirable characters such as red seed color in some countries or excessive plant height as well as the fact that it usually takes between 10 and 15 years to develop and release cultivars with heightened interest in the use of SHW and SBL being relatively recent. Such undesirable characters could be addressed via multiple back-crossing to parents with desirable traits. Results presented above indicate that positive transgression for quantitative traits associated with abiotic stress tolerance is common and that such novel genotypes result from either new D^t or $A^d B^d$ derived alleles or interactions between these new variants and the common wheat genome. Thus, a measure of the impact of SBLs is not exclusive to the either the introgression of the D^t or the $A^d B^d$ variants, but rather from the interactions including epistatic interactions between the genomes.

ACKNOWLEDGMENTS

The authors wish to acknowledge the financial support of Grains Research and Development Corporation, Australia, Generation Challenge Program, CIMMYT, Mexico and BMZ, Germany, global COE program, JSPS, Japan. The authors thank Awais Rasheed to Dr. Michael Baum for critically reading the manuscript and Bikram S. Gill for helpful suggestions. The mentoring support of Dr Maarten van Ginkel is heartily appreciated and gratefully acknowledged.

LITERATURE CITED

Abdalla, O., F. Ogbonnaya, I. Tahir, W. Tadesse, A. Yaljarouka, A. Yahyaoui, K. Nazari, A. Badebo, F. Eticha, S. Gelalcha, B. Hundie, R. Wanyera, and P. Njau. 2011. Breeding for UG99 resistance at ICARDA: Results from regional trials. BGRI Technical Workshop, June 13–16, St. Paul MN.

Alonso, L.C., and G. Kimber. 1984. Use of restitution nuclei to introduce alien genetic variation into hexaploid wheat. Z. Pflanzenzucht. 92:185–189.

Arraiano, L.S., A.J. Worland, C. Ellerbrook, and J.K.M. Brown. 2001. Chromosomal location of a gene for resistance to Septoria tritici blotch (*Mycosphaerella graminicola*) in the hexaploid wheat 'Synthetic 6×'. Theor. Appl. Genet. 103:758–764.

Assefa, S. and H. Fehrmann. 2000. Resistance to wheat leaf rust in *Aegilops tauschii* Coss. and inheritance of resistance in hexaploid wheat. Genet. Resour. Crop Evol. 47:135–140.

Badebo, A. and H. Ferhmann. 2005. Resistance to yellow rust in *Aegilops tauschii* (Coss.), *Triticum durum* and their synthetic amphiploids. Ethiop. J. Agr. Sci. 18:129–135.

Badebo, A., G.H.J. Kema, M.van Ginkel, and C. H.van Silfhout. 1997. Genetics of suppressors of resistance to stripe rust in synthetic wheat hexaploids derived from *Triticum*

turgidum subsp. *dicoccoides* and *Aegilops squarrosa*. African Crop Sci. Conf. Proc. 3:195–202.
Bai, D. and D. R. Knott. 1992. Suppression of rust resistance in bread wheat (*Triticum aestivum* L.) by D-genome chromosomes. Genome 35:276–282.
Bennett, A.M. 1994. Molecular and functional studies of variant HMW glutenins in wheat. Ph.D. thesis. Univ. Sydney, Australia.
Bibi, A., A. Rasheed, A.G. Kazi, T. Mahmood, S. Ajmal, I. Ahmad, and A. Mujeeb-Kazi. 2011. High-molecular-weight (HMW) glutenin subunit composition of the elite-11 synthetic hexaploid wheat sub-set (*Triticum turgidum* × *Aegilops tauschii*; $2n=6x=42$; AABBDD). Plant Genetic Resources: Characterization and Utilization p. 10:1–4.
Braun, H.-J., G. Atlin, and T. Payne. 2010. Multilocation testing as a tool to identify plant response to global climate change. In: M.P. Reynolds (ed.), Climate change and crop production. CABI, London.
Brown-Guedira, G.L., T.S. Cox, B.S. Gill, R.G. Sears, and S. Leath. 1999. Registration of KS96WGRC37 leaf rust-resistant hard red winter wheat germplasm. Crop Sci. 39:596.
Brown-Guedira, G.L., A.K. Fritz, A. Rosa, B.S. Gill, and S. Singh. 2004. Registration of KS00WGRC44 leaf rust-resistant hard red winter wheat germplasm. Crop Sci. 44:702–703.
Brown-Guedira, G.L., M. Guedira, A.K. Fritz, T.J. Martin, O.K. Chung, G.L. Lockhart, B.W. Seabourn, B.S. Gill, and T.S. Cox. 2005a. Notice of release of KS04WGRC49 hard winter wheat germ plasm with unique glutenin and gliadin proteins. Annu. Wheat Newslett. 51:190.
Brown-Guedira, G.L., J.H. Hatchett, X.M. Liu, A.K. Fritz, J.O. Owuoche, B.S. Gill, R.G. Sears, T.S. Cox, and M.S. Chen. 2005b. Registration of KS99WGRC42 Hessian fly resistant hard red winter wheat germplasm. Crop Sci. 45:804–805.
Börner A., E. Schumann, A. Fürste, H. Cöster, B. Leithold, M.S. Röder, W.E. Weber. 2002. Mapping of quantitative trait loci determining agronomic important characters in hexaploid wheat (*Triticum aestivum* L.). Theor. Appl. Genet. 105:921–936.
Cakmak, I., O. Cakmak, S. Eker, A. Ozdemir, N. Watanabe, and H.J. Braun. 1999. Expression of high zinc efficiency of *Aegilops tauschii* and *Triticum monococcum* in synthetic hexaploid wheats. Plant Soil 215:203–209.
Cassman, K. G. 1999. Ecological intensification of cereal production systems: Yield potential, soil quality, and precision agriculture. Proc. Natl. Acad. Sci. (USA) 96:5952–5959.
Chen, F., X.C. Xia, D.S. Wang, M. Lillemo, and Z.H. He. 2006. Detection of allelic variation for Puroindoline alleles in CIMMYT germplasm developed from synthetic wheat crossing with common wheats (in Chinese with an English abstract). Sci. Entia Agr. Sinica 39:440–447.
Cheryan, M., and J. J. Rackis. 1980. Phytic acid interactions in food systems. CRC Crit. Rev. Food Sci. Nutr. 13:297–335.
Chhuneja, P., S. Kaur, R.K. Goel, and H.S. Dhaliwal. 2006. Mapping of leaf rust and stripe rust resistance QTLs in bread' wheat × synthetic RIL population under Indian field conditions. Indian J. Crop Sci. 1:49–54.
Christopher, J.T., A.M. Manschadi, and Y. Dang. 2006. Synthetic hexaploid wheats can exhibit stay-green phenotype and high yield in the Australian northern grains region. p. 33. In: M. Imtiaz, F.C. Ogbonnaya and M.van Ginkel (eds.), 1st Synthetic Wheat Symposium "Synthetics for Wheat Improvement." Sept. 4–6, Horsham, Victoria, Australia. Abstr. book.

Chu, C.-G., J.D. Faris, S.S. Xu, and T.L. Friesen. 2010. Genetic analysis of disease susceptibility contributed by the compatible *Tsn1*–Sn-ToxA and *Snn1*–Sn-Tox1 interactions in the wheat-*Stagonospora nodorum* pathosystem. Theor. Appl. Genet. 120:1451–1459.

Chu, C.-G., T.L. Friesen, S.S. Xu, and J.D. Faris. 2008a. Identification of novel tan spot resistance loci beyond the known host-selective toxin insensitivity genes in wheat. Theor. Appl. Genet. 117:873–881.

Chu, C.-G., T.L. Friesen, S.S. Xu, J.D. Faris, and J.A. Kolmer. 2009. Identification of novel QTLs for seedling and adult leaf rust resistance in a wheat double haploid population. Theor. Appl. Genet. 119:263–269.

Chu, C.-G., S.S. Xu, T.L. Friesen, and J.D. Faris. 2008b. Whole genome mapping in a wheat doubled haploid population using SSRs and TRAPs and the identification of QTL for agronomic traits. Mol. Breed. 22:251–266.

Chu, C.-G., J.D. Faris, T.L. Friesen, and S.S. Xu. 2006. Molecular mapping of hybrid necrosis genes Ne1 and Ne2 in hexaploid wheat using microsatellite markers. Theor. Appl. Genet. 112:1373–1381.

Cox. T.S. 1997. Deepening the wheat gene pool. J. Crop Prod. 1:1–25.

Cox, T.S., R.K. Bequette, R.L. Bowden, and R.G. Sears. 1997a. Grain yield and breadmaking quality of wheat lines with the leaf rust resistance gene *Lr41*. Crop Sci. 37:154–161.

Cox, T.S., W.W. Bockus, B.S. Gill, R.G. Sears, T.L. Harvey, S. Leath, and G.L. Brown-Guedira. 1999. Registration of KS96WGRC40 hard red winter wheat germplasm resistant to wheat curl mite, Stagnospora leaf blotch, and Septoria leaf blotch. Crop Sci. 39:597.

Cox, T.S., W.W. Bockus, B.S. Gill, R.G. Sears, W.F. Heer, J.H. Long, and T.L. Harvey. 1996. KS95WGRC33 Septoria leaf blotch-resistant germplasm released. Ann. Wheat Newslett.42 (http://wheat.pw.usda.gov/ggpages/awn/42/awn42e3.html).

Cox, T.S., L. G. Harrell, P. Chen, and B. S. Gill. 1991a. Reproductive behavior of hexaploid/diploid wheat hybrids. Plant Breed. 107:105–118.

Cox, T.S., J.H. Hatchett, R.G. Sears, and B.S. Gill. 1994a. Registration of KS92WGRC26, Hessian fly-resistant hard red winter wheat germplasm. Crop Sci. 34:1138–1139.

Cox, T.S., J.H. Hatchet, B.S. Gill, W.J. Raupp, and R.G. Sears. 1990a. Agronomic performance of hexaploid wheat lines derived from direct crosses between wheat and *Aegilops squarrosa*. Plant Breed. 105:271–277.

Cox, T.S., T. Hussien, R.G. Sears, and B.S. Gill. 1997b. Registration of KS92WGRC16 winter wheat germplasm resistant to leaf rust. Crop Sci. 37:634.

Cox, T. S., R.G. Sears, and R. K. Bequette. 1995. Use of winter wheat × *Triticum tauschii* backcross populations for germplasm evaluation. Theor. Appl. Genet. 90:571–577.

Cox, T.S., R.G. Sears, and B.S. Gill. 1992. Registration of KS90WGRC10 leaf rust-resistant hard red winter wheat germplasm. Crop Sci. 32:506.

Cox, T.S., R.G. Sears, and B.S. Gill. 1991b. Fall Cereal Conf, Aug. 1–2, Manhattan, KS, p. 18–20.

Cox, T.S., R.G. Sears, B.S. Gill, and E.N. Jellen. 1994b. Registration of KS91WGRC11, KS92WGRC15, and KS92WGRC23 leaf rust-resistant hard red winter wheat germplasms. Crop Sci. 34:546.

Cox, T.S., M.E. Sorrells, G.C. Bergstrom, R.G. Sears, B.S. Gill, E.J. Walsh, S. Leath, and J.P. Murphy. 1994c. Registration of KS92WGRC21 and KS92WGRC22 hard red winter wheat germplasms resistant to wheat streak mosaic virus and powdery mildew. Crop Sci. 34:546.

del Blanco, I.A., S. Rajaram, W.E. Kronstad, and M.P. Reynolds. 2000. Physiological performance of synthetic hexaploid wheat-derived populations. Crop Sci. 40:1257–1263.

Dewey, D.R. 1984. The genomic system of classification as a guide to intergeneric hybridization with the perennial Triticeae. p. 209–279. In: J.P. Gustafsom (ed.), Gene manipulation in plant improvement. Plenuim Press, New York.

Dreccer, M.F., F.C. Ogbonnaya, G. Borgognone, and J. Wilson. 2003. Boron tolerance is present in primary synthetic wheats. p. 1130–1132. In: N. Pogna (ed.), Proceedings of 10th Intl. Wheat Genetics Symposium, Sept. 1–6, 2003, Paestum, Italy.

Dreccer, M.F., F.C. Ogbonnaya, and G. Borgognone. 2004. Sodium exclusion in primary synthetic wheats. p. 118–121. In: Proceedings of 11th Wheat Breeding Assembly, Symposium on Seeding the Future Conference, Sept. 21–24, Canberra, Australia.

Dreccer F., M. Reynolds, F. Ogbonnaya, S. Chapman, R. Trethowan, and G. Borgognone. 2006. Physiology of drought adaptive traits in synthetic wheats. p. 12. In: M. Imtiaz, F.C. Ogbonnaya and M.vanGinkel (eds.), 1st Synthetic Wheat Symp. Synthetics for Wheat Improvement. Sept. 4–6, Horsham, Victoria, Australia. Abstr. book.

Dreccer, M.F., M.G. Borgognone, F.C. Ogbonnaya, R.M. Trethowan, and B. Winter. 2007. CIMMYT-selected derived synthetic bread wheats for rainfed environments: yield evaluation in Mexico and Australia. Field Crops Res. 100:218–228.

Dreisigacker, S. M. J. Lage, Kishii, and M. Warbuton. 2008. Use of synthetic hexaploid wheat to increase diversity for CIMMYT bread wheat improvement. Aust. J. Agr. Res. 59:413–420.

Dubcovsky, J. and J. Dvořák. 2007. Genome plasticity a key factor in the success of polyploidy wheat under domestication. Science 316:1862–1865.

Dubcovsky, J., G. Santa María, E. Epstein, M.-C. Luo, and J. Dvořák. 1996. Mapping of the K^+/Na^+ discrimination locus *Kna*1 in wheat. Theor. Appl. Genet. 92:448–454.

Dudnikov, A.J. and T. Kawahara. 2006. *Aegilops tauschii*: genetic variation in Iran. Genet. Resour. Crop. Evol. 53:579–586.

Dvorak, J., M.C. Luo, Z.L. Yang and H.B. Zhang. 1998. The structure of the *Aegilops* gene pool and evolution of hexaploid wheat. Theor. Appl. Genet. 97:657–670.

Dvořák, J., M.M. Noaman., S. Goyal, and J. Gorham. 1994. Enhancement of the salt tolerance of *Triticum turgidium* L. by the *Kna*1 locus transferred from *Triticum aestivum* L. chromosome 4D by homoeologous recombination. Theor. Appl. Genet. 87:872–877.

Dyck, P.L. and E.R. Kerber. 1970. Inheritance in hexaploid wheat of adult-plant leaf rust resistance derived from *Aegilops squarrosa*. Can. J. Genet. Cytol 12:175–180.

Eastwood, R.F., E.S. Lagudah, R. Appels, M. Hannah, and J.F. Kollmorgen. 1991. *Triticum tauschii*: a novel source of resistance to cereal cyst nematode (*Heterodera avenae*). Aust. J. Agr. Res. 42:69–77.

Eastwood, R.F. 1995. Genetics of resistance to *Heterodera avenae* in *Triticum tauschii* and its transfer to bread wheat. Ph.D. thesis. Univ. Melbourne, Australia.

Eastwood, R., N. Gororo, and E. Martin. 2006. Experiences with synthetic wheat in a commercial wheat breeding program. p. 9. In: M. Imtiaz, F.C. Ogbonnaya and M.van Ginkel (eds.), 1st Synthetic Wheat Symp. Synthetics for Wheat Improvement. Sept. 4–6, Horsham, Victoria, Australia. Abstr. book.

Eastwood, R.F., E.S. Lagudah, and R. Appels. 1994. A directed search for DNA sequences tightly linked to cereal cyst nematode resistance genes in *Triticum tauschii*. Genome 37:311–319.

Eig, A. 1929. Monographisch-kritische Ubersicht der Gattung *Aegilops*. Repertorium Specierum Novarum Regni Vegetabilis, Beihefte 55:1–228.

El Bouhssini, M., F.C. Ogbonnaya, M. Chen, S. Lhaloui, F. Rihawi, and A. Dabbous. 2012. Sources of resistance in primary synthetic hexaploid wheat to insects pests-Hessian fly, Russian wheat aphid and Sunn pest in the Fertile Crescent. Genet. Res. Crop Evol. DOI 10.1007/s10722-012-9861-3.

Farrer, W. 1904. Some notes on the wheat 'Bobs': its peculiarities, economic value and origin. Agr. Gaz. N. S. W. 15:849–854.

Feldman, M., 2001. Origin of cultivated wheat. p. 3–53. In:A.P. Bonjean, and W.J. Angus (eds.), The world wheat book: a history of wheat breeding. Lavoisier Publications Paris.

Flood, R.G., E.S. Lagudah, and G.M. Halloran. 1992. Expression of vernalization requirement and spikelet number in synthetic hexaploid wheats and their *Triticum tauschii* and tetraploid wheat parents. Ann. Bot. 69:213–217.

Friebe, B., J. Jiang, W.J. Raupp, R.A. McIntosh, and B. S. Gill. 1996. Characterization of wheat-alien translocations conferring resistance to diseases and pests: current status. Euphytica 91:59–87.

Friesen, T.L., S.S. Xu, and M.O. Harris. 2008. Stem rust, tan spot, Stagonospora nodorum blotch and Hessian fly resistance in Langdon durum-*Aegilops tauschii* synthetic hexaploid wheat lines. Crop Sci. 48:1062–1070.

Fritz, A.K., T.X. Cox, B.S. Gill, and R.G. Sears. 1995. Molecular marker-facilitated analysis of introgression in winter wheat, *Triticum tauschii* populations. Crop Sci. 35:1691–1695.

Gatford, K.T. 2004. Seed dormancy mechanisms in diploid wheat (*Triticum tauschii* (Coss.) Schmalh.). Ph.D. thesis. Univ. Melbourne, Australia.

Gatford, K.T., P. Hearnden, F. Ogbonnaya, R.F. Eastwood, and G.M. Halloran. 2002. Novel resistance to pre-harvest sprouting in Australian wheat from the wild relative *Triticum tauschii*. Euphytica. 126:67–76.

Garg, M., F. J. El-Haramein, O. Abdalla, and F.C. Ogbonnaya. 2010. End use quality assessment of synthetic derivatives. In: N.I. Dzyubenko (ed.), 8th Intl. Wheat Conf. N.I. Vavilov Research Institute of Plant Industry, June 1–4, 2010. St. Petersburg, Russia.

Gedye, K.R., C.F. Morris, and A.D. Bettge. 2004. Determination and evaluation of the sequence and textural effects of the puroindoline a and puroindoline b genes in a population of synthetic hexaploid wheat. Theor. Appl. Genet. 109:1597–1603.

Genc, Y., and G.K. McDonald. 2004. The potential of synthetic hexaploid wheats to improve zinc efficiency in modern bread wheat. Plant Soil 262:23–32.

Ghaffary, S.M.T., J.D. Faris, T. L. Friesen, R.G.F. Visser, T.A. J. van derLee, O. Robert, G.H.J. Kema. 2011. New broad-spectrum resistance to septoria tritici blotch derived from synthetic hexaploid wheat. doi 10.1007/s00122-011-1692-7.

Ghaffary, S.M.T., J.D. Faris,T. L. Friesen, R.G.F. Visser, T.A. J. van der Lee, O. Robert, and G.H.J. Kema. 2012. New broad-spectrum resistance to septoria tritici blotch derived from synthetic hexaploid wheat. Theor. Appl. Genet. 124:125–142.

Gill, B.S., J.H. Hatchett, T.S. Cox, W.J. Raupp, R.G. Sears, and T.J. Martin. 1986. Registration of KS85WGRC01 Hessian fly resistant hard red winter wheat germplasm. Crop Sci. 26:1266–1267.

Gill, B.S. and W.J. Raupp 1987. Direct genetic transfers from *Aegilops squarrosa* L. to hexaploid wheat. Crop Sci. 27:445–450.

Gill, B.S., W.J. Raupp, L.E. Browder, T.S. Cox, and R.G. Sears. 1991a. Registration of KS89WGRC7 leaf rust-resistant hard red winter wheat germplasm. Crop Sci. 31:246.

Gill, B.S., D.L. Wilson, W.J. Raupp, J.H. Hatchett, T.S. Cox, A. Amri, and R.G. Sears. 1991b. Registration of KS89WGRC3 and KS89WGRC6 Hessian fly-resistant hard red winter wheat germplasm. Crop Sci. 31:245.

Gill, B.S., D.L. Wilson, W.J. Raupp, J.H. Hatchett, T.L. Harvey, T.S. Cox, and R.G. Sears. 1991c. Registration of KS89WGRC4 hard red winter wheat germplasm lines with resistance to Hessian fly, greenbug, and soilborne mosaic virus. Crop Sci. 31:246.

Gill, B.S., B. Friebe, W.J. Raupp, D.L. Wilson, T.S. Cox, R.G. Sears, G.L. Brown-Guedira, and A.K. Fritz. 2006. Wheat genetic resource center: the first 25 years. Adv. Agr. 89:73–136.

Gill, B.S., L. Huang, V. Kuraparthy, W.J. Raupp, D.L. Wilson, and B. Friebe. 2008. Alien genetic resources for wheat leaf rust resistance, cytogenetic transfer and molecular analysis. Aust. J. Agr. Res. 59:197–205.

Godfray H.C.J., J.R. Beddington, I.R. Crute, L. Haddad, D. Lawrence, J.F. Muir, J. Pretty, S. Robinson, S.M. Thomas, and C. Toulmin. 2010. Food security: the challenge of feeding 9 billion people. Science 327:812–818.

Gorham, J., C. Hardy, R.G. Wyn Jones, L.R. Joppa, and C.N. Law. 1987. Chromosomal location of a K/Na discrimination character in the D genome of wheat. Theor. Appl. Genet. 74:584–588.

Gororo, N.N. 1999. Growth, development and yield of wheat lines bearing *Triticum tauschii* germplasm. Ph.D. thesis. Univ. Melbourne, Australia.

Gororo, N.N., H.A. Eagles, R.F. Eastwood, M.E. Nicolas, and R.G. Flood. 2002. Use of *Triticum tauschii* to improve yield of wheat in low-yielding environments. Euphytica 123:241–254.

Gosman, N., H. Jones, R. Horsnell, A. Kowalski, G. Rose, L. Everest, A. Bentley, S. Tha, C. Uauy, D. Novoselovic, R. Simek, B. Kobiljski, A. Kondic-Spika, O. Mitrofanova, Y. Chesnokov, and A. Greenland. 2011. Comparative analysis of D-genome diversity in *Aegilops tauschii*, common bread wheat (*Triticum aestivum*) and synthetic hexaploid wheat. Proc. 21st International Triticeae Mapping Initiative Workshop, Mexico City, pp 31–32.

Harlan, J.R. and J.M.J. deWit. 1971. Toward a rational classification of cultivated plants. Taxon 20:509–517.

Grama, A., and Z.K. Gerechter-Amitai. 1974. Inheritance of resistance to stripe rust (*Puccinia striiformis*) in crosses between wild emmer (*Triticum dicoccoides*) and cultivated tetraploid and hexaploid wheats II. *Triticum aestivum*. Euphytica 23:393–398.

Gu, X.-Y., L. Zhang, K. Glover, C.G. Chu, S.S. Xu, J.D. Faris, T.L. Friesen, and A. Ibrahim. 2010. Genetic variation of seed dormancy in synthetic hexaploid wheat-derived populations. Crop Sci. 50:1318–1324.

Hall, M.D., G. Brown-Guedira, A. Klatt, and A.K. Fritz. 2009. Genetic analysis of resistance to *soil-borne wheat mosaic virus* derived from *Aegilops tauschii*. Euphytica 169:169–176.

Halloran, G.M., F.C. Ogbonnaya, and E.S. Lagudah. 2008. *Triticum* (*Aegilops*) *tauschii* in the natural and artificial synthesis of hexaploid wheat. Aust. J. Agr. Res. 59:475–490.

Hammer, K. 1980. Zur Taxonomie und Nomenklature der Gattung *Aegilops* L. Feddes Rep. 91:225–258.

Hatchett, J.H., T.J. Martin, and R.W. Livers. 1981. Expression and inheritance of resistance to Hessian fly in synthetic hexaploid wheats derived from *Triticum tauschii* (Coss.) Schmal. Crop Sci. 21:731–734.

He, P., B.R. Friebe, B.S. Gill, and J.M. Zhou. 2003. Allopolyploidy alters gene expression in the highly stable hexaploid wheat. Plant Mol. Biol. 52:401–414.

Herde, D., C.D. Percy, T.L. Walters, and S.M. Neate. 2012. Quantitative genes as the basis for phenotypic selection methodology. p.50. In: R.I. S. Brettell and J. M. Nicol (eds.), Proceedings of the 1st International crown rot workshop for wheat improvement, Narrabri, New South Wales, Australia. 22-23 October 2012.

Hsam, S.L.K., R. Kieffer, and F.J. Zeller. 2001. Significance of *Aegilops tauschii* glutenin genes on breadmaking properties of wheat. Cer. Chem. 78:521–525.

Hu, Y.K. and Z.Y. Xin. 2001. RAPD and SSR markers linked to powdery mildew resistance gene in *Triticum durum-Aegilops squarrosa* amphidiploid M53 (in Chinese with English abstract). Acta Agronomica Sinica 27:15–419.

2. SYNTHETIC HEXAPLOIDS

Hu, Y.K., Z.Y. Xin, X. Chen, Z.Y. Zhang, and X.Y. Duan. 2001. Genetic analysis and gene deduction of powdery mildew resistance in *T. durum-Ae. squarrosa* amphiploids (in Chinese with an English abstract). Acta Genetica Sinica 28:152–157.

Hua, W., Z. Liu, J. Zhu, C. Xie, T. Yang, Y. Zhou, X. Duan, Q. Sun, and Z. Liu. 2009. Identification and genetic mapping of *pm42*, a new recessive wheat powdery mildew resistance gene derived from wild emmer (*Triticum turgidum* var. *dicoccoides*). Theor. Appl. Genet. 119:223–230.

Huang, X.Q., H. Cöster, M.W. Ganal, and M.S. Röder. 2003. Advanced backcross QTL analysis for the identification of quantitative trait loci alleles from wild relatives of wheat (*Triticum aestivum* L.). Theor. Appl. Genet. 106:1379–1389.

Huang, X.Q., H. Kempf, M.W. Ganal, and M.S. Röder. 2004. Advanced backcross QTL analysis in progenies derived from a cross between a German elite winter wheat variety and a synthetic wheat (*Triticum aestivum* L.). Theor. Appl. Genet. 109:933–943.

Imtiaz, M., F.C. Ogbonnaya, J. Oman, and M.van Ginkel. 2008. Characterization of quantitative trait loci controlling genetic variation for preharvest sprouting in synthetic backcross-derived wheat lines. Genetics 178:1725–1736.

Imtiaz, M., F.C. Ogbonnaya, and M. van Ginkel (Eds). 2006. Abstract book: 1st Synthetic Wheat Symposium "Synthetics for wheat improvement" September 4–6, Horsham, Victoria, Australia. 39pp.

Innes, R.L. and E.R. Kerber. 1994. Resistance to wheat leaf rust and stem rust in *Triticum tauschii* and inheritance in hexaploid wheat of resistance transferred from *T. tauschii*. Genome 37:813–822.

Jiang, J. B. Friebe, and B. S. Gill. 1994. Recent advances in alien gene transfer in wheat. Euphytica 73:199–212.

Joppa, L.R., R.G. Timian, and N.D. Williams. 1980. Inheritance of resistance to greenbug toxicity in an amphiploid of *Triticum turgidum/T. tauschii*. Crop Sci. 20:343–344.

Joppa, L.R., and N.D. Williams. 1982. Registration of Largo, a greenbug resistant hexaploid wheat. Crop Sci. 22:901–902.

Kaloshian, I.,P.A. Roberts, and I.J. Thomason. 1989. Resistance to Meloidogyne spp. in Allohexaploid Wheat Derived from *Triticum turgidum* and *Aegilops squarrosa*. J. Nematol. 21:42–47.

Kashkush, K., M. Feldman, and A.A. Levy. 2002. Gene loss, silencing, and activation in a newly synthesized wheat allotetraploid. Genetics 160:1651–1659.

Kema, G.H.J., W. Lange, and C.H.van Silfhout. 1995. Differential suppression of stripe rust resistance in synthetic wheat hexaploids derived from *Triticum turgidum* subsp. *dicoccoides* and *Aegilops squarrosa*. Phytopathology 85:425–429.

Kerber, E.R. and G.J. Green. 1980. Suppression of stem rust resistance in hexaploid wheat cv. Canthatch by chromosome 7DL. Can. J. Bot. 58:1347–1350.

Kerber, E.R. 1987. Resistance to leaf rust in hexaploid wheat: *Lr32* a third gene derived from *Triticum tauschii*. Crop Sci. 27:204–206.

Kerber, E.R. and P.L. Dyck, 1979. Resistance to stem and leaf rust of wheat in *Aegilops squarrosa* and transfer of a gene for stem rust resistance to hexaploid wheat. p.358–364. In:S. Ramanujam (Ed.), Proc. 5th Int. Wheat Genet. Symp. February 23–28, 1978. Indian Society of Genetics & Breeding, New Delhi, India.

Khan, I.A., J.D. Procunier, D.G. Humphreys, G. Tranquilli, A.R. Schlatter, S. Marcucci-Poltri, R. Frohberg, and J. Dubcovsky. 2000. Development of PCR based markers for a high grain protein content gene from *Triticum turgidum* ssp. *dicoccoides* transferred to bread wheat. Crop Sci. 40:518–524.

Kihara, H. 1923. On the physical nature of protoplasm of the wheats. (Japanese). J. Soc. Agric. Forest. Sapporo. Year 15, p. 55–68. English abstract in Japanese Jaa. Bot. Vol. 2, p. (8–9).

Kihara, H. 1925. Weitere Untersuchungen uber die pentaploiden *Triticum*-Bastarde, I. Japan. J. Bot. 2:299–305.

Kihara, H. 1982. Wheat Studies-Retrospect and Prospects. Kodansha Ltd., Tokyo.

Kihara, H., K. Yamashita, M. Tanaka, and J. Tabushi. 1957. Some aspects of the new amphidiploids synthesized from the hybrids, Emmer wheats × *Aegilops squarrosa* var. *strangulata*. Wheat Infor. Serv. 6:14–15.

Knott, D.R. 2000. Inheritance of resistance to stem rust in Medea durum wheat and the role of suppressors. Crop Sci. 40:98–102.

Kruse, A. 1967. Intergeneric hybrids between *Hordeum vulgare* L. ssp. *distichum* (v Pallas $2n=14$) and *Secale cereale* L. (v Petkus $2n=14$). p. 82–92. In: Royal Veterinary and Agricultural College Yearbook 1967. Copenhagen.

Kruse, A. 1969. Intergeneric hybrids between *Triticum aestivum* L. (v Koga, $2n=42$) and *Avena sativa* L. (v Stal $2n=42$) with pseudogamous seed formation. p. 188–200. In: Royal Veterinary and Agricultural College Yearbook 1967. Copenhagen.

Kunert, A., A.A. Naz, O. Dedeck, K. Pillen, and J. Léon. 2007. AB-QTL analysis in winter wheat: I. Synthetic hexaploid wheat (*T. turgidum* ssp. *dicoccoides* × *T. tauschii*) as a source of favourable alleles for milling and baking quality traits. Theor. Appl. Genet. 2007. 115:683–695.

Kushnir, U. and G. M. Halloran. 1984. Transfer of high kernel weight and high protein from wild tetraploid wheat (*Triticum turgidum dicoccoides*) to bread wheat (*T. aestivum*) using homologous and homoeologous recombination. Euphytica 33:249–255.

Lage, J. and R.M. Trethowan. 2008. CIMMYT's use of synthetic hexaploid wheat in breeding for adaptation to rainfed environments globally. Aust. J. Agr. Res. 59:461–469.

Lage, J., M.L. Warburton, J. Crossa, B. Skovmand, and S.B. Andersen. 2003. Assessment of genetic diversity in synthetic hexaploid wheats and their *Triticum dicoccum* and *Aegilops tauschii* parents using AFLPs and agronomic traits. Euphytica 134:305–317.

Lagudah E.S., R. Appels, D. McNeil, and D.P. Schachtman. 1993. Exploiting the diploid D genome chromatin for wheat improvement. p. 87–107. In: J.P. Gustafson, R. Appels and P. Raven (eds.), Gene conservation and exploitation. Plenum Press, New York.

Lagudah, E.S., O. Moullet, and R. Appels. 1997. Map-based cloning of a gene sequence encoding a nucleotide-binding domain and a leucine-rich region at the *Cre3* nematode resistance locus of wheat. Genome, 40:659–665.

Lagudah, E.S., F. Macritchie, and G.M. Halloran. 1987. The influence of high-molecular-weight subunits of glutenin from *Triticum tauschii* on flour quality of synthetic hexaploid wheat. J. Cereal Sci. 5:129–138.

Lan, X.J., Y.L. Zheng, X.B. Ren, D.C. Liu, Y.M. Wei, and Z.H. Yan. 2005. Utilization of preharvest sprouting tolerance gene of synthetic wheat RSP (in Chinese with English abstract). J. Plant Genet. Resour. 6:204–209.

Lange, W. and G. Jochemsen. 1992. Use of the gene pools of *Triticum turgidum* ssp. *dicoccoides* and *Aegilops squarrosa* for the breeding of common wheat (*T. aestivum*), through chromosome-doubled hybrids I. Two strategies for the production of the amphiploids. Euphytica 59:197–212.

Langridge, P. and D. Fleury. 2011. Making the most of 'omics' for crop breeding. Trends Biotechnol. 29:33–40.

Lazar, M.D., W.D. Worrall, G.L. Peterson, K.B. Porter, L.W. Rooney, N.A. Tuleen, D.S. Marshall, M. E. McDaniel, and L.R. Nelson. 1997. Registration of 'TAM 110' wheat. Crop Sci. 37:1978–1979.

Lazar, M.D., W.D. Worrall, K.B. Porter, and N.A. Tuleen. 1996. Registration of eight closely related wheat germplasm lines differing in biotype E greenbug resistance. Crop Sci. 36:1419.

Li, G., T. Fang, H. Zhang, C. Xie, H. Li, T. Yang, E. Nevo, T. Fahima, Q. Sun, and Z. Liu. 2009. Molecular identification of a new powdery mildew resistance gene *Pm41* on chromosome 3BL derived from wild emmer (*Triticum turgidum* var. *dicoccoides*). Theor. Appl. Genet. 119:531–539.

Li, G.-Y., X.-C. Xia, Z.-H. He, and Q.-X. Sun. 2007. Allelic variation of puroindoline a and puroindoline b genes in new type synthetic hexaploid wheats from CIMMYT (in Chinese with English abstr.). Acta Agronica Sinica 33:242–249.

Li, J., H.T. Wei, W.Y. Yang, and Z.S. Peng. 2005. Genetic evaluation of a synthetic hexaploid wheat Cerceta/*Aegilops tauschii* 783 with resistances to stripe rust and pre-harvest sprouting by SSR markers (in Chinese with English abstract). Mol. Plant Breed. 3:810–814.

Lillemo, M., F. Chen, X.C. Xia, M. William, R.J. Peña, R. Trethowan, and Z.H. He. 2006. Puroindoline grain hardness alleles in CIMMYT bread wheat germplasm. J. Cereal Sci. 44:86–92.

Liu, S., R. Zhou, Y. Dong, P. Li, and J. Jia. 2006. Development, utilization of introgression lines using a synthetic wheat as donor. Theor. Appl. Genet. 112:1360–1373.

Liu, Z., Q. Sun, Z. Ni, E. Nevo, and T. Yang. 2002. Molecular characterization of a novel powdery mildew resistance gene *Pm30* in wheat originating from wild emmer. Euphytica 123:21–29.

Lopes, M.S. and M. R. Reynolds. 2011. Drought adaptive traits and wide adaptation in elite lines derived from resynthetized hexaploid wheat. Crop Sci. 51:1617–1626.

Loughman R, E.S. Lagudah, M. Trottet, R.E. Wilson, and A. Mathews. 2001. Septoria nodorum blotch resistance in *Aegilops tauschii* and its expression in synthetic amphiploids. Aust. J. Agr. Res. 52:1393–1402.

Lowe, I.,L. Jankuloski, S. Chao, X. Chen, D. See and J. Dubcovsky. 2011. Mapping and validation of QTL which confer partial resistance to broadly virulent post-2000 North American races of stripe rust in hexaploid wheat.

Lu, C.M., W.Y. Yang, and B.R. Lu. 2005b. Differentiation of the high molecular weight glutenin subunit D^tx2.1 of *Aegilops tauschii* indicated by partial sequences of its encoding gene and SSR markers. Euphytica 141:75–83.

Lu, C.M., W.Y. Yang., W.J. Zhang, and B.R. Lu. 2005a. Identification of SNPs and development of allelic specific PCR markers for high molecular weight glutenin subunit $Dx^t1.5$ from *Aegilops tauschii* through sequence characterization. J. Cereal Sci. 41:13–18.

Lu, H., J.C. Rudd, J.D. Burd, and Y. Weng. 2010. Molecular mapping of greenbug resistance genes *Gb2* and *Gb6* in T1AL.1RS wheat-rye translocations. Plant Breed. 129:472–476.

Lutz, J., S.L.K. Hsam, E. Limpert, and F.J. Zeller. 1994. Powdery mildew resistance in *Aegilops tauschii* Coss. and synthetic hexaploid wheats. Genet. Resour. Crop Evol. 41:151–158.

Lutz, J., S.L.K. Hsam, E. Limpert, and F.J. Zeller. 1995. Chromosomal location of powdery mildew resistance genes in *Triticum aestivum* L. (common wheat). 2. Genes *Pm2* and *Pm19* from *Aegilops squarrosa* L. Heredity 74:152–156.

Ma, H., R.P. Singh, and A. Mujeeb-Kazi. 1995. Resistance to stripe rust in *Triticum turgidum, T. tauschii* and their synthetic hexaploids. Euphytica 82:117–124.

Maan, S.S. and T. Sasakuma. 1977. Fertility of amphihaploids in *Triticinae*. J. Hered. 68:87–94.

Mackay, M.C. and K.A. Street. 2004. Focused identification of germplasm strategy – FIGS. Proceedings 11th Wheat Breeding Assembly, Sept. 20–24, 2004, Canberra, Australia.

Mackie, A.M., P.J. Sharp, and E.S. Lagudah. 1996. The nucleotide and derived amino acid sequence of a HMW-glutenin gene from *Triticum tauschii* and comparison with those from the D-genome of bread wheat. J. Cereal Sci. 24:73–78.

Marais, G.F., G.F. Potgieter, and H.S. Roux. 1994. An assessment of the variation for stem rust resistance in the progeny of a cross involving the *Triticum* species *aestivum, turgidum* and *tauschii*. S. Afr. J. Plant Soil 11:15–19.

Marais, G.F., W.G. Wessels, M. Horn. 1998. Association of a stem rust resistance gene (*Sr45*) and two Russian wheat aphid resistance genes (*Dn5* and *Dn7*) with mapped structural loci in common wheat. S. Afr. J. Plant Soil 15:67–71.

Mares, D. and K. Mrva. 2008. Genetic variation for quality traits in synthetic wheat germplasm. Aust. J. Agr. Res. 59:406–412.

Martin, T.J., T.L. Harvey, and J.H. Hatchett. 1982. Registration of greenbug and Hessian fly resistance wheat germplasm. Crop Sci. 22:1089.

Martin, E.M., R.F. Eastwood, and F.C. Ogbonnaya. 2004. Identification of microsatellite markers associated with the cereal cyst nematode resistance gene *Cre3* in wheat. Aust. J. Agr. Res. 55:1205–1211.

Massa, A.N., C.F. Morris, and B.S. Gill. 2004. Sequence diversity of puroindoline-a, puroindoline-b, and the grain softness protein genes in *Aegilops tauschii* Coss. Crop Sci. 44:1808–1816.

Mather, D,H. Wallwork, D. Herde, J. Taylor, E. Vassos, K. Khoo, R. Fox and K. Chalmers. 2012. Small-effect QTL for crown rot resistance on chromosomes 2B and 4B of wheat. p.50. In: R.I. S. Brettell and J. M. Nicol (eds.), Proceedings of the 1st International crown rot workshop for wheat improvement, Narrabri, New South Wales, Australia. 22–23 October 2012.

Matsuoka, Y. and S. Nasuda. 2004. Durum wheat as a candidate for the unknown female progenitor of bread wheat: an empirical study with a highly fertile F_1 hybrid with *Aegilops tauschii* Coss. Theor. Appl. Genet. 109:1710–1717.

Matsuoka, Y., E. Nishioka, T. Kawahara and S. Takumi. 2009. Genealogical analysis of subspecies divergence and spikelet-shape diversification in central Eurasian wild wheat *Aegilops tauschii* Coss. Plant Syst. Evol. 279:233–244.

Matsuoka, Y., S. Takumi, and T. Kawahara. 2007. Natural variation for fertile triploid F_1 hybrid formation in allohexaploid wheat speciation. Theor. Appl. Genet. 115:509–518.

Matsuoka, Y., S. Takumi and T. Kawahara. 2008. Flowering time diversification and dispersalin Central Eurasian wild wheat *Aegilops tauschii* Coss.: Genealogical and ecological framework. PLoS One 3 (9): e3138.

May, C.E. and E.S. Lagudah. 1992. Inheritance in hexaploid wheat of Septoria tritici blotch resistance and other characteristics derived from *Triticum tauschii*. Aust. J. Agr. Res. 43:433–442.

McFadden, E.S. 1930. A successful transfer of emmer characters to vulgare wheat. J. Am. Soc. Agr. 22:1020–1034.

McFadden, E.S. and E.R. Sears. 1944. The artificial synthesis of *Triticum spelta*. Rec. Genet. Soc. Am. 13:26–27.

McFadden, E.S. and E.R. Sears. 1946. The origin of *Triticum spelta* and its free-threshing hexaploid relatives. J. Hered. 37:81–89; 107–116.

McIntosh, R.A. and E.S. Lagudah. 2000. Cytogenetic studies in wheat XVIII. Gene *Yr24* for resistance to stripe rust. Plant Breed. 119:81–83.

McIntosh, R.A., C.R. Wellings, and R.F. Park. 1995. Wheat rusts: an atlas of resistance genes. CSIRO Publications, East Melbourne, Victoria, Australia.

Miranda, L.M., J.P. Murphy, D.S. Marshall, C. Cowger, and S. Leath. 2007. Chromosomal location of *Pm35*, a novel *Aegilops tauschii* derived powdery mildew resistance gene introgressed into common wheat (*Triticum aestivum* L.). Theor. Appl. Genet. 114:1451–1456.

Miranda, L.M., J.P. Murphy, D. Marshall, and S. Leath. 2006. *Pm34*: a new powdery mildew resistance gene transferred from *Aegilops tauschii* Coss. to common wheat (*Triticum aestivum* L.). Theor. Appl. Genet. 113:1497–1504.

Mizuno, N., M. Yamasaki, Y. Matsuoka, T. Kawahara, and S. Takumi. 2010. Population structure of wild wheat D-genome progenitor *Aegilops tauschii* Coss.: implications for intraspecific lineage diversification and evolution of common wheat. Mol. Ecol. 19:999–1013.

Monasterio, J.I. and R.D. Graham. 2000. Breeding for trace minerals in wheat. Food Nut. Bull. 21:393–396.

Mujeeb-Kazi, A. 1993. Interspecific and intergeneric hybridization in the *Triticeae* for wheat improvement. p. 95–102. In: A. Damania (ed.), Biodiversity and wheat improvement. Wiley, Chichester, UK.

Mujeeb-Kazi, A. 2003a. New genetic stocks for durum and bread wheat improvement. p. 772–774. In: 10th Intl. Wheat Genetics Symp., Paestum, Italy.

Mujeeb-Kazi, A. 2003b. Wheat improvement facilitated by novel genetic diversity and in vitro technology. Plant Cell Tissue Organ Cult. 13:179–210.

Mujeeb-Kazi A. and R. Asiedu. 1990. Wide hybridization—potential of alien genetic transfers for *Triticum aestivum* improvement. p. 111–127. In: Y.P.S. Bajaj (ed.), Biotech in agriculture and forestry, Vol. 13, wheat.

Mujeeb-Kazi, A., S. Cano, V. Rosas, A. Cortes, R. Delgado. 2001a. Registration of five synthetic hexaploid wheat and seven bread wheat lines resistant to wheat spot blotch. Crop Sci. 41:1653–1654.

Mujeeb-Kazi, A. and R. Delgado. 2001. A second, elite set of synthetic hexaploid wheats based upon multiple disease resistance. Ann. Wheat Newslett. 47:114–116.

Mujeeb-Kazi, A., R. Delgado, L. Juarez, and S. Cano. 2001b. Scab resistance (type II: spread) in synthetic hexaploid germplasm. Ann. Wheat Newslett. 47:118–120.

Mujeeb-Kazi, A., G. Fuentes-Davila, R. Delgado, V. Rosas, S., Cano A. Cortés, Juarez L, and J. Sanchez. 2000a. Current status of D-genome based, synthetic, hexaploid wheats and the characterization of an elite subset. Ann. Wheat Newslett. 46:76–79.

Mujeeb-Kazi, A., G. Fuentes-Davila, R.L. Villareal, A. Cortés, V. Rosas, and R. Delgado. 2001c. Registration of 10 synthetic hexaploid wheat and six bread wheat germplasms resistant to karnal bunt. Crop Sci. 41:1652–1653.

Mujeeb-Kazi, A., Ll. Gilchrist, R.L. Villareal, and R. Delgado. 2000b. Registration of 10 wheat germplasms resistant to *Septoria tritici* leaf blotch. Crop Sci. 40:590–591.

Mujeeb-Kazi, A., A. Gul, M. Farooq, S. Rizwan, and I. Ahmad. 2008. Rebirth of synthetic hexaploids with global implications for wheat improvement. Austral. J. Agr. Res, 59:391–398.

Mujeeb-Kazi, A. and G. Hettel. 1995. Utilizing wild Grass Biodiversity in Wheat Improvement: 15 years of Wide Cross Research at CIMMYT. p. 1–140.

Mujeeb-Kazi, A. and G. Kimber. 1985. The production, cytology and practicality of wide hybrids in the Triticeae. Cereal Res. Commun. 13:111–124.

Mujeeb-Kazi, A., V. Rosas, and S. Roldan. 1996. Conservation of the genetic variation of *Triticum tauschii* (Coss.) Schmalh. (*Aegilops squarrosa* auct. non L.) in synthetic hexaploid wheats (*T. turgidum* L. s.lat. × *T. tauschii*; $2n = 6x = 42$, AABBDD) and its potential utilization for wheat improvement. Genet. Resour. Crop Evol. 43:129–134.

Mulki, M.A., G. Ye, L.C. Emebiri, D. Moody, and F.C. Ogbonnaya. 2012. Association mapping for soil-borne pathogen resistance in synthetic hexaploid wheat. Mol. Breed. (Accepted).

Murphy, J.P., S. Leath, D. Huynh, R.A. Navarro, and A. Shi. 1998. Registration of NC96BGTD1, NC96BGTD2 and NC96BGTD3 wheat germplasm resistant to powdery mildew. Crop Sci. 38:570–571.

Murphy, J.P., S. Leath, D. Huynh, R.A. Navarro, and A. Shi. 1999. Registration of NC97BGTD7 and NC97BGTD8 wheat germplasms resistant to powdery mildew. Crop Sci. 39:884–885.

Naghavi, M.R. and M. Mardi. 2010. Characterization of genetic variation among accessions of *Aegilops tauschii*. AsPac J. Mol. Biol. Biotechnol. 18:93–96.

Narasimhamoorthy, B., B.S. Gill, A.K. Fritz, J.C. Nelson, and G.L. Brown-Guedira. 2006. Advanced backcross QTL analysis of a hard winter wheat × synthetic wheat population. Theor. Appl. Genet. 112:787–796.

Naz, A.A., A. Kunert, V. Lind, K. Pillen, and J. Léon. 2008. AB-QTL analysis in winter wheat: II. Genetic analysis of seedling and field resistance against leaf rust in a wheat advanced backcross population. Theor. Appl. Genet. 116:1095–1104. 10.1007/s00122-008-0738-y.

Nelson, J.C., C. Andreescu, F. Breseghello, P.L. Finney, D.G. Gualberto, C.J. Bergman, R.J. Peña, M.R. Perretant, P. Leroy, C.O. Qualset, and M.E. Sorrells. 2006. Quantitative trait locus analysis of wheat quality traits. Euphytica 149:145–159.

Nelson, J.C., J.E. Autrique, G. Fuentes-Davila, and M.E. Sorrells. 1998. Chromosomal location of genes for resistance to Karnal bunt in wheat. Crop Sci. 38:231–236.

Nelson, J.C., R.P. Singh, J.E. Autrique, and M.E. Sorrells. 1997. Mapping genes conferring and suppressing leaf rust resistance in wheat. Crop Sci. 37:1928–1935.

Nelson, J.C., A.E. VanDeynze, E. Autrique, M.E. Sorrells, Y.H. Lu, S. Negre, M. Bernard, and P. Leroy. 1995. Molecular mapping of wheat. Homoeologous group 3. Genome 38:525–533.

Nishikawa, K., Y. Furuta, and T. Wada. 1980. Genetic studies on alpha-amylase isozymes in wheat. III. Intraspecifc variation in *Aegilops squarrosa* and birthplace of hexaploid wheat. Jpn. J. Genet. 55:325–336.

Nishikawa, K., T. Mori, N. Takanmi and Y. Furuta. 1974. Mapping of progressive necrosis genes *Ne1* and *Ne2* of common wheat by telocentric method. Jpn. J. Breed. 24:277–281.

Niwa, K., H. Aihara, A. Yamada, and T. Motohashi. 2010. Chromosome number variations in newly synthesized hexaploid wheats spontaneously derived from self-fertilization of *Triticum carthlicum* Nevski / *Aegilops tauschii* Coss. F1 hybrids. Cereal Res. Commu. 38:449–458.

Nkongolo, K.K., J.S. Quick, A.E. Limin, and D.B. Fowler. 1991. Sources and inheritance of resistance to Russian wheat aphid in *Triticum* species amphiploids and, *Triticum tauschii*. Can. J. Plant Sci. 71:703–708.

Ogbonnaya, F.C. 2011. Development, management and utilization of synthetic hexaploids in wheat improvement. p. 823–849. In: A.P. Bonjean W.J. Angus and M. van Ginkel (eds.), The world wheat book: a history of wheat breeding. Vol. 2. Lavoisier, France.

Ogbonnaya, F.C., G.M. Halloran, and E.S. Lagudah. 2005. D genome of wheat-60 years on from Kihara, Sears and McFadden. In: Tsunewaki Koichiro (ed.), Frontiers of Wheat BioScience. Kihara Memorial Yokohama Foundation for the Advancement of Life Sciences, Yokohama, Japan.

Ogbonnaya, F.C., S. Huang, E. Steadman, L. Emebiri, M. Imtiaz, M.F. Dreccer, E.S. Lagudah, R. Munns, and M. van Ginkel. 2008b. Mapping quantitative trait loci associated with salinity tolerance in synthetic derived backcrossed bread lines. 11th International Wheat Genetics Symposium, Brisbane.

Ogbonnaya, F.C., M. Imtiaz, H.S. Bariana, M. McLean, M.M. Shankar, G.J. Hollaway, R.M. Trethowan, E.S. Lagudah, and M.van Ginkel. 2008a. Mining synthetic hexaploids for multiple disease resistance to improve wheat. Aust. J. Agr. Res. 59:421–431.

Ogbonnaya, F.C., M. Imtiaz, and R. DePauw. 2007b. Haplotype diversity of preharvest sprouting QTLs in wheat. Genome 50:107–118.

Ogbonnaya, F.C., E. Steadman, D. Burch, D. Moody, and R. Munns. 2009. Breeding for salinity tolerance using synthetic wheats. Genomics Symp. The Genomics of Salinity. The Grand Chancellor Hotel, South Australia, 16–18 Nov.

Ogbonnaya, F.C., N.C. Subrahmanyam, O. Moullet, J.de Majnik, H.A. Eagles, J.S. Brown, R.F. Eastwood, J. Kollmorgen, R. Appels, and E.S. Lagudah. 2001. Diagnostic DNA markers for cereal cyst nematode resistance in bread wheat. Aust. J. Agr. Res. 52:1367–1374.

Ogbonnaya, F.C., G. Ye, R. Trethowan, F. Dreccer, D. Lush J. Shepperd, and M.van Ginkel. 2007a. Yield of synthetic backcross-derived lines in rainfed environments of Australia. Euphytica 157:321–336.

Ogbonnaya, F.C., M. Imtiaz, P. Hearnden, J. Wilson, R.F. Eastwood, K.T. Gatford and M. van Ginkel.2006.Identification of a novel gene for seed dormancy in wheat. p.634–639 In: C.F. Mercer (ed). Breeding for Success: Diversity in Action.Proceedings of the 13th Australasian Plant Breeding Conference, Christchurch, New Zealand 18–21, April 2006, (CD, ISBN: 978-0-86476-167-8).

Okuno, K., K. Ebana, B. Noov, and H. Yoshida. 1998. Genetic diversity of Central Asian and North Caucasian *Aegilops* as revealed by RAPD markers. Genet. Resour. Crop. Evol. 45:398–394.

Oury, F.-X., F. Leenhardt, C. Rémésy, E. Chanliaud, B. Duperrier, F. Balfourier, and G. Charmet. 2006. Genetic variability and stability of grain magnesium, zinc and iron concentration in bread wheat. Eur. J. Agron. 25:177–185.

Palta, J.A., X. Chen, S.P. Milroy, G.J. Rebetzke, M.F. Dreccer, and M. Watt. 2011. Large root systems: are they useful in adapting wheat in dry environments? Funct. Plant Biol. 38:347–354.

Patterson, F.L., F.B. MaasIII, J.E. Foster, R.H. Ratcliffe, S. Cambron, G. Safranski, P.L. Taylor, and H.W. Ohm. 1994. Registration of eight Hessian fly resistant common winter wheat germplasm lines (Carol, Erin, Flynn, Iris, Joy, Karen, Lola, and Molly). Crop Sci. 34:315–316.

Peña, R.J., J. Zarco-Hernández, and A. Mujeeb-Kazi. 1995. Glutenin subunit compositions and bread-making quality characteristics of synthetic hexaploid wheats derived from *Triticum turgidum* x *Triticum tauschii* (Coss.) Schmal crosses. J. Cereal Sci. 21:15–23.

Pestsova, E.G., A. Börner, and M.S. Röder. 2001. Development of a set of *Triticum aestivum–Aegilops tauschii* introgression lines. Hereditas 135:139–143.

Pestsova, E., V. Korzun, N.P. Goncharov, K. Hammer, M.W. Ganal, and M.S. Roder. 2000. Microsatellite analysis of *Aegilops tauschii* germplasm. Theor. Appl. Genet. 101:100–106.

Porter, K.B., G.L. Peterson, W.D. Worrall, M.E. McDaniel, N.A. Tuleen, D.S. Marshall, and L.R. Nelson. 1989. Registration of TXGH10563B, TXGH10989 and TXGH13622 greenbug-resistant wheat germplasm lines. Crop Sci. 29:1585–1586.

Pritchard, D.J., P.A. Hollington, W.P. Davies, J. Gorham, J.L. Diaz de Leon, and A. Mujeeb-Kazi. 2002. K^+/Na^+ discrimination in synthetic hexaploid wheat lines: Transfer of the trait for K^+/Na^+ discrimination from *Aegilops tauschii* in to a *Triticum turgidum* background. Cer. Res. Commun. 30:261–267.

Qi, L. L., B. Friebe, P. Zhang, and B. S. Gill. 2007. Homoeologous recombination, chromosome engineering and crop improvement. Chromosome Res. 15:3–19.

Ram, S., A. Verma, and S. Sharma. 2010. Large variability exists in phytase levels among Indian wheat varieties and synthetic hexaploids. J. Cereal Sci. 52:486–490.

Rattey, A.R., S.C. Chapman, C.L. McIntyre, and R. Shorter. 2008. Utility of derived synthetic wheat to enhance adaptive traits for yield in the northern region of Australia. In: R. Appels, R. Eastwood, E. Lagudah, P. Langridge, M. Mackay, L. McIntyr (eds.), The 11th Intl. Wheat Genetics Symp. Proce. Sydney Univ. Press (http://hdl.handle.net/2123/3442).

Rattey, A. and R. Shorter. 2010. Evaluation of CIMMYT conventional and synthetic spring wheat germplasm in rainfed sub-tropical environments. I. Grain yield. Field Crops Res. 118:273–281.

Ren, X.-B., X.-J. Lan, D.-C. Liu, J.-L. Wang, and Y.L. Zheng. 2008. Mapping QTLs for pre-harvest sprouting tolerance on chromosome 2D in a synthetic hexaploid wheat x common wheat cross. J. Appl. Genet. 49:333–341.

Rengasamy, P. 2002. Transient salinity and subsoil constraints to dryland farming in Australian sodic soils: an overview. Austral. J. Expt. Agr. 42:351–361.

Reynolds, M., F. Dreccer, and R. Trethowan. 2007. Drought adaptive traits derived from wheat wild relatives and landraces. J. Exp. Bot. 58:177–186.

Rouse, M.N., E.L. Olson, B.S. Gill, M.O. Pumphery, and Y. Jin. 2011. Stem rust resistance in *Aegilops tauschii* germplasm. Crop Sci. 51:2074–2078.

Rudd, J.C., M.A. Lazar, D.W. Worrall, D.S. Marshall, and R.L. Sutton. 2004. Registration of TAM 112 wheat. http://www.ars.usda.gov/research/publications/publications.htm?seq_no_115=174435.

Sadras, V., D. Roget, and G. O'Leary. 2002. On-farm assessment of environmental and management constraints to wheat yield and efficiency in the use of rainfall in the Mallee. Aust. J. Agr. Res. 53:587–598.

Sardesai, N., J.A. Nemacheck, S. Subramanyam, and C.E. Williams. 2005. Identification and mapping of *H32*, a new wheat gene conferring resistance to Hessian fly. Theor. Appl. Genet. 111:1167–1173.

Sax, K. 1918. The behaviour of chromosomes in fertilization. Genetics 3:309–327.

Sax K. 1922a. Sterility in wheat hybrids III. Endosperm development and F2 sterility. Genetics 7:554–559.

Sax K. 1922b. Sterility in wheat hybrids. II. Chromosome behaviour in partially sterile hybrids. Genetics 7:513–552.

Sehgal, S.K., S. Kaur, S. Gupta, A. Sharma, R. Kaur, and N.S. Bains. 2011. A direct hybridization approach to gene transfer from *Aegilops tauschii* Coss. to *Triticum aestivum* L. Plant Breed. 130:98–100.

Shah, S.H., J. Gorham, B.P. Forster, and G.R. Wyn Jones. 1987. Salt tolerance in Triticeae— the contribution of the D genome to cation selectivity in hexaploid wheat. J. Exp. Bot. 38:254–269.

Sharma, H.C. and B.S. Gill. 1983. Current status of wide hybridization in wheat. Euphytica 32:17–31.

Sharma, H.H. 1995. How wide can a wide cross be? Euphytica 82:43–64.

Siedler, H., A. Obst, S.L.K. Hsam, and F.J. Zeller. 1994. Evaluation for resistance to *Pyrenophora tritici-repentis* in *Aegilops tauschii* Coss. and synthetic hexaploid wheat amphiploids. Genet. Resour. Crop Evol. 41:27–34.

Singh, R.P., J.C. Nelson, and M.E. Sorrells. 2000. Mapping *Yr28* and other genes for resistance to stripe rust in wheat. Crop Sci. 40:1148–1155.

Singh, R.P., D.P. Hodson, J. Huerta-Espino, Y. Jin, S. Bhavanai, P. Njau, H. Herrara-Foessel, P.K. Singh, S. Singh, and V. Govindan. 2011a. The emergence of Ug99 races of the stem rust fungus is a threat to world wheat production. Annu. Rev. Phytopathol. 49:465–481.

Singh, R.P., J. Huerta-Espino, S. Bhavanai, SD. Herrara-Foessel, Y. Jin, P. Njau, P.K. Singh, G. Velu, S. Singh, R.J. Pena and J. Crossa. 2011b. p. 98–104. High yielding CIMMYT spring wheats with resistance to Ug99 and other rusts developed through targeted breeding. BGRI Technical Workshop, June 13–16, St. Paul MN.

Smale, M., M.P. Reynolds, M. Warburton, B. Skovmand, R. Trethowan, R.P. Singh, I. Ortiz-Monasterio, and J. Crossa. 2002. Dimensions of diversity in modern spring bread wheat in developing countries from 1965. Crop Sci. 42:1766–1779.

Sohail, Q., T. Inoue, H. Tanaka, A. E. Eltayeb, Y. Matsuoka, and H. Tsujimoto. 2011. Applicability of *Aegilops tauschii* drought tolerance traits to breeding of hexaploid wheat. Breed. Sci. 61:347–357.

Sohail, Q., T. Shehzad, A. Kilian, A. E. Eltayeb, H. Tanaka, and H. Tsujimoto. 2012. Development of diversity array technology (DArT) markers for assessment of population structure and diversity in *Aegilops tauschii*. Breed. Sci. 38–45.

Sorrells, M.E., J.P. Gustafson, D. Somers, S. Chao, D. Benscher, G. Guedira-Brown, E. Huttner, A. Kilian, P.E. McGuire, K. Ross, J. Tanaka, P. Wenzl, K. Williams, and C.O. Qualset. 2011. Reconstruction of the synthetic W9784 × Opata M85 wheat reference population. Genome 54:875–882.

Tadesse, W., M. Schmolke, S.L.K. Hsam, V. Mohler, G. Wenzel, and F.J. Zeller. 2007. Molecular mapping of resistance genes to tan spot (*Pyrenophora tritici-repentis* race 1) in synthetic wheat lines. Theor. Appl. Genet. 114:855–862.

Takumi, S., E. Nishioka, H. Morihiro, T. Kawahara, and Y. Matsuoka. 2009. Natural variation of morphological traits in the wild wheat progenitor *Aegilops tauschii* Coss. Breed. Sci. 59:579–588.

Talbot S. J. 2011. Introgression of genetic material from primary synthetic hexaploids into an Australian bread wheat (*Triticum aestivum* L.). M. Agr. Sci., Univ. Adelaide, Australia.

Tang, Y.-I., W.-Y. Yang, J.-C. Tian, J. Li, and F. Chen. 2008. Effect of HMW-GS 6+8 and 1.5+10 from synthetic hexaploid wheat on wheat quality traits. Agr. Sci. China 7: 1161–1171.

Tanksley, S.D. and J.C. Nelson. 1996. Advanced backcross QTL analysis: a method for the simultaneous discovery and transfer of valuable QTLs from unadapted germplasm into elite breeding lines. Theor. Appl. Genet. 92:191–203.

Thompson, J.P. 2008. Resistance to root-lesion nematodes (*Pratylenchus thornei* and *P. neglectus*) in synthetic hexaploid wheats and their durum and *Aegilops tauschii* parents. Aust. J. Agr. Res. 59:432–446.

Thompson, J.P., P.S. Brennan, T.G. Clewett, J.G. Sheedy, and N.P. Seymour. 1999. Progress in breeding wheat for tolerance and resistance to root-lesion nematode (*Pratylenchus thornei*). Aust. Plant Pathol. 28:45–52.

Trethowan, R.M. and A. Mujeeb-Kazi. 2008. Novel germplasm resources for improving environmental stress tolerance of hexaploid wheat. Crop Sci. 48:1255–1265.

Trethowan, R.M. and M.vanGinkel. 2009. Synthetic wheat—an emerging genetic resource. p. 369–386. In: B. Carver (ed.), Wheat Sci. Trade. Wiley-Blackwell, Ames, IA.

Tsujimoto, H. 2010. Alien genetic resources for dryland wheat breeding. Proceedings of 10th International Conference on Dry Land Development, Dec. 12–15, 2010, Cairo pp 1500–1530.

Uauy C, J. Carlos Brevis, X. Chen, I. Khan, L. Jackson, O. Chicaiza, A. Distelfeld, T. Fahima, and J. Dubcovsky. 2005. High-temperature adult-plant (HTAP) stripe rust resistance gene *Yr36* from *Triticum turgidum* ssp. *dicoccoides* is closely linked to the grain protein content locus *Gpc-B1*. Theor. Appl. Genet. 112:97–105.

Valkoun, J.J. 2001. Wheat pre-breeding using wild progenitors. Euphytica 119:17–23.
Valkoun, J., J. Dostal, and D. Kucerova. 1990. *Triticum* x *Aegilops* hybrids through embryo culture. p. 152–166. In: P.S. Bajaj (ed.), Wheat. Springer Verlag, Berlin, Heidelberg.
van Ginkel, M. and F. Ogbonnaya. 2007. Novel genetic diversity from synthetic wheats in breeding cultivars for changing production conditions. Field Crops Res. 104:86–94.
van Slageren, M.W. 1994. Wild wheats: A monograph of *Aegilops* L. and *Amblyopyrum* (Jaub.et Spach) Eig (Poaceae). ICARDA and Wageningen Agricultural University.
Villareal, R.L., A. Mujeeb-Kazi, and G. Fuentes-Dávila. 1996. Registration of four synthetic hexaploid wheat germplasm lines derived from *Triticum turgidum* × *T. tauschii* crosses and resistant to Karnal bunt. Crop Sci. 36:218.
Villareal R.L., K. Sayre, O. Bañuelos, and A. Mujeeb-Kazi. 2001. Registration of four synthetic hexaploid wheat (*Triticum turgidum/Aegilops tauschii*) germplasm lines tolerant to waterlogging. Crop Sci. 41:274.
Wallwork, H and M. Butt.2012.Breeding for resistance to crown rot in wheat through the accumulation of minor genes. p.51. In:R.I.S. Brettell and J.M. Nicol (eds.),Proceedings of the 1st International crown rot workshop for wheat improvement, Narrabri, New South Wales, Australia. 22-23 October 2012.
Wang, R.R-C. 1989. Intergeneric hybrids involving perennial Triticeae. Genet. (Life Sci. Adv.) 8:57–64.
Wang, T., S.S. Xu, M.O. Harris, J. Hu, L. Liu, and X. Cai. 2006. Genetic characterization and molecular mapping of Hessian fly resistance genes derived from *Aegilops tauschii* in synthetic wheat. Theor. Appl. Genet. 113:611–618.
Waterhouse, W.L. 1933. On the production of fertile hybrids from crosses between vulgare and Khapli emmer wheats. Proceedings of Linnean Society NSW 58:99–104.
Watson, I.A. and D.M. Stewart. 1956. A comparison of the rust reaction of wheat varieties Gabo, Timstein and Lee. Agron. J. 48:514–516.
Weng, Y., W. Li, R.N. Devkota, and J.C. Rudd. 2005. Microsatellite markers associated with two *Aegilops tauschii*-derived greenbug resistance loci in wheat. Theor. Appl. Genet. 110:462–469.
William, M.D.H.M., R.J. Peña, and A. Mujeeb-Kazi. 1993. Seed protein and isozyme variations in *Triticum tauschii* (*Aegilops squarrosa*). Theor. Appl. Genet. 87:257–263.
Williams, N.D., J.D. Miller, and D.L. Klindworth. 1992. Induced mutations of a genetic suppressor of resistance to wheat stem rust. Crop Sci. 32:612–616.
Xie, W.L. and E. Nevo. 2008. Wild emmer: genetic resources, gene mapping and potential for wheat improvement. Euphytica 164:603–614.
Xu, S. and Y. Dong. 1992. Fertility and meiotic mechanisms of hybrids between chromosome autoduplication tetraploid wheats and *Aegilops* species. Genome 35:379–384.
Xu, S.J. and L.R. Joppa. 1995. Mechanisms and inheritance of first division restitution in hybrids of wheat, rye, and *Aegilops squarrosa*. Genome 38:607–615.
Xu, S.J. and L.R. Joppa. 2000. First-division restitution in hybrids of Langdon durum disomic substitution lines with rye and *Aegilops squarrosa*. Plant Breed. 119:33–241.
Xu, S.S., X. Cai, T. Wang, M.O. Harris, and T.L. Friesen. 2006. Registration of two synthetic hexaploid wheat germplasms resistant to Hessian fly. Crop Sci. 46:1401–1402.
Xu, S.S., T.L. Friesen, and A. Mujeeb-Kazi. 2004. Seedling resistance to tan spot and Stagonospora nodorum blotch in synthetic hexaploid wheats. Crop Sci. 44:2238–2245.
Xu, S.S., K. Khan, D.L. Klindworth, and G. Nygard. 2010. Evaluation and characterization of high-molecular weight 1D glutenin subunits from *Aegilops tauschii* in synthetic hexaploid wheats. J. Cereal Sci. 52:333–336.
Xue, S., Z. Zhang, F. Lin, Z. Kong, Y. Cao, C. Li, H. Yi, M. Mei, H. Zhu, J. Wu, H. Xu, D. Zhao, D. Tian, C. Zhang, and Z. Ma. 2008. A high-density intervarietal map of the wheat

genome enriched with markers derived from expressed sequence tags. Theor. Appl. Genet. 117:181–189.
Yamaguchi, T. and E. Blumwald. 2005. Developing salt-tolerant crop plants: challenges and opportunities. Trends Plant Sci. 10:615–620.
Yang, W., J. Li, and Y. Zheng. 2006. Improving wheat disease resistance and yield potential by using CIMMYT synthetic germplasm in Sichuan, China. p. 8. In: M. Imtiaz, F.C. Ogbonnaya and M.vanGinkel (eds.), Abstract book 1st Synthetic Wheat Symp. Synthetics for Wheat Improvement. Sept. 4–6, Horsham, Victoria, Australia.
Yang, W., D. Liu, J. Li, L. Zhang, H. Wei, X. Hu, Y. Zheng, Z. He, and Y. Zou. 2009. Synthetic hexaploid wheat and its utilization for wheat genetic improvement in China. J. Genet. Genom. 36:539–546.
Yang, W.Y., B.R. Lu, Y. Yu, and X.R. Hu. 2001. Genetic evaluation of synthetic hexaploid wheat for resistance to the physiological strain CYR30 and CYR31 of wheat stripe rust in China. J. Genet. Mol. Genet. 12:190–198.
Yu, G.T., X. Cai, M.O. Harris, Y.Q. Gu, M.-C. Luo, and S.S. Xu. 2009. Saturation and comparative mapping of the genomic region harboring Hessian fly resistance gene *H26* in wheat. Theor. Appl. Genet. 118:1589–1599.
Yu, G.T., T. Wang, K.M. Anderson, M.O. Harris, X. Cai, and S.S. Xu. 2012. Evaluation and haplotype analysis of elite synthetic hexaploid wheat lines for resistance to Hessian fly. Crop Sci. 52:752–763.
Yu, G.T., C.E. Williams, M.O. Harris, X. Cai, M. Mergoum, and S.S. Xu. 2010. Development and validation of molecular markers closely linked to *H32* for resistance to Hessian fly in wheat. Crop Sci. 50:1325–1332.
Zadoks, J.C., T.T. Chang and C.F. Konzak. 1974. A decimal code for the growth stages of cereals. *Weed Res.* 14: 415–421.
Zadoks, J.C., T.T. Chang, and C.F. Konzak. 1974. A decimal code for the growth stages of cereals. *Weed Research* 14:415–421.
Zamir, D. 2001. Improving plant breeding with exotic genetic libraries. Nature Rev. Genet. 2:983–989.
Zegeye H., K. Nazari, A. Badebo, W. Denbel, H. Mohammed, and F.C. Ogbonnaya. 2011. Sources of resistance to stripe rust in synthetic hexaploid wheat. Plant Dis. (submitted).
Zevan, A. C. 1981. Determination of the chromosome and its arm carrying the *Ne1*-locus of *Triticum aestivum* L., Chinese Spring and the *Ne1*-expressivity. Wheat Inform. Serv. 33–34:4–6.
Zegeye, H., A. Badebo, A. Mohammed, F.C. Ogbonnaya, W. Denbel, and K. Nazari. 2012. Sources of resistance to stripe rust in synthetic hexaploid wheat., p.53. In: E. Quilligan, P. Kosina, A. Downes, D. Mullen, and B. Nemcova (eds). Wheat for food security conf, Oct. 8–12, Addis Ababa, Ethiopia.
Zhang, L., D. Liu, Z. Yan, X. Lan, Y. Zheng, and Y. Zhou. 2004. Rapid changes of microsatellite flanking sequence in the allopolyploidization of new synthetized hexaploid wheat. Sci. China Ser. C. Life Sci. 47:553–561.
Zhang, P., S. Dreisigacker, A.E. Melchinger, J.C. Reif, A. Mujeeb Kazi, M.vanGinkel, D. Hoisington, and M.L. Warburton. 2005. Quantifying novel sequence variation and selective advantage in synthetic hexaploid wheats and their backcross-derived lines using SSR markers. Mol. Breed. 15:1–10.
Zhou, R., Y. Dong, Z. Zhu, and J. Jia. 2007. GB4, a wheat line had big spike and powdery mildew resistance generated from synthetic wheat. J. Plant Genet. Resour. 8:378.

Zhou, R., Z. Zhu, X. Kong, N. Huo, Q. Tian, P. Li, C. Jin, Y. Dong, and J. Jia. 2005. Development of wheat near-isogenic lines for powdery mildew resistance. Theor. Appl. Genet. 110:640–648.

Zhu, Z., R. Zhou, X. Kong, Y. Dong, and J. Jia. 2005. Microsatellite markers linked to 2 powdery mildew resistance genes introgressed from *Triticum carthlicum* accession PS5 into common wheat. Genome 48:585–590.

Zwart, R.S., J.P. Thompson, and I.D. Godwin. 2004. Genetic analysis of resistance to root-lesion nematode (*Pratylenchus thornei*) in wheat. Plant Breed. 123:09–212.

Zwart, R.S., J.P. Thompson, and I.D. Godwin. 2005. Identification of quantitative trait loci for resistance to two species of root-lesion nematode (*Pratylenchus thornei* and *P. neglectus*) in wheat. Aust. J. Agr. Res. 56:345–352.

Zwart, R.S., J.P. Thompson, A.W. Milgate, U.K. Bansal, P.M. Williamson, H. Raman, and H.S.Bariana. 2010. QTL mapping of multiple foliar disease and root-lesion nematode resistances in wheat. Mol. Breed. 26:107–124.

Zwart, R.S., J.P. Thompson, and J.G. Sheedy, and J.C. Nelson. 2006. Mapping quantitative trait loci for resistance to *Pratylenchus thornei* from synthetic hexaploid wheat in the International Triticeae Mapping Initiative (ITMI) population. Aust. J. Agr. Res. 57:525–530.

3

Breeding Early and Extra-Early Maize for Resistance to Biotic and Abiotic Stresses in Sub-Saharan Africa

B. Badu-Apraku
International Institute of Tropical Agriculture
P.M.B. 5320
Ibadan, Nigeria
c/o L.W. Lambourne & Co., Carolyn House
26 Dingwall Road
Croydon, CR9 3EE UK

M. A. B. Fakorede
Department of Crop Production and Protection
Obafemi Awolowo University
Ile-Ife, Nigeria

ABSTRACT

Maize (*Zea mays*) is an important crop for food, feed, and agroindustrial raw material produced primarily in the savannah but also in other agroecologies of West and Central Africa (WCA). Maize production in the savannah has been constrained by cultivation of inappropriate varieties, biotic and abiotic stresses including parasitism by *Striga hermonthica* (Del.) Benth. low soil N, drought, diseases, and pests. The International Institute of Tropical Agriculture (IITA), in collaboration with member countries of the West and Central Africa Collaborative Maize Network (WECAMAN), has developed early and extra-early maturing germplasm with tolerance or resistance to *Striga*, drought, and low soil N. Tolerance/resistance to the stresses have been increased in the germplasm through population improvement and effective screening methods. Two early populations [TZE-W Pop DT STR (white) and TZE-Y Pop DT STR (yellow)] and two extra-early populations [TZEE-W Pop (white) and TZEE-Y Pop (yellow)] were subjected to four cycles of S_1 recurrent selection for improved grain yield and *Striga* tolerance/resistance. Products of the selection cycles and 14 experimental cultivars derived from them were evaluated in several field experiments. Under *Striga* grain yield improvement was 70.6 kg ha^{-1} (6.3%) cycle^{-1} for TZE-Y Pop DT STR and 352.5 kg ha^{-1} (58.0%) cycle^{-1} for TZE-W Pop DT STR. Under *Striga*-free environments, improvement was 194 kg ha^{-1} (6.6%) cycle^{-1} and 186.5 kg ha^{-1} (6.0%) cycle^{-1}, respectively. For the extra-early yellow population yield improvement was 90% cycle^{-1} and 18.41% cycle^{-1} under *Striga*-infested and *Striga*-free environments, whereas the white extra-early population was improved by 12.7% cycle^{-1} under *Striga* and 12.91% cycle^{-1} in *Striga*-free environments. *Striga* tolerance/resistance improved significantly in response to selection. The derived cultivars from this program performed equally with or better than the best released cultivars in *Striga*-free environments. The breeding program also developed stress-tolerant hybrids and quality protein maize (QPM) cultivars with resistance to multiple stresses. Many stress-tolerant/resistant open-pollinated and hybrid cultivars derived from this program have been released and adopted by farmers in West and Central Africa. For sustained improvement, future efforts should use a multidisciplinary network approach, with establishment of seed companies and development of hybrids as the ultimate goal.

KEYWORDS: biotic and abiotic stresses; drought tolerance; low soil N; maize hybrid; open-pollinated cultivar; quality protein maize; recurrent selection; *Striga hermonthica*

ABBREVIATIONS
  I. INTRODUCTION
    A. Importance of Maize in Sub-Saharan Africa
    B. Constraints to Maize Production and Productivity in Sub-Saharan Africa
      1. *Striga* Parasitism
      2. Drought Stress
      3. Low Soil Nitrogen
      4. Combined Effects of *Striga* Parasitism, Drought, and Low Nitrogen Stresses
  II. DEVELOPMENT OF BREEDING POPULATIONS
    A. Breeding Methodology
    B. Development of Base Populations for Recurrent Selection
      1. Extra-Early Populations
      2. Early Populations
    C. Screening Methodology
      1. *Striga*
      2. Drought
      3. Low Soil Nitrogen
    D. Data Collection and Analysis
  III. S_1 RECURRENT SELECTION PROGRAM FOR *STRIGA* RESISTANCE
    A. Selection Procedure
    B. Evaluation of Progress from Selection
      1. Cycles of Selection
      2. The Early Populations
      3. The Extra-Early Populations

C. Performance of EVs from the Selection Programs
D. Residual Variances, Heritability, and Genetic Correlation
  1. The Early Populations
  2. The Extra-Early Populations
IV. ADAPTATION
  A. Importance of Adaptation to Drought-Prone Environments in Maize Cultivars for West and CentralAfrica
  B. Breeding Methodology for Adaptation to Drought-Prone Environments
V. DEVELOPMENT OF QPM POPULATIONS AND CULTIVARS
  A. Introduction of QPM to West and Central Africa
  B. Genetic Enhancement of Obatanpa GH
  C. Conversion of Normal Endosperm Maize to QPM
  D. Development of *Striga* Resistant QPM Inbred Lines and Hybrids with Enhanced Adaptation to Drought Stress
  E. Evaluation of the Performance of Early and Extra-Early QPM Open-Pollinated and Hybrid Cultivars
VI. BREEDING FOR COMBINED TOLERANCE/RESISTANCE TO MULTIPLE STRESSES IN EARLY AND EXTRA-EARLY MAIZE
VII. INBRED-HYBRID DEVELOPMENT PROGRAM
VIII. TRAITS FOR INDIRECT SELECTION FOR STRESS TOLERANCE/RESISTANCE IN CONTRASTING ENVIRONMENTS
IX. FUTURE CHALLENGES AND PERSPECTIVES
ACKNOWLEDGMENTS
LITERATURE CITED

ABBREVIATIONS

ANOVA	Analysis of variance
ASI	Anthesis-silking interval
CIMMYT	International Center for Maize and Wheat Improvement
CV	Coefficient of variation/variability
DA	Days to 50% anthesis
DS	Days to 50% silking
DAP	Days after planting
DT	Drought tolerance or drought tolerant
EASP	Ear aspect
EPP	Ear number per plant
EV	Experimental variety
GGE	Genotype main effect plus and genotype × environment interaction
GEI	Genotype × environment interaction
GT	Genotype × trait

GTI	Genotype × trait interaction
IARC	International agriculture research center
IITA	International Institute of Tropical Agriculture
MET	Multi environment trial/testing
MSV	Maize streak virus
NARS	National agricultural research system
NARES	National agricultural research and extension system
PASP	Plant aspect
QPM	Quality protein maize
SSA	Sub-Saharan Africa
STR	*Striga* resistant/tolerant
WA	West Africa
WAP	Weeks after planting
WCA	West and Central Africa
WECAMAN	West and Central Africa Collaborative Maize Network

I. INTRODUCTION

A. Importance of Maize in Sub-Saharan Africa

Maize (*Zea mays*) is one of the most important staple foods as well as cash crops in sub-Saharan Africa (SSA) and has the potential to mitigate the present food insecurity in the subregion. Annual maize production in West and Central Africa (WCA) is estimated at about 11 million tonnes. Although the availability of food per person increased worldwide during the last four decades, it went down by about 7% in Africa. The challenge presently facing Africa, therefore, is to feed a population that is growing annually at 2.3% compared with the per capita agricultural GDP of 3.1% (United Nations Statistics Division 2008). Maize has good prospects for rectifying the food deficit in SSA because it is the number-one staple food crop for rural and urban consumers. Furthermore, maize has good potential for use in the feed industry and as well as several other industrial uses. For example, various alcoholic drinks are prepared from maize. It is also important as feed for poultry and other livestock, constituting between 40% and 75% of the ration. The dry grains of appropriate types can be popped. Dry milling of maize grains produces corn meal, corn flour, and corn oil. Cornstarch, obtained from the wet milling process, is used for food, textile and paper sizing, laundry starch, dextrines, and adhesives such as the gums used for stamps and envelopes. Corn syrup, used as a sweetener, is also made from cornstarch. Other industrial products, which are obtained from maize through distillation and fermentation, include ethyl alcohol, butyl alcohol, propyl alcohol, acetaldehyde, acetic acid, acetone, lactic acid, citric acid, glycerol, and whisky. Ethanol is now mixed with gasoline used for running vehicles.

The demand for maize as food, feed, and industrial raw material continues to increase in WCA. This increasing demand is driven by expanding populations and rising incomes in all countries of the subregion. The yearly per capita consumption of maize is greatest in Benin Republic (87 kg), followed by Togo (70 kg), and Ghana (45 kg). The importance of maize in the food basket in SSA has been increasing steadily during the last few decades (Fakorede et al. 2003), and a number of dependable improved technologies (high yielding and stable cultivars with accompanying agronomic practices) are available to farmers (Abalu 2001).

The International Institute of Tropical Agriculture (IITA, Ibadan, Nigeria) in collaboration with the National Agricultural Research Systems (NARS), has developed extra-early and early-maturing maize cultivars that fit the conditions of the savanna agroecology. Availability of the extra-early and early cultivars has made it possible for maize to spread fairly rapidly into the savannas, replacing the traditional crops such as sorghum (*Sorghum bicolor*) and pearl millet (*Pennisetum glaucum*), especially in the Sudan savanna and the northern fringes of the Northern Guinea savanna (NGS) of West Africa, where the short duration of rainfall had hitherto precluded maize cultivation. Extra-early and early-maturing maize cultivars are more responsive to fertilizer application, mature more quickly and can be harvested much earlier in the season than the traditional sorghum and millet crops. The maize cultivars are thus used for filling the hunger gap in July in the West African savanna when all food reserves are depleted after the long dry period and the new crop of the normal growing season is not ready for harvest. There is also a high demand for the early and extra-early maize in the West African forest zone for peri-urban maize consumers because they allow farmers to market the early crop at a premium price in addition to being compatible with cassava (*Manihot esculentum*), cowpea (*Vigna unguiculata*), and soybean (*Glycine max*) for intercropping (IITA 1992). Another important advantage of the early and extra-early maize is that they provide farmers in the various agroecological zones with flexibility in the dates of planting. Early and extra-early maize can be planted when the rains are delayed or used for early plantings when the rainfall distribution is normal.

B. Constraints to Maize Production and Productivity in Sub-Saharan Africa

Although the Guinea savanna of West Africa has the highest potential for increased maize production and productivity due to higher solar radiation, lower night temperatures, and lower incidence of diseases, production is seriously constrained by *Striga hermonthica* parasitism, recurrent drought, and low soil fertility, especially low levels of nitrogen (N). A more rapid adoption of maize in the West African savannas would occur if the available extra-early and early cultivars were resistant or tolerant to *Striga* infestation, drought, and low soil N. A *Striga*-resistant/tolerant maize crop will provide a long-term sustainable option for managing the parasitic weed. Continued use of *Striga* susceptible maize cultivars will lead to continued increase in the *Striga* seed bank until a threshold is reached at which the host plant cannot tolerate it any longer and suffer heavy yield loss.

1. *Striga* Parasitism. *Striga* spp. compete with the maize plant for water and nutrients and have a toxic effect that can cause severe damage to the host even before *Striga* emerges from the soil. *Striga* is prolific in seed setting and may produce up to 500,000 seeds per plant, with most of the seed remaining viable in the soil for about 20 years. *Striga* infestation is therefore extremely difficult to control and is a major threat to the rapid spread of maize into the West African savanna. The levels of infestation are often so high that farmers may be compelled to abandon their fields. Annual yield losses due to *Striga* species in the savanna region have not been estimated in recent years. However, M'Boob (1986) estimated that the losses were about US$7 billion and are detrimental to the lives of over 100 million African people. The *Striga* spp. that parasitize maize are *S. hermonthica*, *S. asiatica*, and *S. aspera*. Of these, *S. hermonthica* is the most important economically and is widely prevalent in SSA (Aggarwal 1991; Lagoke et al. 1991). *Striga* plants impair the photosynthetic efficiency of the host and the major part of the yield loss is caused by the parasite's potent phytotoxic effect (Ransom et al. 1996). The *Striga* problem in SSA is increasing because of crop intensification, which results in deteriorating soil fertility, shortening of the fallow and crop rotation periods, and the expansion of production into marginal lands with little use of external inputs, especially fertilizer (Vogt et al. 1991). As land use intensifies, soil fertility declines, *Striga* seed-bank in the soil increases, and *Striga* infestation becomes severe, greatly reducing crop yields.

The *Striga* plant has an intimate physiological interaction with the host plants thus making it very difficult to develop successful control measures that are compatible with the farming systems and socio-economic conditions of the resource-poor farmers in SSA. A number of control measures have been developed, including hand pulling, crop rotation, trap and catch crops, fertilizer use, fallowing, seed treatments, and host plant resistance or tolerance (Odhiambo and Ransom 1994; Shaxson and Riches 1998). Of these, host plant resistance or tolerance is considered the most affordable and environmentally friendly for resource-poor farmers. Genetic resistance to *Striga* has been reported in several cereal crops including rice (*Oryza sativa*; Bennetzen et al. 2000; Gurney et al. 2006), sorghum (Maiti et al. 1984; Hess et al. 1992; Haussmann et al. 2004), and maize (Adetimirin et al. 2000b; Badu-Apraku et al. 2004a; Gethi and Smith 2004). Resistance to *Striga* refers to the ability of the host plant to stimulate the germination of *Striga* seeds but prevent the attachment of the parasite to its roots, or kill the attached parasite. The resistant genotype supports significantly fewer *Striga* plants and produces a higher yield than a susceptible (converse of resistant) genotype when infested by the weed (Ejeta et al. 1992; Haussmann et al. 2000; Rodenburg et al. 2006). A *Striga*-tolerant genotype supports germination and growth of as many *Striga* plants as the intolerant or, as proposed by DeVries (2000), the sensitive genotype but produces more grain and stover, and shows fewer damage symptoms (Kim 1994; Badu-Apraku and Akinwale 2011). According to Amusan et al. (2008), *Striga* usually penetrates the xylem and shows substantial internal haustorial development on the susceptible or sensitive maize genotype. On the other hand, the haustorial ingress on the resistant genotype is often stopped at the endoderm.

2. Drought Stress. Recurrent drought is the single most important factor limiting the production of maize in SSA where production worth several billion US dollars is lost annually. Global warming and the accompanying increased unpredictability of the intensity and frequency of rainfall patterns call for a more effective improvement of maize yield under drought stress. Abundant sunshine and low night temperatures in most of the savanna agroecology of SSA imply that maize could be produced throughout the year in most areas with adequate rainfall or water supply. Edmeades et al. (1995) reported an estimated 15% annual maize yield loss from drought stress in the West African savannas and indicated that localized losses may be much higher in the marginal areas where the annual rainfall is below 500 mm and soils are sandy or shallow. Grain yield losses can even be greater if drought stress occurs at the most drought-sensitive stages of crop growth, such as the flowering and grain filling periods (Denmead and Shaw 1960; NeSmith and Ritchie 1992).

3. Low Soil Nitrogen. In most of the developing countries of the world, maize production occurs under low N conditions (McCown et al. 1992; Oikeh and Horst 2001) due to limited N-fertilizer use and reduced N-uptake in drought prone environments, high price ratios between fertilizer and grain, nonavailability of fertilizer, or lack of credit for farmers (Bänziger and Lafitte 1997a). The estimated annual loss of maize yield resulting from low N stress environments vary from 10% to 50% (Wolfe et al. 1988). Therefore, the development of maize germplasm with tolerance to low soil N is crucial for an increase in maize productivity (Betrán et al. 2003a). Improvement for drought tolerance also resulted in specific adaptation and improved performance under low N conditions, suggesting that tolerance to either stress involves a common adaptive mechanism (Bänziger et al. 1999b; Badu-Apraku et al. 2011b).

4. Combined Effects of *Striga* Parasitism, Drought, and Low Nitrogen Stresses. Under field conditions, drought, *Striga*, and soil nutrient deficiencies can occur simultaneously and when this happens the combined effect can be devastating (Cechin and Press 1993; Kim and Adetimirin 1997). Badu-Apraku et al. (2004b) compared the performance of 17 early-maturing cultivars subjected to induced drought stress and *Striga* infestation at Ferkéssedougou (Côte d'Ivoire) with their performance without stress (i.e., under well-watered and *Striga*-free conditions). The mean grain yield was reduced by 53% under drought stress and by 42% under *Striga* infestation (relative to a stress-free environment). In another study, Badu-Apraku et al. (2010) obtained yield reduction of 44% under drought stress, 65% under *Striga* infestation, and 40% under low N stress. Kim and Adetimirin (1997) had earlier reported that maize grain yield reduction in plots with both *Striga* and drought stresses ranged between 56% and 77% when compared with the yield reduction in plots with *Striga* as the only stress factor. They concluded that moisture stress increased geometrically the losses caused by *Striga*. Furthermore, drought stress and low soil nutrient status, especially low levels of N, aggravate *S. hermonthica* parasitism on maize (Lagoke et al. 1991; Cechin and Press 1993; Kim and Adetimirin 1995). In the Sudan and NGS where intermittent drought occurs frequently, it is desirable to incorporate drought tolerance into cultivars that have resistance to *Striga* as the two stresses occur at the same time. Farmers in the *Striga* endemic ecologies of SSA are presently demanding for cultivars that combine both stress factors and are unwilling to adopt maize cultivars that do not possess them.

II. DEVELOPMENT OF BREEDING POPULATIONS

For the past three decades, IITA researchers have been working on the improvement of early and extra-early maize populations, cultivars, and inbred lines to combat the constraints indicated above in the savanna agroecologies of SSA. Details of the strategies and methodologies adopted for the development, testing, and promotion of *Striga*-resistant, drought-tolerant, and low soil N-tolerant open-pollinated and hybrid cultivars as well as future perspectives of the improvement activities are presented in the following sections.

The research work on early and extra-early maize has involved the development of several base populations from which inbred lines have been extracted for the development of experimental cultivars, synthetics and hybrids. Recurrent selection for genetic enhancement has been carried out in some specific populations while some populations have been converted to quality protein maize (QPM) along with upgrading for grain yield and other agronomic characters. Emphasis in this review will be focused on the following populations and their derived inbred lines, open-pollinated and hybrid cultivars:

Early populations: TZE-W Pop DT STR, TZE-Y Pop DT STR, DTE-W STR Syn Pop, DTE-Y STR Syn Pop;
Extra-early populatons: TZEE-W Pop DT STR and TZEE-Y Pop DT STR;
QPM populations: GH Pop 63 SR and conversion of other populations to QPM.

A. Breeding Methodology

IITA initiated a breeding program for *Striga* resistance/tolerance about 25 years ago. The first step was to develop artificial infestation and screening methodology for resistance (Kim 1991). Next, the mode of inheritance and heritability were investigated, followed by selection, an important component of breeding. Results of studies conducted at the initial stages of the program showed that resistance to *Striga* in maize is multigenic (Kim 1991; Badu-Apraku et al. 1999). The traits were easily determined, exhibited normal distribution, were moderately heritable, and under the control of predominantly additive gene action, although nonadditive gene action was also important (Kim 1994; Kling et al. 2000; Badu-Apraku and Fakorede 2001). The genetic information generated by IITA suggested that *Striga* resistance was multigenic and amenable to recurrent selection. A breeding program for *Striga* resistance was therefore initiated in Côte d'Ivoire in 1994 by the West and Central Africa Collaborative Maize Network (WECAMAN) with backstopping by IITA. WECAMAN's efforts were to complement those of IITA to combat the threat posed by *S. hermonthica* to maize production in the West African savannas. The objective of this collaborative program was to develop maize populations, cultivars, and inbred lines that combine earliness or extra-earliness with resistance/tolerance to *S. hermonthica*, drought, and low soil nitrogen. The emphasis of the breeding program has been the formation of high-yielding early and extra-early drought and *Striga*-resistant/tolerant populations using drought-tolerant and *Striga*-resistant germplasm from diverse sources identified through several years of extensive testing in WCA. Efforts were concentrated on the introgression of *Striga* resistance/tolerance into the early and extra-early maize populations and cultivars, using inbred lines from IITA (1368 STR and 9450 STR) as the sources of resistance. Backcrossing, inbreeding, and hybridization were adopted in the breeding program. Among the products of the IITA-WECAMAN breeding program are two early-maturing (90–95 days to maturity) populations designated TZE-W Pop DT STR (white) and TZE-Y Pop DT STR (yellow) and two extra-early (80–85 days to maturity) *Striga*-resistant populations, one having white endosperm designated as TZEE-W Pop STR, and the other named TZEE-Y Pop STR with yellow endosperm (Badu-Apraku et al. 1999, 2001). The four breeding populations have been subjected to recurrent selection for stress tolerance and enhancement of grain yield under stress and no-stress conditions.

The ultimate goal of the IITA maize breeding program for biotic and abiotic stress tolerance is improved grain yield of four maize maturity groups, extra-early, early, intermediate, and late maturing maize to ensure food security in SSA. The program in collaboration with the NARS of WCA has focused on developing breeding materials with resistance under five stress factors; that is, resistance/tolerance to *Striga,* drought, and low soil nitrogen for the savanna ecologies of WCA, resistance to major diseases [namely the maize streak virus (MSV), downy mildew, rust, and leaf blight for all ecologies] and borer resistance for the forest ecologies. Four main strategies have been adopted for improvement in the extra-early and early-maturing maize components of the IITA Maize Improvement Program. These are

1. Development of stress-tolerant (*Striga*-resistant, drought, and low N tolerant) maize source populations for recurrent selection.
2. Improvement of source populations using recurrent selection with reliable artificial field infestation and screening methods to increase resistance to relevant stresses in the breeding materials.
3. Extraction of open-pollinated cultivars, inbred lines, and hybrids from source populations.
4. Germplasm enhancement.

B. Development of Base Populations for Recurrent Selection

1. Extra-Early Populations. Efron (1985) indicated that exotic, local landraces, and introduced tropical germplasm are potential sources of unique alleles for broadening and diversifying the genetic base of adapted maize germplasm in WCA to further enhance progress from selection for both grain yield and stability of performance. In line with this, the two broad-based extra-early breeding populations were developed from partial diallel crosses (no reciprocals) involving the most promising materials identified from several years of extensive multi-location testing in WCA (Badu-Apraku et al. 1999; Badu-Apraku and Fakorede 2001). The materials used for the partial diallel crosses were developed by the IITA-facilitated Semi-Arid Food Grain Research and Devlopment (SAFGRAD) project from crosses involving local landraces, exotic and introduced germplasm. They were selected on the basis of high grain yield, resistance to the MSV, and extra-earliness. The most promising extra-early white cultivars selected were TZEE-W-SR BC5, TZEE-W-SR × Gua 314 BC1, Pop 30 × Gua 314 BC1 and Pool 27 × Gua 314 BC1, which were crossed in a diallel fashion and the progeny subjected to two cycles of recombination to develop the white population named TZEE-W Pop. Similarly, the most promising extra-early yellow germplasm identified were CSP-SR BC5, TZEE-Y SR BC5, CSP × Local Raytiri, and TZEF-Y, which were used to form TZEE-Y Pop. Details of the procedure used to develop the two extra-early breeding populations are presented in Table 3.1.

Two inbred lines, TZi 3 (1368 STR) and TZi 25 STR (9450), both with moderate level of tolerance to *S. hermonthica*, were used as sources of resistance and were introgressed into TZEE-W Pop and TZEE-Y Pop,

Table 3.1. Procedure for the development of the *Striga*-resistant extra-early maturing yellow and white maize populations in WestAfrica.

Location year	Activity
Kamboinse	
1993A[z]	Generation of two separate diallel crosses involving four most promising extra-early white and yellow endosperm maize cultivars
1993B[y]	Evaluation of the two sets of diallel crosses of the white and yellow endosperm
Ferkessedougou	
1994A	152 and 142 most promising families selected from the white and yellow diallels, respectively, planted separately in isolated half-sib recombination blocks
	Balanced composite formed by bulking equal number of seed from the families of each grain color used as sole pollen source for the recombination
	Resulting white and yellow extra-early source populations designated as TZEE-W Pop and TZEE-Y Pop, respectively
1994B	Each population advanced to the second cycle of recombination using a balanced composite of selected families of each population with little or no selection pressure imposed
	Concurrently, two inbred lines, TZi3 (1368) and TZi 25 STR (9450), both with moderate levels of *Striga*-resistance/tolerance crossed to TZEE-W Pop and TZEE-Y Pop, respectively to introgress *Striga* resistance into each population
1995B	F_1 crosses involving TZEE-W Pop × 1368STR and TZEE-Y Pop × TZi 25 STR planted in separate blocks and backcrossed to TZEE-W Pop and TZEE-Y Pop, respectively as recurrent parents to recover earliness. Planting of each recurrent parent was delayed for about 3 weeks to ensure pollen-silk synchrony
1995C[x]	Backcrosses planted under artificial infestation of *Striga* and plants with *Striga* resistance, extra-earliness, and other agronomically desirable characters in each population selfed to obtain (TZEE-W Pop × 1368 STR) × TZEE-W Pop S_1 and (TZEE-Y Pop × Tzi 25 STR) × TZEE-Y Pop S_1

[z]Main planting season.
[y]Off season.
[x]Irrigation season.

respectively in an effort to improve the level of *Striga* resistance. In addition to *Striga* tolerance, TZi 3 has good levels of resistance to the major diseases in WCA. Tzi 25 has been found to have low *Striga* seed germination stimulant production. It has also been reported to be tolerant to *S. asiatica* in North Carolina (Efron et al. 1988; Ransom et al. 1990). The primary mechanism of resistance in the two inbreds is tolerance, but the two inbreds, particularly Tzi 25, also support reduced numbers of *Striga* plants. The two *Striga*-tolerant inbred lines were derived from temperate germplasm (Kim et al. 1984) and the tolerance is multigenic. To ensure synchronization of anthesis and silking, the planting of TZEE-W Pop and TZEE-Y Pop was delayed until about 21 days after planting TZi 3 and TZi 25 STR, both of which are later maturing than the populations. Additional rows of the donor parents were planted again 7 days later. The introgression of *Striga* resistance into each source population was followed by backcrossing, generation of S_1 progenies, selection of *Striga*-resistant S_1's from each population and two cycles of recombination under artificial *Striga* infestation. The resulting progenies of the white and yellow populations were designated as TZEE-W Pop STR C_0 and TZEE-Y Pop STR C_0, respectively.

2. Early Populations. The two drought tolerant, *Striga*-resistant early-maturing source populations, TZE-W Pop DT STR and TZE-Y Pop DT STR were also bred from diverse germplasm sources. The program was initiated in 1994 in Côte d'Ivoire. The sources of drought tolerance for TZE-W Pop DT STR were Pool 16 DT, Pool 16 Sequia C_2, DR-W Pool BC_1F_1, TZE Comp 4, and the inbred line 5012. DR-Y Pool BC_2F_2, KU 1414 and 9499 served as the sources of drought tolerance for TZE-Y Pop DT STR (Table 3.2). As with the extra-early populations, TZi 3 and TZi 25 were used as the sources of tolerance to *S. hermonthica* for the development of TZE-W Pop DT STR and TZE-Y Pop DT STR, respectively. The resulting populations after four cycles of random mating were designated TZE-W Pop DT STR C_0 and TZE-Y Pop DT STR C_0 for the white and yellow populations, respectively.

C. Screening Methodology

1. *Striga*. The principal screening and evaluation sites for *Striga* resistance for the four maize populations and derived cultivars were Ferkessedougou and Sinematialli (hereafter called Ferke and Sine, respectively) in Côte d'Ivoire, and Mokwa and Abuja in Nigeria (Table 3.3). The screening method developed by IITA was used (Kim 1991; Kim and Winslow 1991). The *Striga* seeds used for artificial infestation were collected from sorghum fields at the end of the previous growing season, mixed with finely sieved sand in the ratio of 1:99 by weight from which about 5,000 germinable *Striga* seeds were placed in each planting hole made on ridges spaced 75 cm apart with 40 cm between the holes. The sand served as the carrier material and provided adequate volume for rapid and uniform infestation.

Table 3.2. Procedure used for the development of TZE-WPop DT STR C_0 and TZE-Y Pop DT STR C_0 maize populations in West Africa.

Location year	Breeding activity	
Kamboinse		
1990A[z]	Generation of diallel crosses involving seven local and improved drought-tolerant selections of white and yellow endosperm maize	
1990B[y]	Evaluation of diallel crosses using tied and untied ridges	
1991A	Random mating of selected germplasm without reference to grain color to form a mixed grain population. The population was separated into white and yellow fractions. The white version was designated DR-White Pool and the yellow fraction DR-Y Pool. Thereafter, breeding activities proceeded for the two pools separately	
	TZE-W Pop DT STR	TZE-Y Pop DT STR
1991B	DR-W Pool crossed to Pop 30 SR to improve MSV[x] resistance	DR-Y Pool crossed to Pop 31 SR to improve MSV resistance
1992B	DR-W Pool × Pop 30F_1 backcrossed to recurrent parent, DR-W Pool to obtain DR-W Pool BC_1F_1	DR-Y Pool × Pop 31 F_1 backcrossed to recurrent parent, DR-Y Pool to obtain DR-Y Pool BC_1F_1
1993A	Generation of diallel crosses involving selected outstanding early white maize germplasm (Pool 16 DT, 1368 STR, 5012, Pool 16 Sequia × Pool 16 DT, DR-W Pool BC_1F_1, TZE Comp4)	DR-Y Pool BC_1F_1 advanced to DR-Y Pool BC_1F_2 through selfing under artificial streak virus infestation
1993B		
Ferkéssedougou		
1994A	Evaluation of diallel crosses involving selected early white maize germplasm and selection of 2–3 ears from each F_1 cross	Selected families from DR-Y Pool BC_1F_2 crossed to KU1414, 9499 (sources of drought tolerance), and TZi 25 STR (source of *Striga* resistance)
1994B	Recombination of 2–3 ears selected from each F_1 cross involving Pool 16 DT, 1368 STR, 5012, Pool 16 Sequia × Pool 16 DT, DR-W Pool BC_1F_1, TZE Comp4 Generation of backcrosses TZE-W Pop × (1368 STR × Pool 16 DT) and TZE-W Pop × (Pool 16 Sequia × 1368 STR) in an effort to introgress *Striga* resistance into TZE-W Pop	Generation of backcrosses, DR-Y Pool BC_2F_2 × (DR-Y Pool BC_1F_2 × KU1414), (DR-Y Pool BC_1F_2 × 9499) × DR-Y Pool BC_2F_2 and (DR-Y Pool BC_1F_2 × TZi 25 STR) × DR-Y Pool BC_2F_2
1995A	Selfing of backcrosses, TZE-W Pop × (1368 STR × Pool 16 DT) and TZE-W Pop × (Pool 16 Sequia × 1368 STR) to produce S_1's	Evaluation of selected ears from the backcrosses, DR-Y Pool BC_2F_2 × (DR-Y Pool BC_1F_2 × KU1414), (DR-Y Pool BC_1F_2 × 9499) × DR-Y Pool BC_2F_2 for drought tolerance under induced moisture stress

Table 3.2. (*Continued*)

Location year	Breeding activity	
1996A	Introgression of drought-tolerant and *Striga*-resistant white endosperm selections into TZE-W Pop	
Ferké, Siné 1996B–1998A	Four cycles of random mating of TZE-W Pop under alternate *Striga* infestation and induced moisture stress to form TZE-W Pop DT STR C_0	Four cycles of random mating of TZE-Y Pop under alternate *Striga* infestation and induced moisture stress to form TZE-Y Pop DT STR C_0

[z]Main planting season.
[y]Off season.
[x]Maize streak virus.

Table 3.3. Description of the test locations for the regional uniform variety trials (RUVT) of early-maturing varieties conducted in West Africa, 2006–2008.

Country	Location	Code	Agroecological zone[z]	Latitude	Longitude	Altitude (m ASL)	Rainfall during growing season (mm)
Benin	Angaredebou	ANG	SS	11°32′N	3°05′W	297	1,000
	Bagou	BF	SGS	11°28′N,	2°23′W	303	1,125
	Ina	INA	SGS	9°58′N	2°44′W	358	900
Ghana	Ejura	EJ	FT	7°38′N	1°37′E	90	1,460
	Manga	MAN	SS	11°01′N	0°16′W	270	265
	Nyankpala	NYP	NGS	9°25′N	0°58′E	340	611
	Yendi	YD	SGS	9°26′N	0°10′W	157	1,300
Mali	Katibougou	KT	SS	12°50′N	8°09′W	285	700
	Kita	KX	SS	13°05′N,	09°25′W	393	1,000
Nigeria	Bagauda	BG	NGS	12°01′N	8°19′E	520	681
	Ikenne	IKN	FT	6°53′N	3°42′E	60	1,200
	Mokwa	MOK	SGS	9°18′N	5°4′E	457	1,100
	Zaria	ZA	NGS	12°00′N	8°22′E	640	1,120
Togo	Ativeme	ATV	FT	6°25′N	1°06′E	100	1,360
	Sotouboua	SO	SGS	0°59′N	8°34′E	374	1,200

[z]SGS, Southern Guinea savanna; NGS, northern Guinea savanna; FT, forest-savanna transitional zone; SS, Sudan savanna.

Three maize seeds were placed in the same hole with the *Striga* seeds at the same time and the maize plants were thinned to two about 2 weeks after emergence to give a final population density of 66,666 plants ha^{-1}.

Screening of segregating materials derived from the source populations was done using 5 m rows with susceptible checks planted at regular intervals of 10 rows. Both the segregating materials and the susceptible checks were infested. The infested susceptible checks offered an opportunity to monitor the level of *Striga* infestation and to ensure effective selection for STR. *Striga* infestations were conducted in moist soil, resulting from natural rainfall or from sprinkler irrigation applied immediately after planting, thus making preconditioning unnecessary.

Fertilization of the artificially *Striga*-infested maize field was delayed until about 30 days after planting. At this growth stage, 30–50 kg N ha^{-1} were applied as 15-15-15 NPK. The actual quantity of NPK applied was determined by the fertility of the soil. The amount of NPK applied was monitored to minimize escapes due to high levels of fertilizer.

Striga damage rating was used as the index of tolerance while the number of emerged *Striga* plants was used as the index of resistance. Tolerance to *Striga* is quantified by a host damage rating score on a 1–9 scale, where 1 was the most tolerant and 9 was highly intolerant or sensitive (DeVries 2000; Badu-Apraku and Akinwale 2011). The best genotypes under *Striga* infestation are selected using a base index that integrates grain yield, *Striga* emergence counts, *Striga* damage syndrome rating, and ears per plant (EPP) measured under infested conditions (MIP 1996; Menkir and Kling 2007; Badu-Apraku et al. 2011b). The means of the selected traits were expressed in standard deviation units and the base index scores computed as follows:

$$I = [(2 \times YLI) + EPP - (SDR8 + SDR10) - 0.5\,(ESP8 + ESP10)]$$

where YLI was the yield of *Striga*-infested plots, EPP was the number of ears at harvest in the *Striga*-infested plots, SDR8 and SDR10 were *Striga* damage rating at 8 and 10 weeks after planting (WAP), and ESP8 and ESP10 were number of emerged *Stri*ga plants at 8 and 10 WAP. Under artificial *Strga* infestation, each trait was standardized, with a mean of zero and standard deviation of 1 to minimize the effects of different scales, therefore a positive value was an indication of the tolerance of the inbred lines to the particular stress, whereas a negative value was an indication of susceptibility.

2. Drought. At the initial stages, selection for drought tolerance was made under controlled conditions at Ferke and Sinematiali in Côte d'Ivoire, and Kamboinse in Burkina Faso. At the later stages, the screening sites were shifted to Ikenne and Bagauda in Nigeria. Various strategies were adopted at the different sites for improvement of drought tolerance in the maize populations. At Ferke, Sine, and Ikenne the crop was grown under irrigation during the dry season using an overhead sprinkler irrigation system, which applied 12 mm of water per week and induced drought stress was imposed by withdrawing irrigation water from about 2 weeks before anthesis to the end of the season. At Kamboinse, a location in the Sudan savanna zone, tied and untied ridges were used to simulate different levels of drought stress. Another strategy that was often adopted at Sine in the Guinea savanna zone was to use a higher plant density to induce stressed conditions for selection purposes. At Bagauda, the materials were exposed to natural terminal drought, which normally starts from flowering and continues till harvesting. To pre-empt the possibility of making the populations unnecessarily earlier maturing, a restricted selection index consisting of high grain yield, number of EPP, early flowering, shorter anthesis-to-silking interval (ASI), and shorter plants was used in the breeding program to improve the populations for drought tolerance, similar to what was used by Menkir and Akintunde (2001). The best genotypes under induced drought stress were identified, based on the index values estimated using the equation

$$I = [(2 \times \text{Yield}) + \text{EPP} - \text{ASI} - \text{PASP} - \text{EASP} - \text{LD}]$$

where plant aspect (PASP), ear aspect (EASP), and leaf death scores (LD) was the stay green characteristic and was recorded for the drought stressed and low-N experiments at 65 d after planting on a scale of 1 to 10, where 1 almost all leaves green and 10 virtually all leaves dead. Under induced drought stress each trait was standardized, with a mean of zero and standard deviation of 1 to minimize the effects of different scales. Here also, a positive value was an indication of the tolerance of the inbred lines to the particular stress, while a negative value was an indication of susceptibility.

3. Low Soil Nitrogen. Most soils in WCA are characterized by low soil N. For purposes of breeding low N tolerant maize, the genetic materials are exposed to two levels of N fertilizer: 30 and 90 kg N ha^{-1}. The lower level is the testing rate while the higher level serves as the control. The rational for using the two levels is to ensure that selected low N tolerant genotypes were not necessarily mediocre in performance under high N. Soil tests were carried out and inorganic N fertilizer was added to make up the two levels. Three sites were specifically dedicated to low N screening: Ile-Ife in the forest zone, Mokwa in the Southern Guinea savanna (SGS), and Zaria in the NGS. The sites were selected based on the criterion that they had been depleted of inherent soil N as far as possible, as indicated by soil tests. In addition to the specific low N screening sites, the *Striga* screening sites at Mokwa and Abuja also serve as indirect screening sites for low N because only $30\,kg\,N\,ha^{-1}$ was applied to the *Striga*-infested treatment while the noninfested treatment that received normal recommended N rate served as the control.

D. Data Collection and Analysis

Evaluation trials were conducted in the specific screening sites as well as several other sites within each agroecological zone in WCA (Fig. 3.1). The locations shown in Fig. 3.1 have been used in conducting regional uniform variety trials (RUVT) under the supervision of collaborating NARS partners. Two RUVT were conducted—one for the early cultivars and the other for the extra-early. Cultivars bred by this program were submitted for evaluation in the respective trials. The trials were packaged by the WECAMAN coordinator, with standard instructions.

Data collected on genotypes in all screening and evaluation sites included number of days to anthesis (DYA) and silk extrusion from

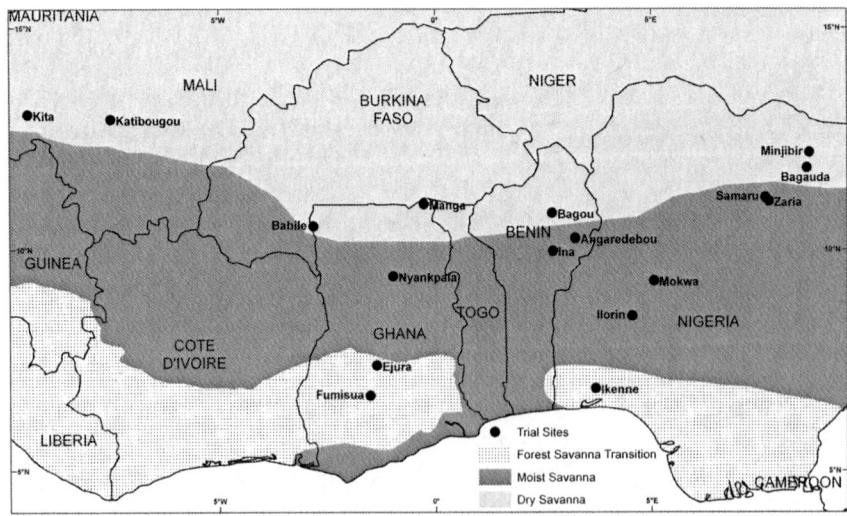

Fig. 3.1. Locations in the major agroecological zones in West Africa used for conducting regional uniform variety trials (RUVT) of WECAMAN from 2006 to 2009.

which anthesis-silking interval (ASI) was determined, plant and ear heights, percent root and stalk lodging, PASP and EASP, husk cover, ear rot, number of EPP, and grain yield. In addition, host plant damage syndrome rating (Kim 1991) and emerged *Striga* counts were made at 8 and 10 WAP (56 and 70 days after planting) in the *Striga*-infested rows, using a 1–9 scale (1 = no damage or indicating normal plant growth and high resistance/tolerance, and 9 = complete collapse or death of the maize plant; that is, highly susceptible or sensitive). *Striga*-tolerant plants normally retain the green leaf and exhibit restricted mild purplish chlorosis while ear and stalk development are little affected by *Striga*. The highly susceptible plants, on the other hand, show greyish leaf color and leaf scorching after initial leaf wilting. These symptoms are usually accompanied by poor development of stalk and ear, resulting in lodging (Kim 1991). Selection was based on grain yield, *Striga* emergence counts, *Striga* damage syndrome, plant and ear heights, and EPP measured under infested conditions.

Genotype × environment interaction (GEI) exists for most traits of maize in WCA thus necessitating extensive testing of cultivars in multiple environments and years before cultivar recommendations for release. Several experiments were conducted to examine the effects of genotype and GEI in early and extra-early cultivar evaluation with particular emphasis on identifying core testing locations in the mega environments of the lowlands of WCA for the early and extra-early maturity groups. In one study, Badu-Apraku et al. (2011c) analyzed grain-yield data of the RUVT-early containing 18 early cultivars evaluated for 3 years in 15 sites representing the dry savanna, moist savanna, and forest/savanna transition zones of WCA (Table 3.3). One of the objectives of the study was to group the sites of the experiment into mega environments.

Yan et al. (2007) proposed that test locations may be classified into three types: (1) locations with low genotype discrimination that should not be selected as test locations; (2) locations with high genotype discrimination, representative of the mega environment as well as close to the ideal mega environment and should, therefore, be chosen for superior genotype selection, when few test locations can be managed due to budget constraints; and (3) locations with high genotype discrimination that do not represent the mega environment, which could be used for unstable genotype evaluation.

The discriminating power of an environment refers to the ability of the environment to identify an ideal test environment while the representativeness refers to the ability of a test location to typify the mega environment. The discriminating power versus representativeness view of GGE biplot analysis of the results of the test locations showed that the test environments in this study could be classified into four mega environments as follows:

First group—Katibougou, Sotouboua, Ejura, and Bagou
Second group—Manga (MAN), Nyankpala (NYP), Bagauda (BG), Yendi (YD), Angaredebou (ANG), Mokwa (MK), Katibougou (KX), and Zaria (ZA)
Third group—Ativeme and Ikenne (IKN)
Fourth group—Ina.

Test locations Ejura, Sotouboua, Bagou, and Katibougou were highly correlated in their ranking of the genotypes, suggesting that a promising early-maturing cultivar selected in anyone of these locations will also be suitable for production in the other locations within the same mega environments in the same or different countries. Similarly, MAN, NYP, BG, YD, ANG, MOK, KX, and ZA were highly correlated in their ranking of the genotypes in the second group and, therefore, a promising cultivar identified in one location will likely be adapted to the other locations in this group. Selecting a cultivar out of these two groups of locations, will likely result in cultivars adapted to IKN and other locations within the same mega environment. Ina stands alone in mega environment 4 in its ranking of the genotypes and was unique in the ranking of the genotypes. Kita was identified as the ideal location while Zaria was close to the ideal location.

The four mega environments stratified in the study did not correspond closely to the maize agroecological zones identified by earlier researchers (Efron 1985; Fajemisin et al. 1985; Menkir 2003; Setimela et al. 2007). The study by Menkir (2003) was based on climatic data obtained from the geographic information system (GIS) for a large number of locations in SSA while those by Efron (1985) and Fajemisin et al. (1985) involved intermediate-to-late maturing cultivars. Furthermore, all of the earlier studies used methods different from those used by Badu-Apraku et al. (unpublished data) who, in addition, sampled fewer locations without any representative location from the mid-altitude agroecology. All of these reasons might have accounted for the differences in the results of the studies. Validation of the discriminating power of the mega environments was done, using repeatability as the indicator. Summarized in Table 3.4 are the repeatability values computed for grain yield and nine other traits, using the data of Badu-Apraku et al. (2011, unpublished data). The output was subjected to factor analysis for the purpose of grouping the 15 locations into factors,

Table 3.4. Repeatability of grain yield and agronomic traits of 18 early-maturing open-pollinated maize cultivars included in the RUVT evaluated for 3 years in 15 sites representing the dry savanna, moist savanna, and forest/savanna transition zones of five WestAfrican countries.

Site	Repeatability											Grain yield	
	Yield	Anthesis	Silk	ASI[z]	PHT	EHT	PASP	EASP	EPP	HUSK	Mean	t ha⁻¹	CV %
Benin													
Anga	0.40	0.58	0.40	−1.37	0.01	0.23	0.21	0.55	−0.07	−0.36	0.058	2.8	21.0
Bagou	−0.06	0.55	0.40	0.50	0.37	0.48	0.00	−0.15	0.27	−1.76	0.060	1.8	30.7
Ina	−0.91	0.50	0.72	−0.43	0.23	0.38	−0.46	−0.53	0.24	−0.11	−0.037	2.9	17.0
Ghana													
Ejura	0.20	0.21	0.31	−0.28	0.36	0.25	−0.08	0.37	−0.53	0.63	0.144	3.7	19.0
Manga	−0.08	0.24	0.68	0.22	0.28	0.59	0.00	−0.23	−0.23	0.49	0.196	1.6	25.2
Nyank	0.44	0.68	0.63	0.09	0.75	0.08	0.17	0.41	−1.42	0.77	0.260	4.2	15.9
Yen	−0.73	0.46	0.23	0.36	0.25	0.07	0.06	0.28	0.38	−0.14	0.122	1.9	57.5
Mali													
Kati	0.69	0.62	0.58	−0.21	0.06	0.31	0.6	0.68	−0.03	0.34	0.364	3.0	21.0
Kita	0.57	0.45	0.61	0.35	0.48	0.61	−0.28	−0.34	−0.27	0.54	0.272	2.5	26.8
Nigeria													
Bagauda	0.51	0.64	0.62	0.06	0.27	0.57	−0.07	0.32	0.64	0.60	0.416	3.9	21.6
Ikenne	0.62	0.59	0.55	0.33	−0.18	0.42	0.13	0.52	0.64	0.50	0.412	2.7	18.8
Mokwa	0.54	0.40	0.44	0.25	0.30	0.26	−0.24	0.81	−0.09	0.19	0.286	3.6	17.2
Zaria	−0.45	0.21	0.39	−0.48	0.58	0.35	−0.58	0.81	−0.16	−0.55	0.012	5.2	18.1
Togo													
Ativeme	0.30	−0.16	0.56	0.64	0.28	0.36	0.45	0.46	0.00	0.67	0.356	2.2	32.4
Sotou	0.41	0.28	0.04	0.29	0.45	0.75	0.55	−0.27	0.18	−0.75	0.193	4.1	19.7
Means	0.16	0.42	0.48	0.02	0.30	0.38	0.03	0.25	−0.03	0.07			

[z]ASI, Anthesis-to-silking interval; PHT, plant height; EHT, ear height; PASP, plant aspect; EASP, ear aspect; EPP, number of ears per plant; HUSK, husk cover rating; CV, coefficient of variation.

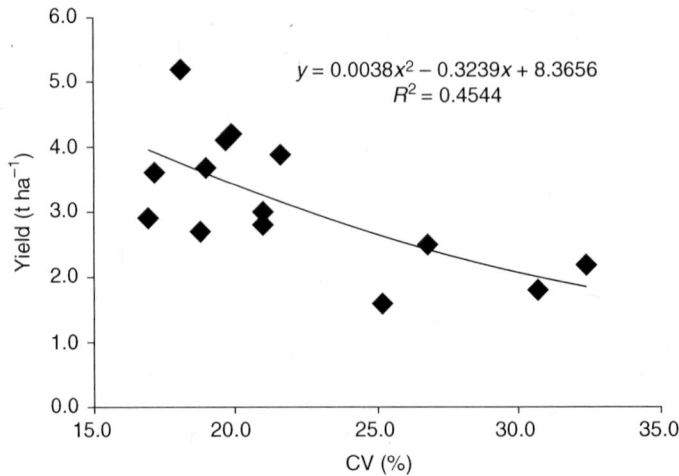

Fig. 3.2. Relationship between grain yield and its CV (%) for 18 early-maturing maize varieties evaluated in 15 locations in West Africa from 2006 to 2008.

which were considered as mega environments. Repeatability for individual traits varied widely among locations. For example, repeatability for grain yield ranged from −0.91 at Ina to 0.64 at IKN. Across sites, most of the traits had low repeatability values, mostly below 0.4. Only DYA and silk had values above 0.4 (Table 3.4). Similarly, across-trait repeatability values were low for most sites; only Ikenne and Bagauda had values of about 0.4, and other sites had values much lower. As had been noted in earlier report on RUVTs (Fakorede et al. 2007), grain yield had a negative relationship with coefficient of variation (CV) also in this study (Fig. 3.2) but grain yield and CV did not influence repeatability.

Five factors (mega environments) were identified and together, they accounted for 87% of the variation among the locations for repeatability (Table 3.5). Although the proportions of the variation accounted for by the factors were not too far apart (about 15%–22%), the number of locations per factor varied from 2 to 5. For this study, the five mega environments for evaluating early-maturing maize germplasm are

1. Kati, Angaradebou, Mokwa, Ejura, and Nyankpala with factor loadings of 0.62–0.82. This group accounted for about 22% of the variability among sites for the repeatability values.
2. Manga and Kita with about equal factor loadings of approximately 0.94 each, accounting for about 19% of the variability among sites.

3. BREEDING EARLY AND EXTRA-EARLY MAIZE

Table 3.5. Factor loadings of repeatability estimates[z] for 15 sites representing the dry savanna, moist savanna, and forest/savanna transition zones of five West African countries as determined for grain yield and agronomic traits of 18 early-maturing open-pollinated maize cultivars included in the RUVT evaluated for 3 years.

Location	Factor loadings				
	1	2	3	4	5
Angarade	**0.794**[y]	−0.107	0.315	0.354	−0.002
Ativeme	−0.088	0.235	**−0.62**	−0.51	−0.205
Bagauda	0.187	0.344	−0.065	**0.885**	0.139
Bagou	−0.031	−0.045	**0.84**	−0.097	0.366
Ejura	**0.7**	0.515	−0.422	−0.089	−0.051
Ikenne	0.086	−0.141	−0.191	**0.846**	−0.145
Ina	−0.054	0.507	0.219	0.351	**0.692**
Kati	**0.817**	−0.074	0.018	0.194	−0.333
Kita	0.079	**0.937**	0.044	0.091	−0.16
Manga	−0.024	**0.939**	−0.095	−0.028	0.178
Mokwa	**0.747**	0.078	−0.194	0.052	0.115
Nyankpala	**0.618**	0.544	−0.243	−0.382	−0.117
Sotou	−0.12	0.058	**0.941**	−0.233	−0.102
Yen	−0.215	−0.148	0.082	−0.071	**0.912**
Zaria	0.619	0.04	0.162	−0.025	**0.704**
Eigen value	4.06	3.06	2.52	1.93	1.47
Percentage variance explained	21.5	18.9	16.5	15.2	14.9
Cumulative variance (%)	21.5	40.4	56.9	72.1	87

[z]Extraction method: principal component analysis. Rotation method: Varimax with Kaiser normalization. Rotation converged in seven iterations.
[y]Values in bold figures on the diagonal indicate locations loaded highly (loading > 0.6) on a factor (or mega environment) and are, therefore, considered as components of the mega environment where a site had values >0.6 on two factors (e.g., Zaria), the larger value was considered.

3. Sotouboua, Bagou, and Ativeme with loadings of 0.94, 0.84, and −0.62, respectively. This group accounted for about 16% of the variability among sites. Ativeme had a negative relationship with this group.
4. Bagauda and Ikenne with loadings of about 0.88 and 0.85, respectively, explaining about 15% of the variation among sites.
5. Yendi, Zaria, and Ina with loadings of 0.69–0.91; also accounted for 15% of the variation among sites.

The mega environments resulting from this analysis were not identical with those from earlier studies but it was probably the most reliable because it took into consideration several traits of the maize plant. Although the GGE biplot analysis used by Badu-Apraku et al.

(unpublished data) and factor analysis both use the principal component analysis approach, factor analysis has the added advantage that the factor loadings may be subjected to rotation to maximize the correlation among locations loaded on the same factor while minimizing the relationship of those between factors. In essence, the mega environments delineated by the factor analysis are orthogonal to each other (Fakorede 1979).

The five factors identified here represent five mega environments. Cultivars that perform well in any of the locations with positive loadings on a particular factor will perform well in other locations loaded on the factor. If a location has negative loading on a factor, as was the case with Ativeme on factor 3, cultivars that perform well in locations with positive loadings on the factor will perform poorly in the location with negative loadings.

Data summarized in Table 3.4 are useful from three perspectives: identification of locations with high repeatability when averaged across traits, traits with high repeatability when averaged across locations, and location × trait interaction for repeatability. Using this interpretation, only DYA and silk were moderately repeatable across locations in this study; repeatability values for all other traits were low. Similarly, apart from Ikenne, Bagauda, Ativeme, and Katibou with repeatability estimates of about 0.4, the aggregate phenotypes of the maize cultivars in this study were poorly repeatable in the different sites. However, repeatability for some traits was high in some locations and low or even zero in some others. It is particularly striking that some of the locations with high grain yield and relatively low CV such as Zaria, had low repeatability for yield and the aggregate phenotype.

III. S_1 RECURRENT SELECTION PROGRAM FOR *STRIGA* RESISTANCE

Recurrent selection is a cyclical scheme designed to increase the frequency of favorable alleles in a population and has been used effectively for improvement of quantitatively inherited traits in maize breeding populations (Sprague and Eberhart 1977; Hallauer and Miranda 1988; Kling et al. 2000; Badu-Apraku et al. 2006a, 2008a; Menkir and Kling 2007) including drought tolerance (Bolanos and Edmeades 1993; Badu-Apraku et al. 1997; Chapman and Edmeades 1999; Monneveux et al. 2006). Although *Striga* resistance or tolerance has been shown to be multigenic (Ejeta et al. 1992; Kim 1994; Lane et al. 1997), a few studies have been conducted on the effectiveness of recurrent selection in improving the level of resistance (Badu-Apraku et al. 2006a, 2008a; Menkir and Kling 2007). It was hypothesized that the use of recurrent selection methods that capitalize on additive gene action in combination with an effective and a reliable artificial method of *Striga* infestation for the screening of segregating families should facilitate the accumulation of *Striga* resistance genes to develop germplasm with multigenic resistance that could be durable over time and effective against the parasitic weed (Berner et al. 1995; Menkir and Kling 2007).

A. Selection Procedure

Recurrent selection for *Striga* resistance was initiated within TZE-W Pop DT STR C_0 and TZE-Y Pop DT STR C_0 populations in 1996, using S_1 lines as the selection unit. That year, progeny trials were conducted at Ferke (under artificial *Striga* infestation), Sine (high yield environment), and Kamboinse (a drought stress environment), and remnant S_1 seed of the top 25%–30%, identified from the across location analysis of the trial data, were intermated in 1998 to complete the first cycle (C1). A base index similar to that indicated above for IITA (MIP 1996) was used to identify the best 25%–30% of the families across locations. In addition, the top 7%–10% of the S_1 families from each population were intermated to form the cultivars EV DT-W 98 and EV DT-Y 98, respectively. Each population has been taken through three additional cycles of S_1 recurrent selection, involving screening under artificial infestation with *S. hermonthica* and under noninfested conditions at Ferke or Abuja and Mokwa. The number of progenies screened in each cycle ranged from 196 to 280, and the top 25%–30% of each population were intermated to reconstitute the respective populations. In addition, the top 10% of the S_1 families of each cycle were intermated to form *Striga*-tolerant experimental varieties (EV) for each population. The cultivars extracted from C_2 and C_3 were EV DT-W 2000 STR, EV DT-Y 2000 STR, TZE-W Pop × 1368 STR, EV DT-W 99 STR EVDT-W 98 C_2, and EVDT-Y 98 C_2.

Similarly, S_1 recurrent selection was initiated within the TZEE-W Pop STR C_0 and TZEE-Y Pop STR C_0 populations during the dry season of 1997. A total of 167 S_1 lines extracted from each population (plus two checks) were tested under *S. hermonthica* infested and noninfested conditions at Sinematialli and Ferke, Côte d'Ivoire, respectively. A 13 × 13 simple lattice design with two replications was used for the evaluation of lines from each population. The first cycle of improvement was completed in 1998 by intermating remnant seed of the top 25%–30% S_1's selected based on the performance across the two testing environments, using the base index mentioned earlier. The top 7%–10% best S_1 progenies from each population were intercrossed to form the cultivars EV 98 TZEE-W and EV 98 TZEE-Y from the TZEE-W-Pop STR C_0 and TZEE-Y Pop STR C_0, respectively. Each population was advanced to cycles 2 and 3 in 1999 and 2001, respectively. The S_1 progenies from each cycle of improvement were screened under artificial infestation with *S. hermonthica* at Ferke and under *Striga*-free conditions at Sinematialli, as described earlier. The number of S_1 lines screened in each cycle ranged from 196 to 256 and 25%–30% selected S_1's were intermated to form the population for the next cycle of selection. Two EVs (2000Syn EE-W and 99 TZEE-Y STR C_1) were formed by intermating the best 10 lines of TZEE-W Pop STR C_2 and TZEE-Y Pop STR C_2, respectively. For the C_3, 355 S_1 progenies derived from TZEE-W Pop STR C_3 (plus five checks) and 249 S_1 lines from TZEE-Y Pop STR C_3 (plus seven checks) were evaluated under artificial infestation with *S. hermonthica* at Abuja and Mokwa, Nigeria in 2003. Based on the across-location data, 30 S_1 lines were selected and recombined to reconstitute cycle 4 of each source population. The best 10 S_1 progenies were recombined separately to form 2004 TZEE-W STR C_4 and 2004 TZEE-Y STR C_4. Each population was taken through another cycle of S_1 family selection under artificial *Striga* infestation in 2008 and the EVs 2008 TZEE-W STR C_5 and 2008 TZEE-Y STRC5 were bred from the white and yellow populations, respectively.

Between 1994 and 2004, 10 EV were bred from the early populations and six from the extra-early populations as follows:

TZE-W Pop DT STR C_0: Six cultivars including EV DT-W 98, EV DT-W 2000, EV DT-W 99 STR, TZE-W Pop × 1368 S6 F_2, EVDT-W 98 C_2, and 2004 TZE-W Pop DT STR C_4.

TZE-Y Pop DT STR C_0: Four cultivars including EV DT-Y 98, EVDT-Y 98 C_2 STR, EV DT-Y 2000 STR, and 2004 TZE-Y Pop DT STR C_4.

TZEE-W Pop STR: Two cultivars including EV 98 TZEE-W and 2000 Syn EE-W, and 2008 TZEE-W STR C_5.

TZEE-Y Pop STR: Two cultivars including EV 98 TZEE-Y and 99 TZEE-Y STR C_1, and 2008 TZEE-Y STR C_5.

B. Evaluation of Progress from Selection

In addition to the time required to complete a cycle of selection, recurrent selection requires input from human labor (skill) and facilities. It is, therefore, expedient to evaluate the progress from selection after two or three cycles have been completed. If progress is not being made, the program may be terminated forthwith or the genetic variability of the base population should be broadened through introgression of some other germplasm. Two approaches were used to evaluate the progress from selection in our program, including evaluation of (i) the cycles of selection *per se* and (ii) the derived cultivars from the selected populations.

1. Cycles of Selection. The progress from four cycles of S_1 selection for grain yield and *Striga* tolerance in the two maturity groups was evaluated in separate field trials conducted under artificial *Striga* infested and noninfested conditions for the early and extra-early populations. The early group study included 13 entries: the C_0, C_2, C_3, and C_4 from TZE-W Pop DT STR; C_0, C_3, and C_4 from TZE-Y Pop DT STR, two derived *Striga*-resistant cultivars of the early white population (2004 TZE-W Pop DT STR C_4 and TZE-W Pop × 1368 S6 F_2); two varieties of the yellow population (EV DT-Y 2000 STR C_1 and 2004 TZE-Y Pop DT STR C_4), an elite *Striga*-resistant cultivar (99 Syn WEC) not from the recurrent selection program; and a *Striga* susceptible check, TZE Comp4. The C_1 from TZE-W Pop DT STR and the C_1 and C_2 from TZE-Y Pop DT STR were not included in the trials because they were lost when the cold room of IITA at Bouaké (Côte d'Ivoire), where the seeds were stored, was looted during the civil war in 2002. The trials were planted at Mokwa and Abuja during the growing seasons of 2005 and 2006.

A total of 17 entries were included in the study for the extra-early group: the original populations (C_0) along with cycles 2, 3, and 4 of TZEE-W Pop STR and cycles 3 and 4 of TZEE-Y Pop STR, two derived varieties from the cycles of selection of the white source population (2000 Syn EE-W, 2004 TZEE-W Pop STR C_4) and two from the yellow (99 TZEE-Y STR C_0, 2004 TZEE-Y Pop STR C_4); three elite *Striga*-resistant cultivars from other selection programs (Sine TZEE-W STR, Ferké TZEE-W STR, 98 TZEE-W STR); and a *Striga* susceptible check, TZEE-W SR BC_5. The 17 entries were planted in *Striga*-infested conditions at Mokwa and Abuja, and *Striga*-free conditions at Mokwa, Abuja, Ikenne, Zaria, and Bagauda, 2005–2007.

2. The Early Populations. Under *Striga* infestation, combined analysis of variance for the early populations showed significant genotype mean squares for grain yield, EPP, and the four *Striga* resistance traits (data not shown). Location and year effects were also significant for grain yield and most other traits, but interaction mean squares were significant for only six of the 40 trait × source of variation cases. Under

Table 3.6. Parameters from the regression of grain yield on four cycles of S_1 recurrent selection in two early (E) and two extra-early (EE) maize populations evaluated in *Striga*-infested and *Striga*-free environments in Nigeria, 2005 and 2006.

Population	Regression equation	Gain cycle^{-1} kg ha^{-1}	%
Striga-infested			
TZEE-W Pop DT STR	$Y = 117.4x + 927; r^2 = 0.493$	117.4	12.7
TZE-W Pop DT STR	$Y = 352.5x + 608; r^2 = 0.559$	352.5	58.0
TZEE-Y Pop DT STR	$Y = 290x + 326.6; r^2 = 0.997$	290.0	90.0
TZE-Y Pop DT STR	$Y = 70.6x + 1115; r^2 = 0.179$	70.6	6.3
Striga-free			
TZEE-W Pop DT STR	$Y = 302.2x + 2336; r^2 = 0.977$	302.2	12.9
TZE-W Pop DT STR	$Y = 186.5x + 3109; r^2 = 0.300$	186.5	6.0
TZEE-Y Pop DT STR	$Y = 350.5x + 1901; r^2 = 0.998$	350.5	18.4
TZE-Y Pop DT STR	$Y = 194.0x + 2963; r^2 = 0.949$	194.0	6.6

Striga-free conditions genotype, year, and location mean squares were significant for 15 of the 18 cases, and in 13 of the 24 cases the interaction mean squares were also significant.

Grain yield response to S_1 recurrent selection in the two early populations under *Striga* infestation differed significantly: 70.6 kg ha^{-1} (6.3%) cycle^{-1} for TZE-Y Pop DT STR and 352.5 kg ha^{-1} (58.0%) cycle^{-1} for TZE-W Pop DT STR (Table 3.6). Corresponding values for the two populations under *Striga*-free environments were similar: 194 kg ha^{-1} (6.6%) cycle^{-1} and 186.5 kg ha^{-1} (6.0%) cycle^{-1}, respectively. *Striga* damage rating at 10 WAP and *Striga* emergence at 8 WAP decreased in the improved cycles of selection in both populations (Table 3.7). Apart from EPP that showed 4%–7% increase per selection cycle, changes in all agronomic traits were not statistically significant under *Striga* infestation. In the *Striga*-free environments, however, stalk lodging of the C_0 was higher than those of the improved cycles in both populations. Similarly in TZE-Y Pop DT STR, ASI decreased in the improved cycles relative to the C_0. Changes in other traits did not show consistent trends in this population.

3. The Extra-Early Populations. Genotype, location, and year significantly affected grain yield, days to silk, ASI, and EPP under both *Striga*-infested and *Striga*-free conditions for the extra-early populations (data not shown). These sources of variation also significantly affected plant height (PLHT) and stalk lodging under *Striga*-free conditions. Genotype mean squares were also significant for stalk lodging as well as *Striga*

Table 3.7. Changes in *Striga* resistance traits under *Striga* infestation, along with agronomic traits associated with four cycles of S_1 recurrent selection in TZE-Y Pop DT STR and TZE-W Pop DT STR maize populations evaluated in *Striga*-infested and *Striga*-free environments in Nigeria, 2005–2007.

Population parameter	Days to silk	ASI[z]	EPP	Plant height (cm)	Stalk lodging (%)	Striga rating at 8 WAP	Striga rating at 10WAP	Striga count at 8 WAP	Striga count at 10 WAP
Striga-infested									
TZE-W Pop DT STR									
Gain per cycle	0.2	0.65	0.02	1.77	0.11	−0.26	−0.22	−7.22	−1.93
Percent response per cycle	0.35	1.25	4.63	1.40	2.29	−4.70	−3.94	−4.88	−1.16
R^2 (%)	35.7	53.5	29.4	96.5	7.7	79.4	41.2	23.5	9.7
Significance of gain	NS	NS	*	NS	NS	**	**	*	NS
TZE-Y Pop DT STR									
Gain per cycle	0.63	0.36	0.04	0.21	−0.30	−0.09	−0.25	−4.75	−3.04
Percent response per cycle	1.13	0.69	6.60	0.16	−5.48	−1.97	−4.76	−3.68	−1.87
R^2 (%)	64.3	66.1	86.8	0.7	55.4	52.2	74.3	7.3	10.9
Significance of gain	ns	NS	*	NS	NS	NS	**	*	NS
Striga-free									
TZE-W Pop DT STR									
Gain per cycle	0.5	0.04	0.01	−1.60	−0.33				
Percent response per cycle	0.92	1.49	0.66	−1.02	−13.86				
r^2 (%)	29.67	1.07	89.54	63.26	96.69				
Significance of gain	*	*	NS	*	*				
TZE-Y Pop DT STR									
Gain cycle^{-1}	0.08	−0.11	0.01	1.10	−0.44				
Percent response per cycle	0.13	−3.02	1.56	0.67	−16.01				
R^2 (%)	3.94	12.29	20.45	12.09	69.89				
Significance of gain	NS	*	NS	NS	**				

[z]ASI, Anthesis-silking interval; EPP, number of ears per plant; WAP, weeks after planting.
Notes: *, **, and NS indicate significant at $P < 0.05$ or $P < 0.01$ and not significant, respectively.

rating, but not *Striga* count at 8 and 10 WAP. Apart from year × location mean squares that were significant for 7 of the 10 traits, interaction effects were not significant in *Striga*-infested environments. In contrast, about 50% of the interaction effects were statistically significant in the *Striga*-free environments, especially genotype × location and year × location sources that were significant for four and five of the six traits, respectively. Four cycles of S_1 recurrent selection improved grain yield by nearly 90% cycle^{-1} in the yellow extra early population when evaluated in *Striga*-infested environments (Table 3.6). This population was improved at the rate of 18.41% cycle^{-1} when evaluated in *Striga*-free environments. Gains from the selection program were about equal for the white extra early population under the two *Striga* environments: about 12.7% cycle^{-1} under *Striga* and 12.91% cycle^{-1} without *Striga* (Table 3.6). *Striga* count at 8 and 10 WAP, along with *Striga* rating at 8 but not 10 WAP were reduced by recurrent selection in TZEE-Y Pop DT STR (Table 3.8). Only *Striga* count at 8 WAP showed significant reduction in TZEE-W Pop STR. Under *Striga* infestation, selection failed to induce significant changes in the agronomic traits of the two populations except EPP, unlike in *Striga*-free environments where significant changes occurred for days to silk, ASI and PLHT in the yellow population as well as EPP and PLHT in the white population (Table 3.8).

C. Performance of EVs from the Selection Programs

An important objective of the recurrent selection program was to develop cultivars from the different cycles of selection. It is important not only to carry out evaluations periodically to determine progress from selection and the efficiency of the recurrent selection method in increasing the frequency of the favorable alleles but also, to assess the performance of the derived cultivars. Therefore, the performance of the derived cultivars from the populations was assessed in several field trials. For the early populations evaluated under *Striga* infestation, ACR 94 TZE Comp5-Y and ACR 94 TZE Comp5-W, which were not from the selection program, were the highest-yielding group: 2,158 and 2,124 kg ha^{-1}, respectively (Table 3.9). The second group comprised six products of the selection program, with grain yield ranging from 1,806 to 1,954 kg ha^{-1}. The third group, with grain yield of 1,498–1,759 kg ha^{-1} contained mostly *Striga* susceptible cultivars and the C_0 of the selection program. Under *Striga*-free conditions, the performance of several cultivars from the selection program was equal to or better than ACR 94 TZE Comp5-Y and ACR 94 TZE Comp5-W. The genotype plus genotype × environment interaction biplot analysis demonstrated that EV DT-Y 2000 STR C_1 and

Table 3.8. Changes in *Striga* resistance traits under *Striga* infestation, along with agronomic traits associated with four cycles of S_1 recurrent selection in TZEE-Y Pop DT STR and TZEE-W Pop DT STR maize populations evaluated in *Striga*-infested and *Striga*-free environments in Nigeria, 2005–2007.

Population parameter	Days to silk	ASI[z]	EPP	Plant height	Stalk lodging	Striga rating at 8 WAP	Striga rating at 10 WAP	Striga count at 8 WAP	Striga count at 10 WAP
Striga-infested									
TZEE-W Pop DT STR									
Gain cycle^{-1}	0.514	−0.46	0.01	−1.51	−0.62	−0.05	−0.09	−11.45	−7.82
Percentage response cycle^{-1}	0.91	−7.62	1.52	−1.11	−9.66	−0.80	−1.57	−11.06	−0.42
R^2 (%)	77	91	2.98	23	98	65	26	0.35	5.61
Significance of gain	NS	NS	**	NS	NS	NS	NS	*	NS
TZEE-Y Pop DT STR									
Gain cycle^{-1}	0.35	−0.27	0.03	−1.00	0.52	−0.11	−0.14	−0.19	−4.12
Percent response cycle^{-1}	0.63	−9.09	7.57	−0.81	11.01	−1.83	−2.15	−0.12	−2.43
R^2 (%)	52	94	55	25	81	35	48	0.01	0.05
Significance of gain	NS	NS	NS	NS	NS	*	NS	**	**
Striga-free									
TZEE-W Pop STR									
Gain per cycle	1.023	−0.257	0.02	2.257	0.283				
Percentage response cycle^{-1}	1.94	−9.08	1.98	1.45	10.14				
R^2 (%)	0.77	0.77	58.16	0.426	0.464				
Significance of gain	NS	NS	**	**	NS				
TZEE-Y Pop STR									
Gain per cycle	0.81	−0.27	4.77	0.24	0.014				
Percentage response cycle^{-1}	1.55	−9.09	3.23	6.45	1.65				
R^2 (%)	94	94	96	74	84				
Significance of gain	**	*	NS	**	NS				

[z] ASI, Anthesis-silking interval; EPP, number of ears per plant; WAP, weeks after planting.
Notes: *, **, and NS indicate significant at $P < 0.05$ or $P < 0.01$ and not significant, respectively.

Table 3.9. Grain yield and other agronomic characters of derived cultivars from S_1 recurrent selection in TZE-Y Pop DT STR and TZE-W Pop DT STR maize populations evaluated under *Striga*-free (noninf.) and *Striga*-infested (inf.) conditions averaged across six locations in West Africa in 2002.

Cultivar	Yield (kg ha^{-1}) Inf.	Yield (kg ha^{-1}) Noninf.	Days to silk Noninf.	Days to silk Inf.	Plant height (cm) Noninf.	Plant height (cm) Inf.	*Striga* 10 WAP[y] Count	*Striga* 10 WAP[y] Rating	Ears harvested Noninf.	Ears harvested Inf.
ACR 94 TZE COMP5-Y	2,158	2,033	52	53	163	137	60	4	24	21
ACR 94 TZE COMP5-W	2,124	2,052	52	53	166	144	67	4	24	20
98 Syn WEC STR C_0	1,954	2,228	52	54	168	143	68	5	25	18
TZE-Y POP DT STR C_3[z]	1,928	2,081	53	55	172	137	64	5	24	16
TZE-W POP DT STR C_3[z]	1,871	2,099	53	54	173	140	67	5	24	18
EV DT-Y 2000 STR C_1[z]	1,831	2,203	53	54	171	139	66	5	24	18
2000 Syn WEC[z]	1,826	2,048	52	54	169	140	66	5	25	17
Kamboinse 88 Pool 16 DT (RE)	1,806	1,881	52	54	169	137	65	5	23	17
EV DT-W 99 STR C_0[z]	1,806	1,803	52	53	159	125	60	5	24	18
EV DT 97 STR C1	1,759	2,110	53	54	166	136	63	4	22	17
ACR 94 POOL 16 DT STR	1,735	1,867	53	55	166	137	68	5	24	15
TZE-W POP X 1368 STR S6 F2[z]	1,721	1,847	53	55	175	133	58	4	23	16
TZE-W POP X 1368 STR C_1[z]	1,689	1,950	51	53	162	130	59	5	25	17
99 Syn WEC	1,541	1,811	55	56	171	187	64	5	24	17
EV DT-W 2000 STR C_0[z]	1,498	1,521	53	52	158	135	67	5	24	15
Grand mean	1,826	1,988	53	54	167	140	64	5	24	173
Coefficient of variation	26	24	5	9	13	42	15	20	15	22
LSD[x] ($P < 0.05$)	269	271	2	3	14	37	6	1	2	2
Standard error of differences	136.7	137.8	0.8	1.5	7.1	18.6	3.0	0.3	1.1	1.1
P stat-test for CV	*	*	*	NS	NS	NS	NS	NS	NS	**
P stat-test for enviroment × CV	**	***	NS	NS	NS	NS	*	***	NS	**

[z]Products from the S_1 recurrent selection program.
[y]WAP, weeks after planting.
[x]LSD, least significant difference.
Notes: *, **, and NS indicate significant at $P < 0.05$, $P < 0.01$, and not significant, respectively.

TZE-W Pop DT STR C_3 from the selection program, along with ACR 94 TZE Comp5-W, had stable grain yield under *Striga* infested and non-infested conditions.

Grain yield of the extra-early genotypes evaluated under *Striga* infestation ranged from 772 kg ha^{-1} for 99 TZEE-Y STR C_0 to 1,588 kg ha^{-1} for 99 Syn EE-W (Table 3.10). The susceptible check, TZEE-W SR BC_5, suffered about 66% yield loss under *Striga* infestation compared with yield when *Striga*-free (783 vs. 2,266 kg ha^{-1}). It also sustained the worst *Striga* damage and was among the genotypes that supported the highest *Striga* emergence (Table 3.10), indicating that the level of infestation was severe in the evaluation trials.

The mean grain yields of the base populations, TZEE-W Pop STR C_0 and TZEE-Y Pop STR C_0, were not significantly different from that of the susceptible check, TZEE-W SR BC_5 under *Striga* infestation. The most promising white-grained genotypes in terms of grain yield, host damage, and level of *Striga* emergence were 99 Syn EE-W and 98 TZEE-W STR from other selection programs, and TZEE-W Pop STR C_4, TZEE-W Pop STR C_3 and 2004 TZEE-W Pop STR from the recurrent selection program. The experimental cultivar 99 Syn EE-W out-yielded the susceptible check by 51% and TZEE-W Pop STR C_4 by 47%. However, there were no significant differences in grain yield among these top ranking white endosperm genotypes. The highest yielding yellow-grained genotype was 2004 TZEE-Y Pop STR C_4 (3,366 ha^{-1}), which was not significantly different in grain yield performance from the derived cultivars, 99 TZEE-Y STR (derived from TZEE-Y Pop STR C_2) and 2004 TZEE-Y STR C_4 (derived from TZEE-Y Pop STR C_4). TZEE-Y Pop STR C_4 was also comparable to the top ranking white endosperm genotypes in terms of grain yield, *Striga* damage, and *Striga* emergence. The improved yield production of the derived cultivars from the advanced cycles of selection in TZEE-Y Pop STR was associated with decreases in *Striga* emergence at 8 and 10 WAP and *Striga* damage at 8 WAP while that of the white population was accompanied by an increase in days to silking (DYS), and EPP as well as a decrease in *Striga* emergence at 8 WAP. Changes in all other traits associated with recurrent selection in both populations were not significant statistically.

D. Residual Variances, Heritability, and Genetic Correlation

Selection and genetic drift can change gene frequencies and genetic variances in the selected population. Information on changes in the genetic variances, heritabilities, and genetic correlations under recurrent selection would be invaluable in determining the modifications in

Table 3.10. Grain yield (kg ha^{-1}) and other traits of derived cultivars from four cycles of S_1 recurrent selection in TZEE-Y Pop DT STR and TZEE-WPop DT STR maize populations evaluated in Striga-infested and Striga-free environments in Nigeria, 2005–2007.

Entry	Grain yield (kg ha^{-1})	Days to silk	ASI[y]	EPP	Plant height (cm)	Stalk lodging (%)	Striga rating 8 WAP	Striga rating 10 WAP	Striga count 8 WAP	Striga count 10 WAP
Striga-infested										
99 Syn EE-W	1,588	56	3	0.6	131.2	5.4	4.7	4.6	136	167
98 TZEE-W STR	1,345	56	4	0.6	128.5	5.8	5.1	4.6	118	154
2004 TZEE-W Pop STR C$_4$[z]	1,235	57	4	0.5	131.5	5.1	5.4	5.2	111	134
Ferke TZEE-W STR	1,106	57	4	0.5	122.7	4.9	5.5	5.2	119	150
99 TZEE-Y STR	1,089	52	4	0.5	113.5	3.9	5.6	5.7	120	126
2000 Syn EE-W[z]	1,049	56	5	0.5	123.8	5.2	5.6	5.4	120	143
Sine TZEE-W STR	1,031	57	4	0.6	122.7	4.2	5.3	5.2	118	143
2004 TZEE-Y Pop STR C$_4$[z]	956	57	5	0.5	130.0	6.2	5.3	5.4	133	170
TZEE-W SR BC5 (RE)	783	52	5	0.4	115.4	5.1	5.9	6.3	189	182
99 TZEE-Y STR C$_0$[z]	772	54	3	0.5	121.2	5.5	5.5	6.0	163	173
LSD[y] ($P < 0.05$)	534.9	2.33	1.93	0.2	15.33	1.9	0.8	1.01	69	68
P stat-test for genotypes	**	**	*	**	NS	*	**	**	NS	NS
Striga-free										
2004 TZEE-W Pop STR C$_4$[z]	3,366	56	2	0.9	165	3.4				
99 Syn EE-W	3,254	56	2	0.9	162	3.3				
2000 Syn EE-W[z]	3,128	54	2	0.9	156	4.0				
Ferke TZEE-W STR	3,002	57	2	0.8	163	3.1				
2004 TZEE-Y Pop STR C$_4$[z]	2,934	55	2	0.8	164	5.6				
98 TZEE-W STR	2,861	56	2	0.9	157	2.8				
99 TZEE-Y STR	2,366	52	2	0.8	146	3.9				
TZEE-W SR BC5 (RE)	2,266	51	2	0.8	152	5.3				
Sine TZEE-W STR	2,134	56	3	0.8	149	2.8				
99 TZEE-Y STR C$_0$[z]	2,067	54	3	0.8	154	4.8				
LSD[y] ($P < 0.05$)	329.7	0.82	0.54	0.06	7.77	2.00				
P stat-test for genotypes	**	**	*	**	**	**				

[z] Products from the S_1 recurrent selection program.
[y] ASI, Anthesis-silking interval; EPP, number of ears per plant; WAP, weeks after planting; LSD, least significant difference.
Notes: *, **, and NS indicate significant at $P < 0.05$, $P < 0.01$, and not significant, respectively.

the breeding methodology and strategies that should be used to ensure continued progress in future cycles of selection in the population improvement programs. Characterization of the genetic variability in the populations after four cycles of S_1 family recurrent selection is important for the evaluation of the effectiveness and progress expected from additional cycles of selection.

Several studies were initiated specifically to (1) determine the relative changes in the mean performance, genetic variances, heritabilities, and genetic correlation coefficients for grain yield, *Striga* resistance, and other agronomic traits under *Striga*-infested and *Striga*-free conditions; (2) estimate the realized gains from the recurrent selection program based on S_1 progenies; and (3) predict possible future gains from S_1 family selection in the populations.

Three hundred full-sib families were extracted from the C_3 of the early populations, TZE-W Pop DT STR and TZE-Y Pop DT STR, and evaluated in separate studies. The full-sib families were produced using the cross-classification (or nested) genetic design (North Carolina Design I) of Comstock and Robison (1948). The full-sib families from TZE-W Pop DT STR were evaluated under *Striga* infestation in Ferke in 2002 and Mokwa in 2003 while those from TZE-Y Pop DT STR were evaluated in Abuja and Mokwa. In addition, 50 S_1 families were extracted by self-pollinating random noninbred (S_0) plants from C_0, C_2, C_3, and C_4 of TZE-Y Pop DT STR and evaluated in Abuja and Mokwa under *Striga* infestation, and in Mokwa and Ikenne under *Striga*-free conditions in 2005 and 2007. Similarly, 50 S_1 families were extracted from C_0, C_2, C_3, and C_4 of TZEE-W Pop STR and TZEE-Y Pop STR. The resulting 200 S_1 families were evaluated for 2 years (2006 and 2007) in Abuja and Mokwa under *Striga* infestation, and in Mokwa and Ikenne under *Striga*-free conditions. Management of the trials and data collection was as earlier described.

Analyses of variance (ANOVA) combined over environments for each population were performed separately for the *Striga*-infested and *Striga*-free environments, using PROC GLM of the statistical analysis systems (SAS). The *Striga* emergence count was transformed to normality using the log transformation $(y+1)$ before the ANOVA. Genetic variance estimates of the population were obtained for each cycle of selection from the combined ANOVA by equating observed mean squares with expected mean squares. Heritability (h^2) was computed as the ratio of the genetic variance (σ_g^2) to the phenotypic variance (σ_p^2) on a progeny-mean basis. Standard errors for σ_g^2 and h^2 estimates were calculated using the method of Hallauer and Miranda (1988). Genotypic correlations between pairs of agronomic traits and the standard errors were computed with the restricted maximum likelihood (REML) method (Holland 2006) using Proc MIXED of SAS. A genetic correlation was considered significant when the standard error was more than twice as large as the genetic correlation.

1. The Early Populations. For the TZE-W Pop DT STR, additive genetic variances (σ_a^2) were moderately large and much larger than the dominance variances (σ_d^2) for most traits (Table 3.11). However, *Striga* emergence count was under the control of σ_d^2. The dominance variance for *Striga* emergence was about twice as large as additive genetic variance at 8 WAP and about four times as large at 10 WAP. Narrow sense heritability (h^2) estimate was 24.5% for grain yield and ranged from 0% to 58% for 12 other traits. Low dominance variance, moderately large additive genetic variances, and high narrow sense heritability were obtained for *Striga* damage scores. Grain yield had a positive additive genetic correlation ($r_a = 0.81$) with EPP, a negative correlation with *Striga* damage ratings at 8 WAP ($r_a = -0.85$) and 10 WAP ($r_a = -0.83$), and negative r_a with flowering traits ($r_a = -0.48$, -0.65 and -0.73 for anthesis, silking and ASI, respectively), and *Striga* emergence count at 10 WAP ($r_a = -0.56$).

For the TZE-Y Pop DT STR, estimates of dominance variances were larger than additive genetic variances for grain yield, PLHT, ear height, number of ears at harvest, and *Striga* damage rating at 8 WAP (Table 3.11). Even though h^2 estimates were generally low for most traits (< 0.4), moderate to large additive genetic variances, and wide ranges were obtained for most traits suggesting that there is adequate genetic variation for improving *Striga* resistance and grain yield in the population. Highly significant correlation coefficients were obtained between grain yield and EPP, PLHT, ear height, DYA and silking, ASI, and *Striga* damage score at 10 WAP (data not shown). Recurrent selection methods that capitalize on both additive and dominance variances would be effective for further improvement of *Striga* resistance and grain yield in this population.

Analysis of the data obtained from the 50 S_1 lines extracted from each cycle of selection in the TZE-Y Pop STR C_0 population revealed that significant improvement in grain yield and *Striga* resistance were associated with recurrent selection (Table 3.12). The advanced cycles of selection significantly out yielded the original cycle in both research environments. However, realized gains from selection for grain yield under *Striga*-infestation (68 kg ha^{-1} cycle^{-1}) and *Striga*-free conditions (169 kg ha^{-1} cycle^{-1}) were significantly lower than the predicted gains (350 and 250 kg ha^{-1} cycle^{-1}, respectively). Under *Striga* infestation, estimates of genetic variances for grain yield, EPP, DYA, PLHT, and *Striga* damage generally increased in the C_4 relative to the C_0 of selection (Table 3.12). In contrast,

Table 3.11. Estimates of additive genetic variance (σ^2_a), phenotypic variance σ^2_p and narrow-sense heritability (h^2) of grain yield and agronomic traits among North Carolina Design I (nested) progenies from TZE-W Pop DT STR C_3 and TZE-Y Pop DT STR C_3 evaluated under artificial *Striga* infestation in Abuja and Mokwa in 2003.

Trait	TZE-W Pop DT STR C_3			TZE-Y Pop DT STR C_3		
	$\sigma^2_a \pm SE$[y]	$\sigma^2_p \pm SE$	$h^2 \pm SE$	$\sigma^2_a \pm SE$	$\sigma^2_p \pm SE$	$h^2 \pm SE$
Anthesis-silking interval	1.18 ± 0.77	4.64	24.64 ± 0.17	0.11 ± 0.32	1.35	8.22 ± 0.48
Days to anthesis	3.54 ± 0.79	6.12	89.55 ± 0.13	6.38 ± 0.36	11.46	55.72 ± 0.13
Days to silking	6.78 ± 1.86	12.99	68.14 ± 0.14	8.57 ± 0.73	13.92	61.58 ± 0.15
Ear height (cm)	39.20 ± 13.23	95.22	47.49 ± 0.14	25.09 ± 7.99	128.53	19.52 ± 0.12
Grain yield (t ha^{-1})	125.7 ± 83.9	473.3	24.51 ± 0.18	46.9 ± 19.0.	274.7	17.07 ± 0.14
No. of ears/plant	0.005 ± 0.00	0.03	19.67 ± 0.12	0.001 ± 0.00	0.03	3.65 ± 0.14
Plant height (cm)	110.73 ± 39.07	309.88	42.98 ± 0.13	63.10 ± 23.14	374.62	17.38 ± 0.12
Root lodging (%)	1.30 ± 1.40	7.01	13.35 ± 0.20	2.85 ± 0.77	10.21	27.90 ± 0.15
Stalk lodging (%)	0.00 ± 1.14	6.27	0.00 ± 0.18	0.62 ± 0.20	1.68	37.01 ± 0.24
Striga rating at 8 WAP	0.25 ± 0.15	0.83	26.13 ± 0.19	0.07 ± 0.04	0.47	14.63 ± 0.16
Striga rating at 10 WAP	0.46 ± 0.17	0.93	44.82 ± 0.18	0.16 ± 0.05	0.39	41.90 ± 0.23
Striga count at 8 WAP	386.47 ± 331.41	2421.65	14.92 ± 0.14	93.40 ± 50.86	451.73	20.68 ± 0.26
Striga count at 10 WAP	212.25 ± 337.10	2325.71	7.79 ± 0.15	96.64 ± 52.63	521.49	18.53 ± 0.20

[z]Not computed because of estimated zero values for additive genetic variance.
[y]SE, standard error for testing if any estimates of variance or heritability is different from 0.

Table 3.12. Estimates of genetic variance (±SE and broad sense heritability (±SE for measured traits of 50 S_1 lines derived from four cycles of selection in the TZE-Y Pop STR maize population tested under artificial *Striga* infestation at Mokwa and Abuja, and *Striga*-free conditions at Ikenne and Mokwa, Nigeria in 2005 and 2007.

Trait	Genetic variance		Broad-sense heritability	
	C_0	C_4	C_0	C_4
Striga-infested environment				
Anthesis-silking interval	3.13 ± 0.6917	2.31 ± 0.70	72.11 ± 0.16	60.17 ± 0.18
Days to anthesis	1.90 ± 0.45	2.76 ± 0.60	69.53 ± 0.17	78.75 ± 0.17
Days to silking	4.66 ± 1.06	4.37 ± 1.17	69.20 ± 0.16	70.70 ± 0.19
Ear aspect	0.08 ± 0.0338	0.13 ± 0.046	41.08 ± 0.17	49.20 ± 0.17
Ear per plant	0.01 ± 0.0022	0.01 ± 0.003	50.94 ± 0.18	68.14 ± 0.18
Grain yield (t ha^{-1})	12.8 ± 4.2	48.8. ± 12.7	49.88 ± 0.16	62.88 ± 0.16
Plant height (cm)	59.60 ± 22.0725	89.72 ± 25.93	50.44 ± 0.19	58.97 ± 0.17
Striga rating 8	0.17 ± 0.0538	0.39 ± 0.102	59.46 ± 0.18	71.28 ± 0.18
Striga rating 10	0.14 ± 0.0484	0.26 ± 0.082	51.89 ± 0.18	55.15 ± 0.17
Striga count 8	172.91 ± 35.6703	108.74 ± 27.45	74.92 ± 0.18	64.96 ± 0.16
Striga count 10	291.61 ± 58.9121	137.74 ± 35.54	76.74 ± 0.16	64.90 ± 0.17
Striga-free environment				
Anthesis-silking interval	0.06 ± 0.06	+^{z}	18.75 ± 0.17	++
Days to anthesis	0.94 ± 0.27**	0.13 ± 0.13	54.08 ± 0.16	17.49 ± 0.17
Days to silking	1.16 ± 0.32*	0.31 ± 0.17	52.65 ± 0.15	29.53 ± 0.16
Ear aspect	0.01 ± 0.01	0.01 ± 0.01*	21.13 ± 0.19	7.86 ± 0.19
Ear height (cm)	3.36 ± 5.00	14.72 ± 7.70	12.52 ± 0.19	37.57 ± 0.20
Ear per plant	+	0.004 ± 0.02	++^{y}	4.80 ± 0.22
Grain yield (t ha^{-1})	43.4 ± 12.5**	36.1 ± 15.7**	58.92 ± 0.17	43.23 ± 0.19
Plant aspect	0.01 ± 0.006*	0.01 ± 0.006	13.12 ± 0.20	27.08 ± 0.22
Plant height (cm)	8.86 ± 10.71	25.24 ± 15.37	14.31 ± 0.17	32.85 ± 0.20
Stalk lodging (%)	+	0.23 ± 0.15	++	31.06 ± 0.20

$^{z+}$ Negative estimates of genetic variances were equated to zero.
$^{y++}$ Heritability was not computed because of negative estimates of genetic variances.
Notes: * and ** are significantly different from zero at 0.05 and 0.01 levels of probability, respectively.

the genetic variances for days to silk, ASI, EASP, and number of emerged *Striga* plants decreased with selection. Heritability for grain yield, *Striga* damage, and number of emerged *Striga* plants were significantly greater than zero. Under *Striga*-free conditions, the genetic variances for grain yield, DYA, silking, ASI, and EASP generally decreased as selection progressed. On the other hand, increases were generally detected for PLHT, EPP, and stalk lodging. Genetic variances for the traits were generally higher when *Striga*-infested than when *Striga*-free.

Changes in the broad-sense heritability for grain yield and other traits with selection followed trends similar to that of the genetic variances and were all significantly greater than zero in both research environments (Table 3.12). In general, heritability estimates of measured traits were higher when *Striga* infested than when *Striga* free. Genetic correlation between grain yield and most other traits were not significant under *Striga*-free conditions (Table 3.13). Under *Striga* infestation, grain yield had highly significant genetic correlation with ears plant^{-1}, EASP, and *Striga* damage at 8 and 10 WAP for all or most cycles of selection. This study provided further evidence that adequate genetic variability exists in cycle 4 of the scheme to ensure future gains from selection.

Table 3.13. Genetic correlations between selected pairs of traits of S_1 lines derived from four cycles of selection in TZE-Y Pop STR evaluated under *Striga*-free environments at Ikenne and Mokwa, and *Striga*-infested environments at Mokwa and Abuja in 2005 and 2007.

Trait	Genetic correlation			
	C_0	C_2	C_3	C_4
Striga-free environments				
Grain yield versus anthesis-silking interval (ASI)	−1.00 ± 2.70	−1.00 ± 2.72	−0.34 ± 0.97	+
Versus days to silk	−0.30 ± 0.30	0.44 ± 0.47	−0.06 ± 0.86	−0.31 ± 0.50
Versus ear aspect	−0.40 ± 0.48	−1.00 ± 1.30	−0.76 ± 1.02	−1.00 ± 1.05
Versus ear height	1.00 ± 1.16	1.00 ± 1.16	1.00 ± 1.16	0.73 ± 0.30*
Versus ears per plant	1.00 ± 5.15	0.87 ± 0.32*	−0.03 ± 0.91	0.75 ± 1.17
Versus plant aspect	−0.73 ± 0.53	−0.88 ± 0.37*	−1 ± 1.24	−0.57 ± 0.26*
Versus plant height	+^{z}	1.00 ± 0.98	0.3 ± 0.75	0.52 ± 0.31
Plant height versus ear height	0.78 ± 0.96	0.83 ± 0.19**	−1.00 ± 1.99	0.89 ± 0.89
Striga-infested environments				
Anthesis silking interval (ASI)				
Versus *Striga* rating at 8 WAP	0.47 ± 0.20*	0.31 ± 0.23	0.43 ± 0.27	0.17 ± 0.22
Versus *Striga* rating at 10 WAP	0.45 ± 0.23	0.37 ± 0.28	0.57 ± 0.43	0.04 ± 0.26
Ear aspect				
Versus *Striga* rating at 8 WAP	0.89 ± 0.19**	0.76 ± 0.17**	0.32 ± 0.32	1.00 ± 0.08**
Versus *Striga* rating at 10 WAP	1.00 ± 0.18**	0.66 ± 0.23*	0.22 ± 0.54	0.92 ± 0.098*

(*continued*)

Table 3.13. (*Continued*)

Trait	Genetic correlation			
	C_0	C_2	C_3	C_4
Ears per plant versus plant height	−0.09 ± 0.43	−0.09 ± 0.43	1.00 ± 2.69	1.00 ± 0.62
Versus *Striga* count at 8 WAP	−0.39 ± 0.50	−0.003 ± 0.45	+	−0.43 ± 0.29
Versus *Striga* count at 10 WAP	0.01 ± 0.36	0.01 ± 0.38	+	−0.46 ± 0.31
Versus *Striga* rating at 8 WAP	−0.96 ± 0.19	−0.96 ± 0.18**	−1.00 ± 0.42*	−0.73 ± 0.22**
Versus *Striga* rating at 10 WAP	+	+	−0.54 ± 0.04**	−0.88 ± 0.17**
Grain yield versus anthesis-silking interval	−0.04 ± 0.28	−0.05 ± 0.37	−0.79 ± 0.37*	−0.24 ± 0.24
Versus days to silk	−0.39 ± 0.26	−0.43 ± 0.31	−0.46 ± 0.35	−0.35 ± 0.19
Versus ear aspect	−0.93 ± 0.15**	−0.99 ± 0.22**	−0.79 ± 0.24**	−0.89 ± 0.09**
Versus ears per plant	± 0.21**	1.00 ± 0.27**	−0.002 ± 0.78	0.93 ± 0.08**
Versus plant height	−0.11 ± 0.36	0.57 ± 0.43	0.58 ± 0.45	0.18 ± 0.24
Versus *Striga* count at 8 WAP	0.47 ± 0.76	0.47 ± 0.76	−0.68 ± 0.60	−0.33 ± 0.37
Versus *Striga* count at 10 WAP	0.47 ± 0.63	0.47 ± 0.62	−0.36 ± 0.64	−0.47 ± 0.38
Versus *Striga* rating at 8 WAP[y]	−0.64 ± 0.21**	−0.79 ± 0.23**	−0.56 ± 0.04**	−1.00 ± 0.06**
Versus *Striga* rating at 10 WAP	−0.78 ± 0.19**	−0.81 ± 0.21**	−0.22 ± 0.78	−0.99 ± 0.08**
Striga rating at 8 WAP versus *Striga* count at 8 WAP	−0.11 ± 0.24	0.31 ± 0.22	0.05 ± 0.29	0.42 ± 0.19**
Versus *Striga* rating at 10 WAP	0.89 ± 0.08**	1.00 ± 0.08**	1.00 ± 0.26**	0.95 ± 0.06**
Versus *Striga* count at 8 WAP	0.01 ± 0.23	0.19 ± 0.23	−0.08 ± 0.29	0.26 ± 0.21
Versus *Striga* rating at 10 WAP	1.00 ± 0.02**	1.00 ± 0.03**	0.98 ± 0.05**	1.00 ± 0.03**
Striga rating at 10 WAP versus *Striga* count at 8 WAP	−0.05 ± 0.26	0.51 ± 0.24*	−0.06 ± 0.46	0.49 ± 0.23*
Versus *Striga* count at 10 WAP	0.17 ± 0.26	0.35 ± 0.26	−0.22 ± 0.47	0.32 ± 0.25

[z,+]Correlations not estimable because one or both variances involved were estimated to be zero.
[y]Weeks after planting.
Notes: * and ** are significantly different at 0.05 and 0.01 levels of probability, respectively.

2. The Extra-Early Populations. Results of the analysis of the 50 S_1 families extracted from each of C_0, C_2, C_3, and C_4 of TZEE-W Pop STR population showed gain in grain yield was 26% cycle^{-1} under *Striga* infestation and 16.4% ha^{-1} when *Striga*-free (Table 3.14). Under *Striga* infestation, genetic variances decreased with selection for *Striga*

Table 3.14. Estimates of genetic variance (±standard error) and broad sense heritability on an entry mean-basis (±standard error) for grain yield and other traits of S_1 families derived from the C0 and C4 of S1 recurrent selection in TZE-W Pop DT STR population tested under artificial *Striga hermonthica* infestation in Mokwa and Abuja, and under *Striga*-free conditions in Mokwa and Ikenne, Nigeria in 2006 and 2007.

	Genetic variances		Broadsense heritability	
Trait	C_0	C_4	C_0	C_4
Striga-infested environments				
Anthesis-silking interval	0.283 ± 0.39	0.15 ± 0.29	0.13 ± 0.19	0.12 ± 0.23
Days to anthesis	0.40 ± 0.27	0.61 ± 0.19*	0.29 ± 0.20	0.48 ± 0.15*
Days to silk	0.56 ± 0.55	1.14 ± 0.57*	0.18 ± 0.17	0.37 ± 0.18*
Grain yield (t ha^{-1})	2.1 ± 6.4	37.3 ± 18.6*	0.07 ± 0.21	0.37 ± 0.18*
Ear aspect	0^z	0.07 ± 0.03*	0^y	0.30 ± 0.18
Ears per plant	0.004 ± 0.001*	0.005 ± 0.002*	0.44 ± 0.17*	0.47 ± 0.16*
Plant height	22.82 ± 14.59	17.53 ± 11.31	0.31 ± 0.20	0.28 ± 0.18
Stalk lodging	0.19 ± 0.14	0.29 ± 0.16	0.25 ± 0.18	0.32 ± 0.18
Striga damage rating				
8 WAP^x	0.08 ± 0.04	0.08 ± 0.04	0.34 ± 0.18	0.32 ± 0.18
10 WAP	0.004 ± 0.02	0.05 ± 0.04	0.02 ± 0.19	0.23 ± 0.20
Striga emergence count				
8 WAP	37.35 ± 14.79*	41.87 ± 15.65*	0.39 ± 0.15*	0.43 ± 0.16*
10 WAP	56.72 ± 17.57**	52.54 ± 19.03*	0.48 ± 0.14**	0.41 ± 0.15*
Striga-free environments				
Anthesis-silking interval	0.025 ± 0.04	0.030 ± 0.03	0.11 ± 0.18	0.15 ± 0.19
Days to silking	0.61 ± 0.20*	0.47 ± 0.17**	0.43 ± 0.15*	0.47 ± 0.17*
Days to anthesis	0.56 ± 0.19*	0.27 ± 0.12*	0.45 ± 0.15*	0.37 ± 0.18*
Ear aspect	0.001 ± 0.001	0.0210 ± 0.010*	0.26 ± 0.17	0.39 ± 0.19*
Ear height	2.85 ± 5.29	3.08 ± 6.83	0.10 ± 0.18	0.10 ± 0.22
Ears/plant	0.0015 ± 0.002	0.0010 ± 0.002	0.12 ± 0.19	0.16 ± 0.20
Grain yield (t ha^{-1})	12.5 ± 15.2	37.2 ± 18.3*	0.16 ± 0.18	0.37 ± 0.17*
Plant aspect	0.001 ± 0.010	0.0001 ± 0.001	0.04 ± 0.18	0.03 ± 0.21
Plant height	16.58 ± 14.75	5.97 ± 14.39	0.21 ± 0.18	0.08 ± 0.20
Stalk lodging	0.08 ± 0.10	0.03 ± 0.07	0.17 ± 0.21	0.07 ± 0.23

^{z}Negative estimates of genetic variances were equated to zero.
^{y}Heritabilty was not computed because of negative estimates of genetic variances.
^{x}WAP, weeks after planting.
Notes: * and ** are significantly different at 0.05 and 0.01 levels of probability, respectively.

emergence and EPP (Table 3.14). Under *Striga* free conditions, genetic variability also decreased for flowering traits. Genetic variances were significant for *Striga* emergence in all cycles and for EPP, in C_0 and C_4. Response to selection for improved *Striga* emergence, EPP and grain yield are expected in subsequent cycles.

When *Striga*-free, the genetic correlations between grain yield, the primary selection trait, and DYS, PLHT, EPP, and ASI of the four cycles of selection were not significant (Table 3.15). Under *Striga* infestation, none of the four possible genetic correlation coefficients between *Striga* traits and grain yield was significant at the C_0, increasing to one, two, and three at C_2, C_3, and C_4, respectively. In every case, the significant coefficients were negative. This implied that high yielding families had low *Striga* counts and *Striga* ratings. Hence, there were more resistant families at the C_4 than the C_0 of the selection program in this population. For most other traits, there were also more significant genetic correlation coefficients at the C_4 than the earlier selection cycles of this population.

The studies led to the following conclusions: (1) recurrent selection was effective for improving *Striga* resistance traits and grain yield, which were characterized by low-to-medium heritability estimates in two early and two extra-early maize populations; (2) residual genetic variances were significant for *Striga* emergence and rating, grain yield, EPP and several other traits of the populations; and (3) response to selection for improved *Striga* emergence, EPP and grain yield are expected in subsequent cycles of selection in these populations.

IV. ADAPTATION

A. Importance of Adaptation to Drought-Prone Environments in Maize Cultivars for West and Central Africa

The maize crop has good prospects for rectifying the food deficit in WCA because there are a number of dependable improved technologies available to farmers. The Guinea savanna has the greatest potential for increased maize production due to high solar radiation, low night temperatures and low incidence of diseases. Unfortunately, maize is plagued by two major production constraints in the Guinea savanna, namely drought and *S. hermonthica* infestation. In a study to compare the effects of drought stress and *S. hermonthica* on maize under field conditions, Badu-Apraku et al. (2004b) reported that grain

Table 3.15. Genetic correlation estimates between selected pairs of traits of S_1 lines derived from four cycles of selection in TZEE-W Pop STR under *Striga* free (Mokwa and Ikenne) and *Striga*-infested environments (Abuja and Mokwa) in 2006 and 2007.

Trait pair	C_0	C_1	C_2	C_3	C_4
Striga-free environment					
Anthesis-silking interval (ASI) versus grain yield	−0.80 ± 0.001**	†^{z}		−0.89 ± 0.89	−0.20 ± 0.14
Days to silk versus grain yield	0.28 ± 1.15	−0.51 ± 0.42		0.33 ± 0.10**	0.71 ± 0.26**
Ear aspect versus grain yield	−1.00 ± 0.001**	−0.29 ± 0.001**		−0.39 ± 0.17*	−0.58 ± 0.69
Ear height versus grain yield	0.89 ± 0.001**	1.00 ± 0.7		†	0.90 ± 0.85
Ears per plant versus grain yield	0.68 ± 1.17	0.24 ± 0.29		0.91 ± 0.43*	0.005 ± 0.01*
Plant aspect versus grain yield	−0.72 ± 0.001**	−0.86 ± 0.001**		−0.11 ± 0.59	−0.65 ± 0.85
Plant height versus grain yield	0.45 ± 0.39	0.78 ± 0.55		0.37 ± 0.97	1.00 ± 2.31
Root lodging versus grain yield	0.59 ± 0.001**	0.06 ± 0.03		−0.15 ± 0.83	−0.39 ± 0.82
Stalk lodging versus grain yield	−0.79 ± 0.16**	0.03 ± 0.02		−1.00 ± 3.45	−0.56 ± 0.81
Striga-infested environment					
ASI versus ears per plant	−0.91 ± 0.62	−1.00 ± 0.67		†	1.00 ± 0.35*
ASI versus grain yield	−1.00 ± 0.93	−0.02 ± 0.85		−0.15 ± 0.85	−0.10 ± 0.44
Days to silk versus grain yield	−1.00 ± 1.02	−0.38 ± 0.57		−0.26 ± 0.46	0.14 ± 0.45
Ear aspect versus ASI	†	0.04 ± 0.82		−0.87 ± 0.72	1.00 ± 0.82
Ear aspect versus grain yield	†	−0.76 ± 0.36*		−0.84 ± 0.40*	−0.81 ± 0.27**
Ears per plant versus grain yield	−0.13 ± 0.63	1.00 ± 0.25**		1.00 ± 7.91	0.94 ± 0.12**
Plant height versus grain yield	0.08 ± 0.55	−0.09 ± 0.59		0.51 ± 0.49	1.00 ± 0.48*
Stalk lodging versus grain yield	−0.11 ± 0.85	0.49 ± 0.77		0.09 ± 0.59	−1.00 ± 0.52
Striga rating at 8 WAP^y versus grain yield	−0.59 ± 0.43	−1.00 ± 1.25		−0.30 ± 0.51	−0.87 ± 0.23**
10 WAP versus grain yield	1.00 ± 2.19	−0.86 ± 0.54		−0.58 ± 0.35	−0.94 ± 0.24**
Striga count at 8 WAP versus grain yield	0.59 ± 0.71	−0.87 ± 0.02**		−0.65 ± 0.02**	−0.39 ± 0.39
10 WAP versus grain yield	0.43 ± 0.57	−0.90 ± 0.84		−0.36 ± 0.01**	−0.41 ± 0.01**
Striga count 8 WAP versus *Striga* count 10 WAP	1.00 ± 0.03**	1.00 ± 0.03**		0.98 ± 0.05**	1.00 ± 0.03**

(*continued*)

Table 3.15. (*Continued*)

Trait pair	C_0	C_2	C_3	C_4
Striga rating at 8 WAP versus ASI	0.94 ± 0.70	†	-0.12 ± 0.78	$1.00 \pm 0.19^{**}$
Versus ear aspect	†	-1.00 ± 6.23	$0.79 \pm 0.38^{*}$	$0.79 \pm 0.36^{*}$
Versus ears per plant	$-0.96 \pm 0.18^{**}$	$-1.00 \pm 0.36^{*}$	†	$-0.73 \pm 0.21^{**}$
Versus *Striga* rating at 10 WAP	$1.00 \pm 0.25^{**}$	1.00 ± 7.83	$0.96 \pm 0.13^{**}$	$1.00 \pm 0.18^{**}$
Versus *Striga* count at 8 WAP	-0.14 ± 0.52	-1.00 ± 3.77	0.61 ± 0.52	0.29 ± 0.42
Versus *Striga* count at 10 WAP	-0.12 ± 0.44	-1.00 ± 7.54	0.50 ± 0.56	0.31 ± 0.45
Striga rating at 10 WAP versus ASI	†	0.44 ± 1.02	-0.36 ± 0.74	$1.00 \pm 0.43^{*}$
Versus ear aspect	†	-0.38 ± 1.03	$1.00 \pm 0.43^{*}$	$1.00 \pm 0.25^{**}$
Versus ears per plant	†	-0.74 ± 0.46	$-1.00 \pm 0.42^{*}$	$-0.88 \pm 0.17^{**}$
Versus *Striga* count at 8 WAP	-0.59 ± 0.43	-1.00 ± 0.92	0.98 ± 0.50	$0.69 \pm 0.30^{*}$
Versus *Striga* count at 10 WAP	-1.00 ± 1.34	-1.00 ± 1.01	0.81 ± 0.41	$0.70 \pm 0.34^{*}$

[z]†Genetic correlations could not be estimated because variance estimate was less than or equal to zero.
[y]WAP, weeks after planting.
Notes: * and ** are significantly different at 0.05 and 0.01 levels of probability, respectively.

yield was reduced by 53% under drought stress and by 42% under *Striga* infestation. Furthermore, drought stress and low soil nutrient status, especially N aggravates *S. hermonthica* parasitism on maize (Lagoke et al. 1991; Cechin and Press 1993; Mumera and Below 1993; Kim and Adetimirin 1995). Therefore, in the Sudan and NGSs where intermittent drought occurs frequently, it is desirable to incorporate drought tolerance into cultivars that have resistance to *Striga* as the two stresses occur at together. Presently, farmers in *Striga* endemic ecologies of WCA are demanding for cultivars that combine *Striga* resistance with drought tolerance and are unwilling to adopt maize cultivars that do not possess both adaptation to drought-prone environments and *Striga* resistance. During the past decade, IITA and the national programs of WCA have released a limited number of improved early, extra-early, intermediate and late maturing cultivars and inbred lines that combine *Striga* resistance with adaption to drought-prone environments to increase productivity on-farm. There is a tremendous opportunity for improving the overall performance and suitability of these cultivars by incorporating higher levels of adaptation to drought-prone environments. Several alleles govern the expression of adaptation to drought-prone environments in maize. Therefore, a major strategy of the IITA maize program has been to screen maize inbred lines with adaptation to drought-prone environments from diverse sources. The promising inbred lines with enhanced adaptation to drought-prone environments are also screened for *Striga* resistance under artificial infestation. The promising inbreds with both better adaptation to drought-prone environments and *Striga* resistance are then evaluated in hybrid combinations for better adaptation to drought-prone environments as well as adaptive traits in selected screening sites. The selected lines are used as sources of genes for withstanding both stresses, which are used for further introgression into breeding populations undergoing S_1 family recurrent selection. Further improvement of the early populations under controlled drought stress using recurrent selection can generate new productive cultivars that combine enhanced levels of adaptation to drought-prone environments with improved levels of resistance to *Striga*. The *Striga*-resistant hybrids with enhanced adaptation to drought-prone environments are evaluated in regional and farmer participatory on-farm trials using the mother-baby approach at several contrasting environments in WCA in collaboration with national maize programs and then released for production by farmers.

B. Breeding Methodology for Adaptation to Drought-Prone Environments

The goal of our program is to develop open-pollinated and hybrid maize cultivars adapted to the different forms of climatic variation prevalent in WCA with emphasis on drought stress. Specifically, the naturally available drought escape and drought tolerance mechanisms in the maize germplasm and the prevailing production environments in WCA were exploited to develop cultivars with enhanced adaptation to stressful environments.

Drought escape occurs when the plant completes critical physiological processes before drought sets in. This trait is quite desirable in cultivars to be released to farmers in the areas of WCA where terminal drought is most prevalent. Adaptation to drought-prone environments, on the other hand, is under genetic control and it indicates the presence of physiological mechanisms to minimize or withstand the adverse effects of drought if and when it occurs. Cultivars with enhanced adaptation to drought-prone environments are useful where drought occurs randomly and at any growth stage of the maize crop. This is quite relevant in WCA where drought occurrence is erratic, with varying intensity and timing.

Two strategies have been adopted for developing cultivars adapted to drought-prone environments in WCA:

1. Develop early and extra-early maturing cultivars that complete their life cycles before severe moisture deficit occurs.
2. Breed cultivars (DT) with better adaptation to drought-prone environments under controlled stress.

Selection for earliness/extra-earliness was carried out in the savanna agroecological zones of the subregion, and several cultivars have been bred, some of which have been released to the farmers after extensive testing in the different countries of the subregion. Induced drought stress is achieved by withdrawing irrigation water 21 and 28 days after planting until maturity for extra-early and early genotypes, respectively. Promising inbred lines selected for drought tolerance have been used to develop early and extra-early maturing open-pollinated and hybrid cultivars with enhanced adaptation to drought-prone environments. The selected lines are also used as sources of tolerance genes for introgression into early breeding populations that are undergoing S_1 family recurrent selection. Using this strategy, several early or extra-early and *Striga*-resistant cultivars with enhanced adaptation to drought-prone environments have been bred. Grain yield under drought stress has strong positive correlation with EPP, kernels per ear and strong negative correlation with ASI. A restricted selection index (grain yield, shorter ASI, EPP, and rate of leaf senescence) was used to improve adaptation to drought-prone environments. Recurrent selection methods such as S_1 family selection and full-sib family selection are used to increase the frequency of alleles enhancing adaptation to drought stress. Inbreeding is also adopted to fix favourable alleles. Our focus is on screening of diverse maize inbred lines with resistance to *Striga,* and other adaptive traits for adaptation to drought under induced moisture stress.

Evaluation of cultivars developed by both strategies showed considerable improvement in yield under moisture stress without any trade-off under nonstress conditions. In a study to evaluate the performance of 17 early-maturing cultivars in 11 WCA locations, Badu-Apraku et al. (2008b) obtained mean grain yield of $4\,t\,ha^{-1}$ with a range of 3.7–$4.4\,t\,ha^{-1}$ (Table 3.16). In similar studies carried out at Ile-Ife, a rain forest location in Nigeria, several high-yielding cultivars, under stress, maintained their relative ranks under nonstress conditions (Table 3.17).

V. DEVELOPMENT OF QPM POPULATIONS AND CULTIVARS

A. Introduction of QPM to West and Central Africa

Maize plays a critical nutritional role in the SSA diets because it is one of the three most important staple food crops, the others are cassava and rice. Traditionally, maize is consumed as a starchy base in a variety of forms such as gruels, porridge, and pastes. It is also widely fed as porridge to weaning children (2–3 months old, until the children are completely weaned at the age of 15–24 months old) and preschool children (3–5 years old) without protein supplements. The normal maize has a major nutritional constraint as human food because its protein, which is about 10% of the grain weight, is deficient in two essential amino acids, lysine and tryptophan, which humans and nonruminant animals cannot synthesize. Infants fed on normal maize without any balanced protein supplements suffer from malnutrition and develop diseases such as kwashiorkor, a fatal syndrome characterized by initial growth failure, irritability, skin lesions, edema, and fatty liver. The higher lysine content of QPM improves the absorption of Zn and Fe in the human digestive system and may thus contribute to improved micronutrient status. The first QPM cultivar released in WCA was Obatanpa GH. This cultivar has been widely adopted by farmers and consumers in Ghana. Presently, it covers

Table 3.16. Grain yield, days to silk, plant height, and ear height of 17 early-maturing maize cultivars evaluated in 11 West Africa locations, 2002.

Cultivar	Grain yield (kg ha^{-1})	Days to silk	Plant height (cm)	Ear height (cm)
HP 97 TZE COMP 3X4	4,403	52	153	80
ACR 97 TZE COMP 3X4	4,192	53	153	77
ACR 94 TZE COMP5-Y	4,171	52	141	75
ACR 94 TZE COMP5-W	4,162	53	146	75
TZE COMP 3X4 C2 F2	4,140	54	151	80
EV DT 97 STR C_1	4,125	53	151	82
TZE-Y POP DT STR C_3	4,103	54	161	87
Local check	4,097	54	158	87
AK 9331-DMRSR	4,094	53	147	77
TZE-W POP DT STR C_3	4,085	54	154	85
KAMBOINSE 88 POOL 16 DT (RE)	4,081	53	147	79
EV FN 9190 DWDP	4,078	55	152	81
TZE-W POP x 1368 STR C_1	4,058	53	146	76
98 Syn WEC STR C_0	4,020	54	150	83
AC 95 TZE COMP4 C_3 F3	3,982	54	151	77
2000 Syn WEC	3,689	53	149	85
99 Syn WEC	3,678	56	144	80
Grand mean	4,068	54	150	80
Range	3,678–4,403	52–56	141–161	75–87
Coefficient of variation (%)	17	3	8	13
Least significant difference ($P < 0.05$)	277	1	5	4
Standard error of differences	147.4	0.3	2.7	2.3
F-test for cultivar	NS	**	**	**
F-test for environment × cultivar interaction	**	**	**	**

Notes: ** and NS indicate significant at $P < 0.01$ and not significant, respectively.
Source: Badu-Apraku et al. (2008).

more than 50% of the maize hectarage (650,000 ha) in Ghana (Dankyi et al. 2005). It has also been released formally or informally in several other African countries including Benin (as Faaba), Burkina Faso, Cameroon, Côte d'Ivoire, Ethiopia, Guinea, Malawi, Mali (as Debunyuman), Mozambique (Susuma), Nigeria (as SAMMAZ 14), Senegal, South Africa, Swaziland, Togo, Uganda, and Zimbabwe (Badu-Apraku et al. 2006c). Obatanpa GH also serves as a source of inbred lines for the development of QPM hybrids and synthetic cultivars in several maize breeding programs in Africa.

Table 3.17. Mean grain yield (t ha^{-1}) of 17 open-pollinated maize cultivars under natural drought stress in the late season and nonstress conditions in the early season at the T & R Farm of Obafemi Awolowo University, Ile-Ife, Nigeria 2002–2007.

Cultivar	Stressed Yield	Stressed Rank	Nonstressed Yield	Nonstressed Rank	Across Yield	Across Rank
TZECOMP3x4C$_2$	4.0	1	6.2	3	5.1	1
TZECOMP3DT	3.8	2	4.6	11	4.2	8
TZEE-WSRBC$_5$	3.8	2	5.9	4	4.9	3
HEI 97TZE COMP4C$_3$	3.5	4	5.8	5	4.7	5
Sin 9432	3.3	5	4.5	12	3.9	11
EV 8435-SR	3.2	6	5.5	7	4.4	6
ACR 95TZE COMP4C$_3$	3.1	7	4.2	13	3.7	12
TZE-WPOPx1368STRC1	3.1	7	6.6	2	4.9	3
AK95DMR-ESRW	3.1	7	5.0	10	4.1	10
BAG 97TZECOMP3x4	3.0	10	5.8	5	4.4	6
ACR 90POOL 16-DT	2.9	11	7.1	1	5.0	2
ACR 9931-DMRSR	2.8	12	5.5	7	4.2	8
TZEE-SRxDamascus	2.3	13	5.1	9	3.7	12
DMR-ESRW C1F2	2.3	13	3.5	14	2.9	14
EV 32-SR	2.2	5	2.9	16	2.6	15
DMR-ESRY C1F2	1.5	16	3.3	15	2.4	16
TZECOMP4C$_2$	1.2	17	1.6	17	1.4	17
Mean	2.9		4.9		3.9	
SE	0.19		0.35		0.25	

B. Genetic Enhancement of Obatanpa GH

Obatanpa GH has good levels of resistance to MSV, lowland rust (caused by *Puccinia polysora* Underw.), and moderate levels of resistance to blight (caused by *Bipolaris maydis*). Obatanpa GH was derived from Population 63 SR, a white dent QPM, adapted to the lowland tropics. Population 63 SR is a composite of intermediate maturing tropical maize germplasm originally developed by International Center for Maize and Wheat Improvement (CIMMYT) in Mexico. IITA incorporated resistance to MSV into the population. Following multi-location testing of Pop 63 SR in Ghana between 1987 and 1989 (Badu-Apraku et al. 2006b), the population was identified as a promising source for new QPM cultivars. At that time, the major defect of the population as a source of QPM was the low level of MSV resistance, poor husk cover, the presence of high percentage of kernels with soft chalky endosperm, and low grain yield potential. In 1989, while a visiting scientist in IITA-Ibadan, the senior author initiated a breeding program to breed MSV resistant, high yielding QPM cultivars with improved husk cover, appropriate hard endosperm modification, as well as elevated levels of lysine and tryptophan (Badu-Apraku et al. 2006b). A bulk of Pop 63 SR was planted at Ibadan and S_0 plants were infested with viruliferous leaf hoppers (*Cicadulina* spp.) about 9 days after planting. Two weeks after MSV infestation, MSV susceptible plants were rogued out and MSV resistant S0 plants were selfed. At harvest, about 500 ears from agronomically desirable S_1 plants with good husk cover were selected and screened under a light box for kernels with the desirable endosperm modification. Kernels with good endosperm modification selected from 250 ears were planted ear-to-row in a recombination block in the field at Ibadan. The S_1 plants were advanced to the second cycle of recombination under artificial infestation with viruliferous leaf hoppers. This breeding step was followed by screening of the selected ears for desirable kernel modification under the light box. The selected kernels of each ear were planted ear-to-row and advanced to the S_1 stage by selfing under artificial MSV infection. At harvest, about 250 S_1 ears selected from plants with good husk cover and other desirable agronomic traits were planted ear-to-row in a recombination block at Fumesua in Ghana to constitute the QPM population designated GH Pop 63 SR. In addition, about 30 kernels from each S_1 ear were sent to CIMMYT in Mexico for tryptophan and lysine analyses. Based on the results from laboratory analyses, 80 S_1 lines with higher levels of the two essential amino acids were advanced to two cycles of recombination with selection for improved husk cover, grain yield, and desirable kernel characteristics. The resulting cultivars were designated as Obatanpa, which in Ghanaian language means "good nursing mother." Results of multilocation field tests showed that Obatanpa was superior or comparable in grain yield and other agronomic characters to the top improved intermediate and late maturing normal endosperm maize cultivars in Ghana (Sallah et al. 1997; Twumasi-Afriyie et al. 1997). Furthermore, results of feeding trials with piglets and chicken showed that Obatanpa had higher nutritional value and could be used as a replacement for normal endosperm maize in animal feeds with economic advantage (Okai et al. 1994; Osei et al. 1994). Because of its superior performance and the elevated levels of lysine and tryptophan, Obatanpa was released for production by farmers in Ghana in 1992. As an open-pollinated cultivar, it has been necessary to upgrade the genetic purity of Obatanpa periodically since its release in 1992. For instance, in 2001, the lysine and tryptophan levels of the grains of Obatanpa were found to be low and some plants were observed to be susceptible to the MSV. To upgrade the lysine and tryptophan content of this open-pollinated cultivars, 277 half-sib families selected from Obatanpa were analyzed for the two essential amino acids in the QPM laboratory at CIMMYT in Mexico. Based on the results of the laboratory analysis, 40 families with high levels of lysine and tryptophan were selected and intermated to reconstitute the cultivar during the off-season of 2001 in Ghana. A program also was initiated during the major season of 2002 to upgrade the level of MSV resistance of the reconstituted cultivar. More than 500 families selected from the reconstituted Obatanpa were planted under artificial infestation with viruliferous leaf hoppers at Ferkessedougou in Côte d'Ivoire. The streak susceptible plants were rogued 3 WAP. At flowering, MSV-resistant plants with agronomically desirable characteristics were selfed. At harvest, the selected S_1 ears were recombined under artificial MSV infestation in Ibadan during the off-season of 2003 to reconstitute the new version of Obatanpa designated, Obatanpa GH. Evaluation trials involving five QPM hybrids, six open-pollinated QPM cultivars including the new and old versions of Obatanpa, and three normal endosperm varieties were conducted at seven locations in the forest and forest-savanna agroecologies of Ghana during the major and minor planting seasons of 2004. The results showed that Obatanpa GH was the highest yielding open-pollinated cultivar with a grain yield of $4.96\,t\,ha^{-1}$ compared with $3.56\,t\,ha^{-1}$ for the normal endosperm local check. Obatanpa GH silked at 55 DAP with a PLHT of 205 cm. In 19 on-farm trials conducted in the NGS zone of the Republic of Benin in 2004, Obatanpa averaged $3.67\,t\,ha^{-1}$. This yield level was comparable to that of EV 97 IWDT STR ($3.37\,t\,ha^{-1}$), a popular *Striga*-tolerant normal endosperm cultivar with enhanced adaption to drought stress. Obatanpa GH outyielded the improved normal endosperm cultivar used as local check ($2.66\,t\,ha^{-1}$). Laboratory analyses conducted in 2005 at IITA showed that Obatanpa GH contains 10% total protein in the grain with 0.88% tryptophan in the protein. In contrast, the normal endosperm check had 9.6% total protein with 0.49% tryptophan in the grain protein.

One important lesson learnt in our QPM breeding program is that there is a need for periodic upgrading of the level of streak resistance in released QPM cultivars in order to maintain the recessive QPM trait in open-pollinated cultivars that can be grown by small farmers who save seed from their farms after harvesting each year for planting the following season.

C. Conversion of Normal Endosperm Maize to QPM

To mitigate the effects of the two major constraints on maize production and productivity in WCA, *S. hermonthica*, and drought, IITA has since 1980, developed several high yielding early and extra-early *Striga*-resistant normal endosperm populations, open-pollinated and hybrid cultivars, and inbred lines with enhanced adaptation to drought-prone environments. Inbreeding, hybridization, and recurrent selection have all been used in the program. To increase the level of quality protein intake by maize consumers, it was necessary to convert elite normal endosperm maize to QPM. Our strategy in this regard has been to focus primarily on crossing elite populations, open pollinanted cultivars and inbred lines to QPM donor sources followed by selection of genetic modifiers, which are minor genes that exert their influence mainly by intensifying or diminishing the expression of major genes. The modifiers stimulate the *opaque-2* (o_2) gene to produce kernels with desirable characteristics. The sources of the QPM trait used in the conversion program for the normal endosperm white, early-maturing populations, cultivars, and inbred lines were Pool 15 SR QPM and DMR-ESR-W QPM (both white grained) and Pool 18 SR QPM (yellow grained) for the yellow materials. The donor for the conversion of the normal endosperm, extra-early maturing, white populations, cultivars, and inbred lines to QPM was EV 99 QPM. Seven early (TZE-W Pop DT STR C_3, EV DT-W 99 STR C_1, 98 Syn WEC STR C_0, TZE-W Pop x 1368 STR C_1, TZE-Y Pop DT STR C_3, TZE-Y Pop C_0 S_6, and EV DT-Y 2000 STR C_0) and four extra-early (TZEE-W Pop C_3, 2000 Syn EE-W, TZEE-Y Pop STR C_3, and 99 TZEE-Y STR) *Striga*-resistant, elite, normal endosperm maize cultivars, and populations with enhanced adaptation to drought stress were crossed in 2002 to QPM donors for partial conversion of the populations and varieties to QPM.

Available information indicates maternal endosperm hardness of F_1 kernels exist in F_2 kernels. Therefore, endosperm hardness in F_1 is usually higher when hard endosperm parents are used as females. Endosperm hardness and lysine content in F_1 kernels are usually significantly less than both parents when hard × hard endosperm combinations were analyzed. Hence, combinations involving both soft and hard endosperm parents or between semihard endosperm parents is a more practical way to obtain higher lysine content and keep the endosperm hardness. Consequently, in all our conversions, hard endosperm normal maize parents were used as the females in F_1 crosses while the donor parent had semihard QPM endosperm. The F_1 crosses were advanced to the F_2 stage and screened under the light box in 2003.

The F_2 kernels with good endosperm modification were selected and backcrossed to the respective recurrent parents to obtain the BC_1F_1. After one generation of backcrossing to the respective recurrent parents, more than 350 S_1 plants were extracted from each backcross population and the kernels from selected ears were screened under the light box for desirable endosperm modification. The selection for the appropriate endosperm modification during the conversion program was based on a rating 1–5 scale (where 1 = kernels completely translucent with no opaqueness; 2 = 25% opaqueness; 3 = 50% opaqueness; 4 = 75% opaqueness; and 5 = 100% opaqueness).

The S_1 kernels with a 2 or 3 score were selected and advanced to the S_2 stage by selfing. Kernels with score of 1 were not selected because while there was the probability of having the o_2 gene in the homozygous recessive condition, (o_2o_2), kernels could also be heterozygous (O_2o_2) or homozygous dominant (O_2O_2) in which case the kernel will be lower in lysine and tryptophan. Kernels with score 4 were not selected because even though the presence of o_2o_2 was guaranteed, the probability of obtaining well modified kernels in advanced generations was lower. The type 5 kernels were rejected as they had soft endosperm with no modifiers (Vivek et al. 2008). The S_2 kernels of each ear were screened under the light box and those with a score of 2–3 were again selected, grown ear-to-row, and the agronomically desirable plants were selfed to obtain about 250 S_3 ears at harvest. At this stage, only the kernels of the S_3 ears with a score of 2 were selected under the light box and planted in isolation blocks for recombination to obtain the F_1 generations of each partially converted variety/population. The F_1 seed of each recombined backcross material was planted and taken through the second cycle of recombination resulting in the extra-early and early QPM cultivars and populations. Seed samples of each F_2 population or cultivar were analyzed for tryptophan content in the laboratory. There was no conscious effort to select for *Striga* resistance or enhanced adaptation to drought-prone environments during the QPM conversions. However, we have used artificial *Striga* infestation to maintain the levels of *Striga* resistance in germplasm converted to QPM. Similarly, artificial infestation with viruliferous leaf hoppers has been used to maintain good levels of MSV resistance in converted germplasm.

D. Development of *Striga* Resistant QPM Inbred Lines and Hybrids with Enhanced Adaptation to Drought Stress

A total of 22 early white, 15 early yellow, and two extra-early white normal endosperm elite *Striga*-resistant inbred lines were crossed to

QPM donor sources (Pool 15 SR for the white and Pool 18 SR for the yellow) in 2003 in an effort to convert them to QPM. The F_1 crosses were backcrossed to the recurrent parents during the major season of 2005 to obtain BC_1. The BC_1 ears were screened under the light box and the kernels with desirable endosperm modification were selected and advanced to the BC_2 stage during the dry season of 2005. The BC_2 kernels with the desirable endosperm modification were selected for planting and selfing during the dry season of 2006. One hundred BC_2S_0 kernels of the early and extra-early QPM inbred lines in the conversion program with the desirable endosperm modification were selected and advanced to the BC_2S_1 and BC_2S_2 stages during the dry season of 2006 and rainy season of 2007, respectively. The selected BC_2S_2 of the early and extra-early inbreds with appropriate endosperm modification were planted in Ibadan in June 2008 and the inbreds with agronomically desirable characteristics were advanced to the BC_2S_3 stage. The BC_2S_3 inbreds were screened under the light box and those with appropriate endosperm modification were selected. The BC_2S_3 lines were planted at Ikenne at the end of November 2008 for screening for enhanced adaptation to drought stress. The S_3 lines with enhanced adaptation to drought-prone environment were recombined to form QPM synthetics. Five hundred BC_1S_3 early-maturing lines in the QPM inbred line conversion program were planted at Ikenne during the 2008–2009 dry season for screening for drought tolerance. Based on the results, 270 drought-tolerant lines were selected and evaluated under *Striga* infestation at Mokwa during the 2009 growing season. In addition, 80 BC_1S_2 extra-early QPM inbreds in the conversion program were evaluated under artificial *Striga* infestation at Abuja in 2009. During the *Striga* evaluations at Mokwa and Abuja, the BC_1S_3 of the early QPM inbreds, and the BC_1S_2 lines of the extra-early QPM were advanced to the BC_1S_4 and BC_1S_3 stages, respectively. Based on the results of the *Striga* evaluations, the best lines of each maturity group were selected and recombined in a diallel fashion to form synthetic cultivars for each maturity group and grain color. Furthermore, the selected S_3 lines of the extra-early and S_4 lines of the early QPM lines were advanced to the S_4 and S_5, respectively. Based on the results of the *Striga* evaluations, the best 30 lines of the early QPM inbreds and best 23 lines of the extra-early QPM were selected and advanced to the S_6 and S_5, respectively during the 2010 growing season. Ninety-three of the early QPM lines at S_6 stage comprising of 71 white grained and 22 yellow grained color were given TZEQ designations and analyzed for lysine and tryptophan contents in the laboratory in August 2010. Based on the analysis, the best 14 yellow endosperm QPM lines were planted and 91 single cross hybrids were generated using a diallel mating scheme.

The diallel crosses were evaluated under induced drought stress and well-watered conditions at Ikenne during the 2010–2011 dry season. Also, the best 30 white endosperm early maturing QPM inbreds were selected and single cross hybrids were made using North Carolina Design II (factorial) mating scheme to determine the performance of their crosses under drought stress and well-watered conditions, examine the combining abilities and inheritance patterns of the inbred lines, and identify the best testers for use in our breeding program. Preliminary results showed that the highest yielding single-cross QPM white hybrid out-yielded the normal endosperm check (TZEI 3 × TZEI 26) by 42% while the best single-cross yellow hybrid out-yielded the best yellow endosperm single-cross hybrid by 18%.

E. Evaluation of the Performance of Early and Extra-Early QPM Open-Pollinated and Hybrid Cultivars

There are several reports indicating that certain QPM cultivars produce grain yield equal to conventional cultivars currently grown in the developing world, and that several QPM experimental cultivars performed better than the checks in several regions of the world (National Research Council 1988). We did not have sufficient basis to make such assertions in WCA. There was, therefore, a need to assess how the available QPM cultivars bred by IITA and partners perform versus their normal counterparts, and to carry out extensive testing of the available QPM cultivars to promote their adoption by farmers in WCA. Sixteen early- and nine extra-early-maturing QPM and normal cultivars were evaluated under *S. hermonthica* infestation and *Striga*-free environments in Nigeria from 2006 to 2008 to identify stable and high yielding cultivars. The extra-early normal maize cultivars, 2000 Syn EE-W and 99 TZEF-Y STR, were similar in yield to the QPM versions under both research conditions. While TZEE-Y Pop STR C_4 was superior in grain yield to its QPM version only under *Striga*-free conditions, TZEE-W Pop STR C_4 significantly outyielded the QPM version, under both test conditions and was superior in *Striga* resistance. In the early maturity group, TZE-W DT STR C_4 outyielded the QPM version by 21% under *Striga* infestation and by 10% when *Striga*-free. In contrast, the QPM cultivar, 98 Syn WEC STR QPM C_0, outyielded the normal endosperm version by 31% when *Striga*-infested. GGE-biplot analysis demonstrated that two extra-early and three early-maturing cultivars had outstanding performance in both environments. TZEE-W Pop STR QPM C_0 and EV DT-W 99 STR QPM C_0 were high-yielding and stable when *Striga*-infested while TZE-Y Pop DT STR C_4, TZE-W Pop DT STR

C_4, and TZE-Y Pop STR QPM C_0 were superior when grown in *Striga*-free environments.

At the initiation of the QPM conversion program at IITA in 2002, normal endosperm extra-early and early-maturing populations and elite cultivars were selected for partial conversion to QPM based on high yield potential, resistance to *Striga* and enhanced adaptation to drought stress. During the conversion to QPM, no conscious effort was made to select for *Striga* resistance or enhanced adaptation to drought-prone environments. It is therefore striking that some products of the conversion program were superior or comparable to the normal endosperm versions, not only in terms of grain yield but also low *Striga* damage and emerged *Striga* plants (2000 Syn EE-W STR vs. 2000syn EE-W STR QPM C_0; 98 Syn WEC vs. 98 Syn WEC QPM C_0; EV DT-W 99 STR C_1 vs. EV DT-W 99 STR QPM C_0). Results of a trial involving 20 early-maturing cultivars with enhanced adaptation evaluated in 2008 at nine locations in drought prone environments in Northern Nigeria also revealed the cultivars Tillering Early DT, EV DT-W 99 STR QPM C_o, EV DT-Y 2000 STR QPM C_o, and Pool 18-SR/AK 94 DMESRY as outstanding because they outyielded the best local normal maize check by 8%–51% (DTMA 2008). Based on these results, EV DT-W 99 STR QPM C_0 and EV DT-Y 2000 STR QPM C_0, which consistently showed superior performance across six locations, were tested extensively on-farm in 2009 in the drought-prone and *Striga* endemic zones of the NGS of Nigeria. Also, EV DT-W 99 STR QPM C_0 is undergoing extensive on-farm testing in Benin while it is at the release stage in Ghana. In a similar trial involving extra-early cultivars evaluated at two locations in the Upper West Region of Ghana, the highest yielding QPM cultivar TZEE-W Pop STR QPM C_0 had 34% higher grain yield than the QPM check (Buah et al. 2009). Based on the superior performance of TZEE-W Pop STR QPM C_0, 2000 Syn EE-W QPM C_0, and TZEE-Y Pop STR QPM C_0 in WA, the three cultivars are presently undergoing extensive testing in Ghana and Mali using the mother-baby on-farm testing approach (S. Buah, SARI; J. Kambiok, SARI; N. Coulibaly, IER; personal communications 2009). TZEE-W Pop STR QPM C_0 and TZEE-Y Pop STR QPM C_0 were released in Ghana in 2010.

The outstanding performance of the QPM cultivars may be attributed to the large population sizes sampled during QPM conversion, which might have ensured that the favorable *Striga*-resistant and drought adaptation alleles were maintained in the respective populations during selection for high grain yield, desirable agronomic characteristics, and appropriate endosperm modification. This finding indicates the effectiveness of the backcross, inbreeding, and hybridization methods adopted in our conversion program.

VI. BREEDING FOR COMBINED TOLERANCE/RESISTANCE TO MULTIPLE STRESSES IN EARLY AND EXTRA-EARLY MAIZE

Maize cultivars targeted to the *Striga* prone areas of WCA must also be adapted to drought- and low N-prone environments. Consequently, the ultimate goal of maize breeding for biotic and abiotic stress tolerance in WCA is improved grain yield under three specific stress factors; that is, low soil nitrogen, drought, and *S. hermonthica* infestation. During the past decade, IITA in collaboration with the National Agricultural Research and Extension Systems (NARES) of WCA has used the S_1 recurrent selection method, improved artificial field infestation with *S. hermonthica*, and screening under drought stress as strategies to develop two early-maturing source populations—TZE-W Pop DT STR (white) and TZE-Y Pop DT STR (yellow)—and several early-maturing cultivars and inbred lines, These populations combined tolerance to drought with moderate levels of resistance to *S. hermonthica* and MSV. Inbreeding, hybridization, and recurrent selection have all been used in this breeding undertaking.

Badu-Apraku et al. (2011c) conducted three studies for 3 years (2007–2009) at five locations in Nigeria to evaluate the performance of selected early-maturing cultivars under contrasting environments; that is, drought stress versus well-watered; *Striga* infested versus *Striga* free; and in low N (30 kg ha^{-1}) versus high N (90 kg ha^{-1}) environments. Each trial contained 15–20 cultivars with three replications and 14 cultivars common to all environments. The objectives of the study were to use the GGE biplot methodology to (i) determine the best performing cultivar in each of four environments: under *Striga* infestation, induced drought stress, low N, and nonstress (ii) compare any two cultivars in individual environments, and identify the most suitable environment for each cultivar, and (iii) identify the most stable genotypes for each environment.

Drought stress reduced grain yield by 44%, *Striga* infestation by 65%, and low N by 40%. GGE biplot analysis showed that TZE-W DT STR C_4, Tillering Early DT, TZE-W DT STR QPM C_0, and TZE-Y DT STR C_4 performed relatively well in all study environments. TZE-W DT STR C_4 and TZE Comp3 C_1F_2 were outstanding under drought, TZE-W DT STR C_4, EVDT-W 99 STR QPM C_0, and TZE-W DT STR QPM C_0 under *Striga* infestation, and Tillering Early DT, EVDT 97 STR C_1, TZE-W DT STR C_4, and TZE Comp3 C_3 under N deficiency. Maize productivity can be significantly improved by promoting cultivation of genotypes that combine high resistance/tolerance to *Striga* and enhanced adaptation to drought with improved N-use efficiency. The high yielding and stable cultivars identified in this study (TZE-W DT STR C_4, Tillering Early DT, TZE-W DT STR QPM C_0, and TZE-Y

DT STRC$_4$) with tolerance to the three stresses and outstanding performance under optimal growing conditions are available for adoption by farmers in WCA.

VII. INBRED-HYBRID DEVELOPMENT PROGRAM

The predominant maize types cultivated by farmers in WCA are open-pollinated cultivars. Hybrid maize accounts for only about 5% of the total area under production (Abdoulaye et al. 2009). IITA has devoted considerable attention and resources in the last decade to develop genetically enhanced maize germplasm with higher yield potential and stability across varying levels of water availability and growing conditions. Strategies for developing germplasm useful for African farmers included the following: backcrossing, inbreeding, hybridization, the S_1 recurrent selection method, improved artificial field infestation with *S. hermonthica*, and screening under managed drought stress. Through these strategies, several early and extra-early maturing white grained source populations, inbred lines, open-pollinated cultivars, and hybrids have been developed, several of which combine tolerance to drought and resistance to *S. hermonthica* and MSV. Until recently, hybrids were grown mainly in Nigeria, which had a few seed companies. However, during the last 2–5 years, more seed companies have emerged in Nigeria, as well as in Ghana and Mali, setting the stage for large-scale hybrid production. The challenge for the NARES of WCA is to develop, test, and make available to farmers productive hybrids with tolerance to the important biotic and abiotic stresses including *Striga*, drought, and low soil N. Maize hybrids tolerant of these three stresses are vital to the sustainable productivity of the maize-based farming systems and the survival of the emerging seed companies in WCA.

An inbred-line development program was initiated in 1997 by IITA at Côte d'Ivoire to extract inbred lines from several broad-based *Striga* and MSV resistant early and extra-early populations (TZEE-W Pop STR C$_0$ and TZEE-Y Pop STR C$_0$) and crosses (TZEE-W SR BC$_5$ × 1368 STR, TZEE-W Pop STR × LD, and TZEF-Y SR BC$_1$ × 9450 STR). S_1 lines extracted from each population were evaluated at Ferkessedougou (mean annual rainfall of 1,400 mm) and Sinematialli (mean annual rainfall of 1,200 mm) during the rainy season of 1997. At Ferkessedougou, the lines were evaluated under artificial *Striga* infestation (about 5,000 germinable *Striga* seeds per maize hill (Kling et al. 2000)) and outstanding S_1 lines selected from each population were taken through six cycles of pedigree inbreeding and selection under artificial *Striga* infestation. In the S_4, 250–300 lines derived from each population were crossed to the corresponding base population as the tester. The S_4 lines *per se* and the testcrosses were evaluated at Sinematialli under *Striga*-free conditions and at Ferkessedougou under artificial infestation with *S. hermonthica* seeds collected from sorghum fields near each testing site. The yield performance of the lines *per se*, their combining abilities for grain yield, *Striga* damage rating, *Striga* emergence count, ear number, and other desirable agronomic characters across the two locations were used as criteria for selecting 90–100 S_4 lines, which were advanced to S_6. A total of 81 lines selected from all the source populations were evaluated during the dry and rainy seasons of 2001 under artificial infestation with *S. hermonthica* and under noninfested conditions at Ferkessedougou. In the dry season trial, the inbred lines were irrigated up to physiological maturity using an overhead sprinkler irrigation system, which applied 12 mm of water per week. Irrigation water was withdrawn based on the results of the progeny yield trials of 1997 rainy season. The selected S_1 lines from each population had undergone six cycles of pedigree inbreeding and selection under artificial *Striga* infestation. At the S_4 stage of inbreeding and selection, 250–300 lines derived from each population were evaluated *per se* and in testcrosses at Ferkessedougou and Sinematialli for general combining ability (GCA), with the same population as the tester. Based on the combining ability for grain yield, desirable agronomic characters, *Striga* resistance and the yield performance of the lines *per se* across the two locations, 8–10 S_4 lines from each population were selected and separately randomated to form synthetic cultivars. Furthermore, 90–100 S_4 lines were selected based on the test performance and advanced to the S_6 using pedigree selection under artificial *Striga* infestation. Through this program, several S_6 inbred lines and synthetic cultivars have been bred from each population. A large number of *Striga*-resistant early and extra-early maturing inbred lines have been developed and evaluated extensively as inbred lines *per se* in WCA countries by WECAMAN (Badu-Apraku and Fakorede 2001). The inbred lines were classified into groups of similar phenotypes based on agronomic and morphological traits, using multivariate statistical analyses (Badu-Apraku et al. 2006b).

Agbaje et al. (2008) conducted field trials to confirm the genotypic groupings of 77 (35 yellow and 42 white) lines on the basis of their heterotic patterns with a view to identifying lines that could be used in the development of high yielding, early maturing, *Striga* and disease-resistant heterotic populations, as well as hybrid and synthetic cultivars. The 42 white endosperm lines were crossed to two intermediate maturing, white endosperm tester lines (1368 and 9071) to generate 84 testcrosses. Similarly, the 35 yellow endosperm lines were crossed to two yellow endosperm testers (KU1414 and 4001) to generate 70 testcrosses. The 84 white testcrosses and four checks and the 70 yellow testcrosses and seven checks were evaluated separately in *Striga*-infested environments at Mokwa and Abuja, and *Striga*-free environments at Mokwa and Ile-Ife in 2005. Significant GCA and specific combining ability (SCA) effects for grain yield were estimated. Combining ability effects and grain yields of the testcrosses were used to classify the lines into heterotic groups. To belong into a heterotic group, the line must have significant ($P < 0.05$) SCA effects with one of the testers and significant ($P < 0.05$) negative SCA effects with the other, along with a mean yield equal to or greater than 1 SE above the grand mean of all testcrosses involving the positive SCA tester. Lines that had zero SCA effects were not classified into either heterotic group. Only 13 of the 42 white endosperm lines could be classified into heterotic groups based on the SCA effects and testcross mean grain yield in *Striga*-free environments (Table 3.18).

Under *Striga*-infested conditions, 12 white lines were classified into heterotic groups. Four of the lines (TZE-W Pop C_0 S_6 Inb50-2-4, TZE-W Pop × LD S_6 Inb 6 from the 1368 group and TZE-W Pop C_0 S_6 Inb12-2-2, TZE-W Pop C_0 S_6 Inb50-3-4 from the 9071 group) maintained their heterotic groups in both *Striga*-free and *Striga*-infested environments. Although none of the yellow lines could be classified into heterotic groups under any of the evaluation environments, high yielding hybrids were identified for both white and yellow types. The study led to the conclusion that the testers used did not discriminate effectively among all the yellow and most of the white inbred lines to allow classification into heterotic groups. Therefore, additional efforts were put in place to identify inbred lines in this maturity group that could be used as testers in future studies.

During the last two decades, IITA in collaboration with the National Agricultural Research Systems (NARS) of WCA has developed a wide range of high yielding, early and extra-early populations (white and yellow endosperm), inbred lines, and cultivars that show escape or better adaptation to drought in an effort to combat the threat posed by recurrent drought and low N in the savannas of WCA. The extra-early populations from which the inbred lines and cultivars were derived were formed from crosses of local (landraces) with exotic and introduced germplasm identified through extensive multilocation trials in WCA (Badu-Apraku and Fakorede 2001). These materials were selected on the basis of high grain yield, earliness, resistance to MSV and, above all, adaptation to the heat and drought stresses characteristic of the

3. BREEDING EARLY AND EXTRA-EARLY MAIZE

Table 3.18. Mean grain yield, general (GCA) and specific (SCA) combining ability effects, and respective standard errors (SE) of white inbred lines classified into two heterotic groups in testcrosses with two testers following evaluation in two *Striga*-free and two *Striga*-infested environments in Nigeria, 2005.

Pedigree	Testcross yield (t ha^{-1})		Effect (t ha^{-1})		Heterotic group
	1368^y	9071^y	SCA	GCA	
Striga-free environments					
TZE-W Pop C$_0$ S$_6$ Inb151-1-2	5.19	2.78	0.62	0.24	1368
TZE-W Pop C$_0$ S$_6$ Inb50-2-4^z	4.65	3.26	0.36	0.23	1368
TZE-W Pop STR S$_7$ Inb141-2-2	4.17	3.88	0.09	0.27	1368
TZE-W Pop C$_0$ S$_6$ Inb50-3-4^z	4.12	4.67	−0.12	0.45	9071
TZE-W Pop × LD S$_6$ Inb6^z	4.10	3.22	0.24	0.08	1368
TZE-W Pop C$_0$ S$_6$ Inb74^z	4.00	3.71	0.09	0.18	1368
TZE-W Pop × LD S$_6$ Inb12	3.82	4.65	−0.19	0.34	9071
TZE-W Pop C$_0$ S$_6$ Inb12-1-2	3.59	4.29	−0.16	0.22	9071
TZE-W Pop × 1368 STR S$_7$ Inb5	3.55	4.37	−0.19	0.23	9071
TZE-W Pop C$_0$ S$_6$ Inb12-2-2^z	3.28	4.49	−0.29	0.20	9071
TZE-W Pop C$_0$ S$_6$ Inb143-3-3	3.26	4.28	−0.24	0.14	9071
TZE-W Pop C$_0$ S$_6$ Inb1-2-4	3.19	4.52	−0.32	0.18	9071
TZE-W Pop × 1368 STR S$_7$ Inb9	3.00	4.58	−0.38	0.15	9071
Grand mean	3.46	3.53	0.00	0.00	
SE	0.58	0.64	0.52	0.37	
Striga-infested environments					
TZE-W Pop C$_0$ S$_6$ Inb75-1-3	2.10	2.75	−0.07	0.58	9071
TZE-W Pop C$_0$ S$_6$ Inb20	1.96	1.02	0.32	0.11	1368
TZE-W Pop STR S$_7$ Inb34-2-3	1.76	2.28	−0.04	0.37	9071
TZE-W Pop C$_0$ S$_6$ Inb34-1-3	1.74	1.40	0.17	0.15	1368
TZE-W Pop × LD S$_6$ Inb6^z	1.68	0.92	0.28	0.02	1368
TZE-W Pop C$_0$ S$_6$ Inb74^z	1.60	2.54	−0.15	0.40	9071
TZE-W Pop C$_0$ S$_6$ Inb50-2-4^z	1.57	1.80	0.03	0.21	1368
TZE-W Pop × 1368 STR S$_7$ Inb10	1.51	1.31	0.14	0.07	1368
TZE-W Pop × 1368 STR S$_7$ Inb2	1.17	2.27	−0.19	0.23	9071
TZE-W Pop C$_0$ S$_6$ Inb50-3-4^z	0.92	2.32	−0.26	0.18	9071
TZE-W Pop × LD S$_6$ Inb4	0.92	1.97	−0.17	0.09	9071
TZE-W Pop C$_0$ S$_6$ Inb12-2-2^z	0.48	2.06	−0.31	0.00	9071
Grand mean	1.09	1.44	0.00	0.00	
SE	0.41	0.50	0.48	0.34	

^{z}Classified into the same heterotic group under the two evaluation environments. Inbreds within the same heterotic group will not show hybrid vigor (heterosis) when crossed but will show hybrid vigor when crossed to inbreds of opposing heterotic group.
^{y}Tester.

Sudan savannas in Burkina Faso, Mali, Mauritania, Ghana, and Nigeria during both the rainy and dry seasons. It was hypothesized that, after several years of cultivation in these environments, the extra-early germplasm should have adaptive traits to these stresses where they had survived. We observed that some of the extra-early IITA inbred lines would not only escape drought stress but also seemed to possess genes for adaptation to drought stress. The inbreds should therefore be able to withstand the mid-season drought that occurs during the flowering and grain filling periods in the savannas of WCA as have been found in early, intermediate, and late maturing cultivars. Badu-Apraku et al. (2011a) conducted three experiments between 2007 and 2010 in Nigeria to identify extra-early inbreds for adaptation to low N or drought stress at the flowering and grain-filling periods, and to determine the potential of the inbreds for hybrid production as well as source germplasm for the improvement of breeding populations. In the first two experiments, 90 extra-early maturing maize inbred lines bred by IITA were evaluated in Nigeria under managed drought stress during the dry seasons of 2007–2008 and 2008–2009 and in well-watered environments at Ikenne, and in low N (30 kg ha^{-1}) and high N (90 kg ha^{-1}) plots at Mokwa, during the planting seasons of 2008 and 2009. The GT biplot was used to identify from the 90 extra-early inbreds, low N extra-early inbreds with enhanced adaptation to drought as candidate parents with favorable alleles for introgression into maize breeding populations and for the production of extra-early hybrids with enhanced adaptation to drought at the flowering and grain-filling periods. Results of the analysis of variance revealed significant genotypic mean squares for grain yield and most other traits of the inbreds under drought or low N stresses indicating wide genetic variability for both traits. TZEEI 6 was closest to the ideal inbred and was thus the best under drought. TZEEI 4, TZEEI 36, and TZEEI 38 were identified as ideal inbreds under drought. Under low N conditions, TZEEI 19 was ranked closest to the ideal genotype and TZEEI 96 and TZEEI 45 were also top ranking. TZEEI 19, TZEEI 29, TZEEI 56, TZEEI 38, and TZEEI 79 showed enhanced adaptation to both drought and low N stresses. A total of 18 single-cross hybrids produced above-average yields across drought and well-watered environments. A total of 18 of the 36 hybrids produced above-average yields across environments with four hybrids identified as highly stable. TZEEI 29 × TZEEI 21 was the closest to the ideal genotype because it combined large mean performance with high yield stability. TZEEI 29 × TZEEI 21 was also one of the top 10 extra-early hybrids under *Striga* infestation (Table 3.19). Similarly, several high-yielding QPM hybrids were identified in a preliminary trial conducted under drought stress at Ikenne

Table 3.19. Grain yield and other traits of extra-early maturing hybrids (the best 10 and the worst 10 based on the base index) and checks evaluated under *Striga* infestation in Nigeria in 2008 and 2009.

Hybrid	Yield (kg ha⁻¹)	Days to silk	Days to anthesis	ASI^z	PHT (cm)	Ear aspect	Ears/plant	Striga damage at 8 WAP	Striga damage at 10 WAP	Striga emergence at 8 WAP	Striga emergence at 10 WAP	Base index
TZEEI 29 × TZEEI 14	3,385	56	55	0.4	142	4.6	1.0	3.7	4.4	20	26	10.6
TZEEI 39 × TZEEI 13	3,317	53	52	1.2	158	3.8	0.8	3.6	3.9	21	25	9.8
TZEEI 4 × TZEEI 13	3,246	56	54	0.4	150	4.7	0.8	3.2	3.9	18	30	11.4
TZEEI 4 × TZEEI 14	3,127	55	55	1.5	145	4.0	0.9	3.1	3.5	17	22	11.6
TZEEI 6 × TZEEI 14	2,865	53	52	2.0	132	4.2	0.8	3.8	4.3	12	14	8.3
TZEEI 21 × TZEEI 13	2,862	54	52	1.6	152	4.4	1.1	3.2	4.3	11	12	10.8
TZEEI 14 × TZEEI 54	2,673	56	54	2.0	121	5.0	0.9	3.8	4.2	19	24	8.2
TZEEI 13 × TZEEI 38	2,666	55	54	2.5	138	4.7	0.8	4.1	4.2	16	25	8.2
TZEEI 6 × TZEEI 13	2,640	54	53	2.4	158	4.2	0.9	3.5	3.9	20	24	8.7
TZEEI 29 × TZEEI 21	2,628	57	55	3.9	139	4.9	0.8	3.2	3.9	18	29	8.7
TZEEI 6 × TZEEI 90	1,199	51	54	5.2	133	5.8	0.5	5.5	7.8	58	59	-9.4
TZEEI 32 × TZEEI 55	1,107	53	56	4.3	128	5.1	0.4	5.6	6.2	39	42	-7.4
TZEEI 37 × TZEEI 54	1,104	54	58	5.7	124	5.8	0.5	5.5	6.3	49	52	-8.0
TZEEI 90 × TZEEI 37	1,104	56	54	1.6	121	5.6	0.4	5.8	6.7	50	54	-8.6
TZEEI 46 × TZEEI 39	1,078	52	54	4.1	128	5.9	0.4	5.8	7.2	43	48	-9.3
TZEEI 32 × TZEEI 39	955	56	55	3.3	120	5.5	0.4	5.6	6.7	25	30	-7.7
TZEEI 32 × TZEEI 3	838	57	57	5.6	125	5.9	0.5	5.3	6.0	35	45	-7.2
TZEEI 3 × TZEEI 37	756	55	57	3.7	108	6.2	0.4	6.1	6.4	36	34	-9.5
TZEEI 13 × TZEEI 54	752	53	55	7.1	117	6.0	0.4	6.3	6.8	47	51	-10.2
TZEEI 3 × TZEEI 49	424	51	50	4.1	110	6.3	0.3	7.2	8.4	32	36	-15.7
Mean	1,758	54	54	3	130	5.1	0.67	4.7	5.50	25.0	30.0	
LSD	1,119	2.9	3.4	3.2	22.5	1	0.2	1.3	1.4	18	20	
P stat for genotype (G)	**	**	**	**	**	**	**	**	**	*	*	
P stat for environ (E)	*	**	**	NS	**	**	**	NS	*	**	**	
P stat for G × E	NS	NS	NS	**	NS	NS	NS	NS	NS	NS	NS	

^{z}ASI, Anthesis-silking interval; PHT, plant height; WAP, weeks after planting.
Notes: *, **, and NS are significant F-test at 0.05 and 0.01 levels of probability, or nonsignificant, respectively.

during the 2010–2011 dry season (Table 3.20). In general, hybrid yield under drought had large positive correlation with grain yield under well-watered environments. Selection for inbred traits such as DYS and ASI under drought predicted fairly accurately hybrid yield under well-watered environments. It was concluded that extra-early inbreds and hybrids are not only drought escaping but also possess genes for adaptation to drought and low N stresses.

Two other studies were conducted from 2007 to 2010 with the objective of determining the combining ability of seven extra-early yellow-grained inbreds, place them into heterotic groups, and identify best testers and superior single-cross hybrids under *Striga*-infested, drought, optimal environments, and across environments. GCA and SCAmean squares were significant for grain yield across environments. Similarly, GCA and SCA mean squares for all measured traits were significant except GCA mean squares of *Striga* damage at 8 WAP and number of emerged *Striga* plants, and SCA mean squares of emerged *Striga* plants and *Striga* damage. The GCA mean squares of grain yield and other traits were larger than those of SCA in all environments, indicating that additive gene action is more important for these traits. Three testers (TZEEI 79, TZEEI 76, and TZEEI 63) and opposing heterotic groups were identified across environments. TZEEI 79 had the highest positive GCA effects across environments. The hybrid TZEEI 79 × TZEEI 76 was the most stable and high yielding across environments and compared favorably in yield to the best open-pollinated early check EV DT-Y 200 STR C_0. In another study, 36 diallel crosses derived from nine early-maturing white maize inbreds were evaluated under drought, well-watered, *Striga*-infested, and *Striga*-free conditions at five Nigeria locations in 2007 and 2009. The objective was to evaluate the combining ability, performance and stability of the early inbreds, and to identify the heterotic groups. Results showed that additive and nonadditive gene actions were important in the genetic control of adaptation to drought and *Striga* resistance in the inbreds. GCA mean squares of grain yield and other traits were larger than SCA, indicating that additive gene action was more important in the inheritance of adaptation to drought-prone environments (Fig. 3.3). The GCA mean squares for *Striga* damage rating at 8 and 10 WAP were about six and five times greater than those of the SCA, indicating that additive gene action played a major role in host plant resistance to *Striga*. The GGE biplot analysis revealed that TZEI 4 and TZEI 5 were the most promising inbreds in yield performance and stability across the test environments. Two heterotic groups were identified [TZEI 7, TZEI 19, TZEI 2, and TZEI 4] and [TZEI 5 and TZEI 3]. Entry TZEI 7 had the

Table 3.20. Grain yield and other agronomic traits of selected early white QPM hybrids evaluated under drought stress at Ikenne during 2010/2011 dry season.

Hybrid	Grain yield (kg ha⁻¹)	Days to anthesis	Anthesis-silking interval	Plant height (cm)	Plant aspect	Ear aspect	Stay green characteristic	Ears/plant
TZEQI 34 × TZEQI 35	3,455	53	0	175	2.3	2.0	3.3	0.8
TZEQI 39 × TZEQI 14	3,352	54	4	151	3.0	2.8	4.7	0.7
TZEQI 24 × TZEQI 5	3,106	53	4	129	2.9	2.9	3.6	0.7
TZEQI 23 × TZEQI 45	3,082	52	3	159	2.7	2.5	5.2	0.7
TZEQI 29 × TZEQI 45	3,042	53	2	144	2.6	2.5	4.2	0.9
TZEQI 44 × TZEQI 24	2,993	51	3	149	2.7	2.7	4.8	0.8
TZEQI 24 × TZEQI 25	2,819	53	3	147	2.4	2.7	4.7	0.7
TZEQI 39 × TZEQI 23	2,814	51	4	143	2.9	2.8	4.8	0.7
TZEQI 49 × TZEQI 6	2,779	54	2	129	2.3	2.5	3.7	0.8
TZEQI 13 × TZEQI 44	2,753	53	2	136	2.5	2.6	4.2	0.9
TZEQI 34 × TZEQI 12	2,726	55	6	146	2.7	3.0	4.2	0.7
TZEQI 59 × TZEQI 29	2,593	54	2	134	2.8	2.6	4.0	0.8
TZEQI 39 × TZEQI 44	2,487	51	3	136	2.6	2.6	5.1	0.7
TZEQI 35 × TZEQI 39	2,487	54	2	147	2.5	2.8	3.7	0.7
TZEQI 59 × TZEQI 14	2,470	54	1	151	2.7	2.4	4.4	0.9
Check 5—								
TZEI 3 × TZEI 26	2,020	53	3	172	2.9	3.1	4.9	0.7
TZEQI 4 × TZEQI 49	2,009	54	6	133	3.1	3.1	4.1	0.6
Check 3—								
TZEI 2 × TZEI 87	1,977	54	3	131	2.9	2.9	3.9	0.7
TZEQI 56 × TZEQI 59	1,975	56	3	146	2.8	2.9	4.4	0.7
Mean	1,364	54	7	133	3.2	3.3	4.9	0.5
SE	463	1	3	10	0.2	0.3	0.5	0.1
F-statistic for genotype	**	**	**	**	**	**	**	**

Notes: ** is significant F-test at 0.01 levels of probability.

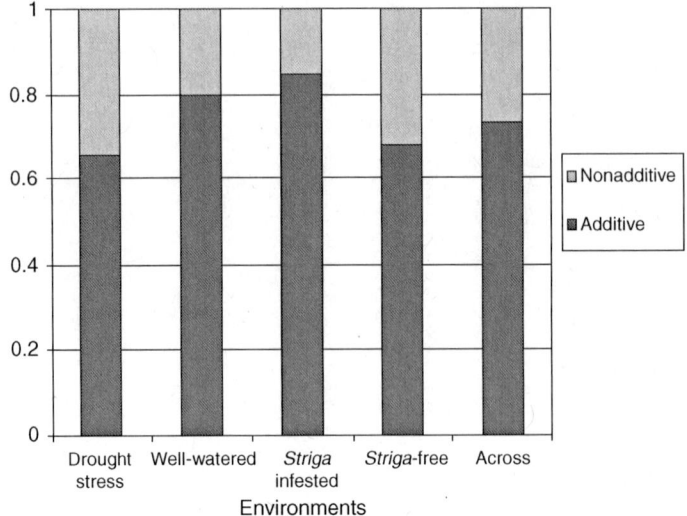

Fig. 3.3. Proportion of additive (lower bar) and nonadditive (upper bar) genetic variance for grain yield at four stress and nonstress environments and across environments for a 9 × 9 diallel cross among nine white-grained early maize inbred lines.

highest GCA effects and TZEI 2, the lowest. Tester TZEI 3 was the closest to the ideal tester.

VIII. TRAITS FOR INDIRECT SELECTION FOR STRESS TOLERANCE/RESISTANCE IN CONTRASTING ENVIRONMENTS

The effect of selection under stress on yield performance of genotypes under optimal conditions and *vice versa* has been an ongoing debate among plant breeders for decades. Because of the presence of GEI, selection in one type of environment may not be repeatable in another environment. GEI, which results from the varied response of genotypes to different environments, reduces the correlation between phenotypic and genotypic values (Comstock and Moll 1963) under stress environments. Furthermore, selection for grain yield under severe drought stress or low N has often been considered inefficient because the estimates of heritability of grain yield has been observed to decline with reduced yield levels (Bolaños and Edmeades 1993). Under these conditions, secondary traits may increase selection efficiency provided they have adaptive value, relatively high heritability, significant genetic correlation with grain yield, and are easy to measure (Falconer 1989; Bolanos and Edmeades 1996). Secondary traits can improve the precision with which drought-tolerant genotypes are identified, compared with measuring only grain yield under drought stress. Studies have shown that, under stress, the estimates of heritability of grain yield usually decreases, whereas the heritability of some secondary traits remains high, while at the same time, the genetic correlation between grain yield and those traits remains about the same or increases sharply (Bolanos and Edmeades 1996; Bänziger and Lafitte 1997a; Badu-Apraku et al. 2004a). Therefore, secondary traits have been proposed as a means to select for drought tolerance. Furthermore, Bänziger et al. (1999a,b) found that improvement for drought tolerance also resulted in specific adaptation and improved performance under low N conditions. This is desirable, and if further studies confirm this finding, the implication is that tolerance to either stress involves a common adaptive mechanism. Secondary traits such as EPP, stay green characteristics, and ASI possess strong correlations with grain yield under drought and have been used to select for higher levels of tolerance to drought and low N in maize (Lafitte and Edmeades 1994; Bänziger and Lafitte 1997a,b; Edmeades et al. 1998). Therefore, a base index that integrates grain yield under drought stress, ASI, EPP, stay green characteristics, PASP and EASP is used by CIMMYT for selecting maize germplasm for drought-prone environments. The index maintains constant DYA and grain yield under well-watered conditions while selecting for improved grain yield, EASP and PASP, decreased ASI, and good stay-green characteristic under drought. At IITA, a base index similar to that of CIMMYT, which integrates increased grain yield under drought stress and well-watered environment with a short ASI, increased number of EPP, good stay green characteristics, PASP and EASP scores under drought stress has been used to select for early, intermediate and late maturing maize germplasm with enhanced adaption to drought stress (Menkir and Akintunde 2001; Badu-Apraku et al. 2004b). Furthermore, the base index has been used in IITA to characterize intermediate/late maturing maize (Menkir et al. 2003; Meseka et al. 2006) and early and extra-early-maturing maize germplasm (Badu-Apraku 2010) for tolerance to low N in WCA. The secondary traits were identified based on phenotypic and genotypic correlations between the traits and grain yield as well as on their heritability. It is very important to ensure that greater progress is achieved using grain yield and selected secondary traits in selection than by using grain yield alone. Thus, not only must secondary traits be identified but their value in breeding must also be highly correlated with main traits. This may be achieved through determination of phenotypic and genetic correlations and heritability among progenies of a single population; selection indices (Fukai and Cooper 1995); divergent selection; analysis of physiological and morphological changes in cultivars that have been consistently selected for performance under drought; simulation models (Bänziger et al. 2000); path-coefficient analysis (Wright 1921) and the genotype × trait (GT) biplot proposed by Yan and Kang (2003). Apart from the GT biplot, all of these methods have the common disadvantage of not being capable of identifying genotypes with specific desirable traits that could be used in a selection program. As noted by Fakorede (1979), correlation measures the mutual association between a pair of variables independently of all other variables across all genotypes. Regression analysis, including stepwise multiple regression and path analysis, which is a special case of partial multiple regression analysis as well as multivariate technique, examines the association among traits measured on a set of genotypes without identifying individual genotypes superior for specific traits (Fakorede 1979; Badu-Apraku and Akinwale 2011). On the other hand, the GT biplot has proven to be a very effective statistical tool for evaluating cultivars based on multiple traits and for identifying superior genotypes for use as parents in a breeding program. It allows genetic correlations among traits to be visualized (Yan and Rajcan 2002; Lee et al. 2003) and multitrait evaluation of genotypes (Yan and Rajcan 2002; Yan and Kang 2003; Morris et al. 2004; Ober et al. 2005). It also provides information on the suitability of cultivars for production and helps detect the less important (redundant) traits, and identify those that are appropriate for indirect selection purposes.

One aspect of the IITA program focuses on the development of extra-early and early maturing, inbred lines, open-pollinated and hybrid cultivars and their performance in contrasting environments created by imposing specific stresses and comparing with nonstresss conditions. While several studies have been conducted to assess the value of the secondary traits used in selecting genotypes under low soil N and drought stress by CIMMYT in Mexico or in Eastern and Southern Africa (Bolanos and Edmeades. 1993; Lafitte and Edmeades 1994; Edmeades et al. 1998; Bänziger et al. 2000) the effectiveness of the base index in selecting for adaption to drought or low N stresses, particularly in the early and extra-early maturing germplasm bred by IITA has not been adequately assessed in WCA. Badu-Apraku et al. (2010) observed a low correlation between grain yield and stay green characteristics of extra-early inbreds under drought stress and low N environments. There is, therefore, a need to assess the influence of these contrasting environmental effects on the effectiveness of the index in selecting germplasm with enhanced adaptation to low N and drought-prone environments. A study was conducted to identify secondary traits for indirect selection for yield improvement under drought or low N stresses. Two sets of experiments were conducted at four locations in Nigeria for 2 years. In the first experiment, cultivars were evaluated under low ($30\,kg\,ha^{-1}$) and high ($90\,kg\,ha^{-1}$) N and in the second experiment, under drought and well-watered environments. The most reliable traits for selection for yield under drought were EASP, PASP, ASI, and number of EPP. DYA and DYS were the most reliable under well-watered environments. Under low N, most reliable traits were DYA, DYS, stay green characteristic (LDTH), ASI, PLHT, EPP, EASP, and PASP whereas PLHT, EPP, PASP, and EASP were most reliable traits under high-N environments. The ASI, EPP, EASP, and PASP were identified as most reliable traits for simultaneous selection of genotypes with enhanced grain yield performance under both drought stress and low N conditions. Furthermore, the GT biplot analysis revealed that selecting for EPP, PLHT, EASP, and PASP under low N environments will improve grain yield under N-stress and nonstress environments. The inclusion of staygreen characteristic in the selection index for selection for yield improvement under drought stress was not justified in this study. It was concluded that breeders can make faster progress in selecting for improved grain yield under drought and low N by selecting tolerant genotypes using a selection index that integrates EPP, EASP, PASP, and ASI under either drought stress or low N environment.

Maize cultivars that combine improved grain yield with *Striga* resistance are desirable in SSA to ensure food security. Breeding for high yielding maize cultivars with resistance *Striga* requires suitable selection measures for both characteristics. In breeding maize for tolerance or resistance to *Striga*, appropriate tolerance or resistance indicator-traits can improve the precision with which resistant genotypes are identified. For the stress tolerance aspect of its research activities, IITA focuses on four maize maturity groups: late, intermediate, early, and extra-early. At the initial stages of the stress-tolerance/resistance research work, IITA concentrated on late and intermediate maturity groups and used a base index, which combines grain yield under *Striga* infestation, *Striga* damage rating, *Striga* emergence, and EPP to select for high grain yield measured under *Striga*-infested and noninfested conditions (MIP 1996; Menkir and Kling 2007). At the time research on improvement of the breeding populations in the extra-early and early maturity groups started in 1994, the early-maturity component adopted the base index used for the improvement of the late/intermediate maturity groups but found inconsistent results, depending on traits used and type of germplasm subjected to selection.

The primary traits of interest in selecting for tolerance or resistance and high grain yield under *Striga* infestation are host-plant damage rating (*Striga* damage) and *Striga* emergence count (number of emerged *Striga* plants). There are contradictory reports on the importance of *Striga* emergence count as a reliable trait for selecting for *Striga* resistance and improved grain yield under artificial *Striga* infestation. For instance, contrary to the results obtained for the late and intermediate maturity groups (Kim and Adetimirin 1995; Gethi and Smith 2004; Menkir and Kling 2007; Yallou et al. 2009), Badu-Apraku et al. (2005, 2006b, 2007), reported weak phenotypic and genotypic correlations between grain yield and *Striga* emergence count in early germplasm. In contrast, Badu-Apraku (2010) studied the relative changes in genetic variances, heritabilities, and genetic correlations following four cycles of S_1 family selection in the extra-early white population and reported that under *Striga* infestation, yield was not correlated with other traits at C_0, but was significantly correlated with EPP, *Striga* damage and emerged *Striga* plants in advanced cycles. Therefore, the value of the traits that are used in the base index by IITA for selecting for *Striga*-tolerant and resistant genotypes required assessment and confirmation to determine whether or not they were appropriate for selection purposes in the early and extra-early maturity groups. Consequently, the GT biplot analysis was used to assess the appropriateness of these traits in the base index. Results revealed that EPP, STRA1, STRA2, and EASP were the most reliable of the eleven measured traits for selection for *Striga* resistance thus justifying the use of EPP, STRA1, and STRA2 in the base index. It was not surprising that EPP was identified as one of the most reliable traits for selection for *Striga* resistance. Badu-Apraku et al. (2008a) reported EPP to be a major component of the increased grain yield associated with recurrent selection programs under drought stress and *Striga* infestation. Similar results were also reported under drought stress by Bolanos and Edmeades (1993), Chapman and Edmeades (1999), and Monneveux et al. (2006). The result, therefore, justified the inclusion of EPP in the selection index for yield improvement in *Striga* prone environments as earlier reported by Adetimirin et al. (2000a), Badu-Apraku (2007), and Badu-Apraku et al. (2008a). Similarly, Badu-Apraku et al. (2007) reported high negative genetic and phenotypic correlations between grain yield and host plant damage rating and concluded similarly to Kim and Adetimirin (1995) that *Striga* damage rating is an appropriate trait for the assessment of tolerance under *Striga* infestation. In contrast, STC1 and STC2 were among the traits that had weak correlation with yield, suggesting that they do not qualify to be included in the base index. This finding is supported by Badu-Apraku et al. (2007) who reported weak phenotypic and genotypic correlations between grain yield and *Striga* emergence count, indicating that it is not a reliable trait for detecting *Striga* resistance. However, it was argued that the result could also mean that grain yield and *Striga* emergence count were genetically independent (no linkage or pleiotropy) and can be effectively selected for simultaneously using an appropriate index. Contrary to this result, Badu-Apraku (2007) in a study of the genetic variances and correlations in an early white maize population, reported a negative genetic correlation ($r_g = -0.56$) between grain yield and *Striga* emergence count at 10 WAP. This result was further supported by Badu-Apraku (2010) who reported that under *Striga* infestation, yield was not correlated with other traits at C_0, but was strongly correlated with EPP, *Striga* damage, and emerged *Striga* plants in advanced cycles of the extra-early white population. Similar results were also reported by Menkir and Kling (2007) for *Striga* emergence count at 8 ($r_p = -0.78$) and 10 ($r_p = -0.72$) WAP and EASP ($r_p = -0.97$) under *Striga* infestation for a late maturing tropical maize population. In this study, further analysis using the stepwise multiple regression analysis revealed that *Striga* emergence count at 8 WAP was among the five traits identified as important yield determinants. Similarly, based on a study using the path coefficient analysis, Badu-Apraku et al. (2010), found a high direct effect of STC1 and STC2 on grain yield, indicating that the two traits were useful selection criteria for selecting for improved grain yield under artificial *Striga* infestation and thus justifying their inclusion in the base index. EASP had a consistently high correlation with grain yield under *Striga* infestation suggesting the need for its inclusion in the base index. This result was consistent with the findings of a similar study with early-maturing inbred lines by Badu-Apraku et al. (2011a).

IX. FUTURE CHALLENGES AND PERSPECTIVES

The role of maize in the nutrition of the people and the economy of WCA has increased tremendously during the last four decades. Consequently, research to improve the yield potential has been of high priority on the agricultural agenda in the subregion. The potential of the Guinea savanna as the maize belt of WCA had long been recognized but could not be tapped because of several production constraints, which were too formidable for individual countries to overcome. The constraints included nonavailability of appropriate (early/extra-early maturing) cultivars to fit into the relatively short cropping season, parasitism by *S. hermonthica*, low levels of soil nutrients, unpredictable recurrent drought, and attack by insect pests and disease organisms.

The establishment of IITA in Nigeria in 1967 provided an avenue for systematic collaborative research in WCA to study and proffer solution to the constraints. Mandatorily, IITA must work in collaboration with NARS thus making it possible for the Institute, within a short time of its existence to identify the research strengths and weaknesses of the NARS in the subregion. The scientists, in collaboration with NARS researchers in WCA prioritized maize production constraints in the different agroecological zones and tackled them with the best possible facilities and methodology. For example, IITA started maize research in the Nigeria savanna zones in 1980 and clearly targeted maize varieties of different maturities to the ecology. Intermediate-to-late maturing varieties were available for the lowland savannas, but much of the savannas needed early and extra-early cultivars while the mid-altitude agroecology needed specific cultivars all of which, unfortunately, were not available. Therefore, breeding efforts were initiated along these lines and it became necessary to cover the whole of WCA.

In 1977, Foreign Ministers of WCA met to discuss and proffer solution to the problem of recurrent drought, which was plaguing the subregion. One of the outcomes of the meeting was the establishment of the SAFGRAD project composed of several commodity networks, including maize. Research on early and extra-early maize was devolved on the maize network of SAFGRAD. In 1987, this maize network became autonomous and was named WECAMAN. The Network has served as an effective mechanism for all stakeholders in maize production and productivity to tackle the regional constraints. These include national and international scientists, extension workers, farmers, seed technologists, industrialists, and policymakers.

Over the years, WECAMAN has developed and transferred improved maize production technologies to farmers in the subregion. Much of the research and development efforts included in this report were initiated under the auspices of WECAMAN. Funding support for WECAMAN's activities ended in 2007, but maize research and development activities have come to stay in WCA. A few subregional maize research and development programs are still in place in WCA. They support maize research and development activities in a few countries following WECAMAN's approach to research. Some of the open-pollinated and hybrid cultivars released to WCA farmers by IITA in recent years are

Table 3.21. Drought-tolerant and *Striga*-resistant extra-early and early-maturing open-pollinated maize cultivars formally released and adopted in four countries of West Africa (2007–2011).

Country	No. of cultivars	Cultivars name[z]
Benin	3	2000 Syn EE-W, TZEE-Y Pop STR and Bag 97 TZE Comp. 3 DT
Ghana	3	TZE W Pop. STR QPM C4, EV DT W99 STR QPM C0 and TZEE-Y Pop. STR QPM C0
Mali	2	Jerobana and TZEE-
Nigeria	4	EV DT 99 DT STR, TZE Comp3 DT, 2000Syn EE-W and 99 TZEE-Y STR

[z]DT, drought tolerant; STR, *Striga* resistant/tolerant; TZE, early (90–95 days to maturity); TZEE, extra-early (80–85 days to maturity).

presented in Table 3.21 while those being processed for release are listed in Table 3.22. The rate of adoption of improved early and extra-early maize cultivars in the savannas has gone far beyond the expectation of WECAMAN partners and has revolutionized maize production in WCA

Table 3.22. Drought tolerant and *Striga*-resistant hybrids and open-pollinated (OP) maize cultivars in the pipeline for release in WestAfrica as at 2011.

Country	No. of cultivars	Cultivar name[z]
Benin Republic	5 OP	EV DT- Y 2000 STR QPM C0, TZEE-W POP STR C_4, EV DT-Y 2000 STR QPM C0, TZE-W DT STR C4, TZE-Y DT STR C4,
	2 Hybrids	TZE-W Pop DT STR × TZEI 22, (TZEI 135 × TZEI 17) × TZEI 157
Ghana	3 OP	2004 TZE-W Pop STR C_4, 2004 TZEE-W Pop STR C_4, TZE COMP 3 DT C2F2
	2 Hybrids	TZE-Y Pop DT STR × TZEI 11, (TZEI 129 × TZEI 17) × TZEI 157
Mali	6 OP	TZE Comp 3 DT C2F2, EV DT-Y 2000 STR QPM C0, 2000 Syn EE-W QPM, TZEE- W Pop STR QPM C0, TZEE-W POP STR C_4, DT Syn-1 W, DT SR-W C0 F2
	1 Hybrid	TZE-W Pop DT STR × TZEI 4
Nigeria	7 OP	EV DT-W 2008 STR EV, DT-Y 2008 STR, TZE-W DT STR C4, TZE Y DT STR QPM C0, EV DT 2000-Y STR QPM, 2008 DTMA – Y STR, DTE W STR SYN
	3 Hybrids	TZE-Y Pop DT STR × TZEI 11, (TZEI 17 × TZEI 157) × TZEI 129, (TZEI 135 × TZEI 17) × (TZEI 16 × TZEI 157)

[z]DT, drought tolerant; STR, *Striga* resistant/tolerant; TZE, early (90–95 days to maturity); TZEE, extra-early (80–85 days to maturity).

(Fakorede et al. 2003; Onyibe et al. 2003). With the availability of these cultivars, green maize is now produced throughout the year in most of WCA, using hydromorphic soils or irrigation during the dry season.

WECAMAN's approach to maize research and development has been unique and is worthy of adoption in future endeavours for maize improvement in WCA. Some of the lessons learned from WECAMAN are as follows:

1. Use of NARS that were relatively stronger than others in specific subject-matter areas as Lead Centers (LCs). Technologies emanating from the research conducted in the LCs were made available to other member countries through RUVT, onfarm tests, and demonstrations.
2. Constraints to maize production were identified in a participatory manner by all stakeholders.
3. An *ad hoc* Research Committee screened proposals and allocated funds to the LCs for research projects to address the constraints.

The experience of WECAMAN firmly established the networking approach to maize research as the most effective for WCA. Future maize improvement efforts should therefore use this approach to continue solving the production constraints. Although considerable progress has been made in solving the maize production constraints in WCA, they have not been completely eradicated. The problem of recurrent drought in the savannas, as well as some parts of the forest zone, is far from solved. Rather, it has been aggravated by climate change, which is now negatively impacting maize production in WCA (Fakorede and Akinyemiju 2003). In addition to earliness and extra-earliness, which are drought-escaping mechanisms, maize breeders in WCA must continue to develop open-pollinated cultivars and hybrids showing enhanced adataption to water scarcity primarily for the savannas. Necessarily, all maize open-pollinated and hybrid cultivars must be able to perform very well under low soil N, and those targeted to *Striga*-endemic areas must be *Striga* tolerant/resistant.

A major strategy of WECAMAN has been to emphasize technology transfer and improve the research and development capacities of collaborators in all member countries. The age-long approach of leaving technology transfer in the hands of extension agents alone was not effective, at least for early and extra-early maize. On-farm trials along with field days were organized by WECAMAN as avenues for technology transfer. This approach, which strengthened the research-extension-farmer linkage was very effective and should be continued in future. One bottle-neck of technology transfer is availability of seeds to the farmers. This is particularly acute with early/extra-early cultivars that are usually of low priority for the private seed sector. Future development activities should focus on this aspect by involving seed companies and entrepreneurs in the activities.

In summary, the following are the important strategies that must be sustained:

- Incorporation of multiple stress resistance has been, and must continue to be a major enhancement strategy of the germplasm for release to farmers in SSA.
- Genes controlling host plant resistance to biotic stresses and enhanced adaption to abiotic stresses need to be stacked in the cultivars using genomic-led approaches.
- Future maize improvement activities in WCA must intensify efforts on the development of hybrids.
- Strengthening the capacity and capability of NARS researchers has been an important strategy for maize improvement in the SSA and it must continue.

ACKNOWLEDGMENTS

The authors are grateful to the IITA staff working on maize research at its headquarters in Ibadan for their technical support and the colleagues of the National Agricultural Research Systems of West and Central Africa for their collaboration. The financial support of the United Stated Agency for International Development, Bill and Melinda Gates Foundation, Howard Buffet Foundation, Nippon Foundation, IFAD, Rockefeller Foundation, as well as the backstopping of IITA and CIMMYT for this research is also gratefully acknowledged.

LITERATURE CITED

Abalu, G.I. 2001. Policy issues in maize research and development in sub-Saharan Africa in the next millennium. In: B. Badu-Apraku, M.A.B. Fakorede, M. Ouedraogo and M. Quin (eds.), Impact, challenges and prospects of maize research and development in West and Central Africa. Proceedings of Regional Maize Workshop, May 4–7, 1999. WECAMAN/IITA-Cotonou, Benin Republic.

Abdoulaye, T., D. Sanogo, A. Langyintuo, S.A. Bamire, and A. Olanrewaju. 2009. Assessing the constraints affecting production and deployment of maize seeds in DTMA countries of West Africa. IITA, Ibadan, Nigeria.

Adetimirin, V.O., L. Akanvou, and S.K. Kim. 2000a. Effects of *Striga hermonthica* on yield components in maize. J. Agr. Sci. 135:185–191.

Adetimirin, V.O., S.K. Kim, and M.E. Aken'Ova. 2000b. Expression of mature plant resistance to *Striga hermonthica* in maize. Euphytica 115:149–158.

Agbaje, S., B. Badu-Apraku, and M.A.B. Fakorede. 2008. Heterotic patterns of early maturing maize inbred lines in *Striga*-free and *Striga*-infested environments. Maydica 53:87–96.

Aggarwal, Pawan K. 1991. *"Identification of Non-Filer Potential Income Tax Payers."* Asia. Pacific Tax and Investment Research Centre Bulletin, 9 (6):217–224.

Amusan, I.O., P.J. Rich, A. Menkir, T. Housley, and G. Ejeta. 2008. Resistance to *Striga hermonthica* in a maize inbred line derived from *Zea diploperennis*. New Phytol. 178:157–166.

Badu-Apraku, B. 2007. Genetic variances and correlations in early tropicalwhite maize population after three cycles of recurrent selection for *Striga* resistance. Maydica 52:205–217.

Badu-Apraku, B. 2010. Effects of recurrent selection for grain yield and *Striga* resistance in an extra-early maize population. Crop Sci. 50:1735–1743.

Badu-Apraku, B., and R.O. Akinwale. 2011. Cultivar evaluation and trait analysis of tropical early maturing maize under *Striga*-infested and *Striga*-free environments. Field Crop Res. 121:186–194.

Badu-Apraku, B., A.F. Lum, R.O. Akinwale, and M. Oyekunle. 2011a. Biplot analysis of diallel crosses of early maturing tropical yellow maize inbreds in stress and nonstress environments. Crop Sci. 51:173–188.

Badu-Apraku, B., M. Oyekunle, and R.O. Akinwale. 2011b. Combining ability of early-maturing white maize inbreds under stress and nonstress environments. Agron. J. 103:544–557.

Badu-Apraku, B., R.O. Akinwale, S.O. Ajala, A. Menkir, M.A. B. Fakorede, and M. Oyekunle. 2011c. Relationships among traits of tropical early maize cultivars in contrasting environments. Agron. J. 103:717–729.

Badu-Apraku B., M.A.B. Fakorede, M. Ouedraogo, and M. Quin. 1999. Strategy for sustainable maize production in West and Central Africa. Proceedings of Regional Maize Workshop, April 21–25, 1997, IITA-Cotonou, Benin Republic.WECAMAN/IITA, Ibadan, Nigeria.

Badu-Apraku B., and M.A.B. Fakorede. 2001. Progress in breeding for *Striga hermonthica* resistant early and extra-early maize varieties. p. 147–162. In: B. Badu-Apraku, M.A.B. Fakorede, M. Ouedraogo and M. Quin (eds.), Impact, challenges and prospects of maize research and development in West and Central Africa. Proceedings of Regional Maize Workshop, May 4–7, 1999. WECAMAN/IITA-Cotonou, Benin Republic.

Badu-Apraku, B., M.A.B. Fakorede, and A.F. Lum. 2006. Evaluation of experimental varieties from recurrent selection for *Striga* resistance in two extra-early maize populations in the savannas of West and Central Africa. Expl. Agric. 43:183–200.

Badu-Apraku, B., M.A.B. Fakorede, and A.F. Lum. 2008. S_1 family selection in early maturing maize population in *Striga*-infested and *Striga*-free environments. Crop Sci. 48:1984–1994.

Badu-Apraku, B., A. Menkir, and A.F. Lum. 2005. Assessment of genetic diversity in extra-early *Striga* resistant tropical inbred lines using multivariate analyses of agronomic data. J. Genet. Breed. 59:67–80.

Badu-Apraku, B., A. Menkir, and A.F. Lum. 2007. Genetic variability for grain yield and components in an early tropical yellow maize population under *Striga hermonthica* infestation. Crop Improv. 20:107–122.

Badu-Apraku, B., M.A.B. Fakorede, A. Menkir, A.Y. Kamara, and A. Adam. 2004a. Effects of drought-screening methodology on genetic variances and covariances in Pool 16 DT maize population. J. Agr. Sci. 142:445–452.

Badu-Apraku, B., M.A.B. Fakorede, A. Menkir, A.Y. Kamara, L. Akanvou, and Y. Chabi. 2004b. Response of early maturing maize to multiple stresses in the Guinea savanna of west and central Africa. J. Genet. Breed. 58:119–130.

Badu-Apraku, B., I. Hema, C. The, N. Coulibaly, and G. Mellon. 1999. Making improved maize seed available to farmers in West and Central Africa- the contributionof of WECAMAN. p. 138–149. In: Badu-Apraku B., M.A.B. Fakorede, M. Ouedraogo and M. Quin (eds.), Strategy for sustainable maize production in West and Central Africa. Proceedings of Regional Maize Workshop, April 21–25, 1997, IITA-Cotonou, Benin Republic. WECAMAN/IITA, Ibadan, Nigeria.

Badu-Apraku, B., M.A.B. Fakorede, M. Ouedraogo, R.J. Carsky, and A. Menkir (eds), 2003. Maize Revolution in West and Central Africa. Proceedings of a Regional Maize Workshop, IITA-Cotonou, Benin Republic, 14–18 May, 2001. WECAMAN/IITA.

Badu-Apraku, B., A.O. Diallo, J.M. Fajemisin, and M.A.B. Fakorede. 1997. Progress in breeding for drought tolerance in tropical early maturing Maize for the semi-arid zone of west and central Africa. In: G.O. Edmeades, M. Banzinger, H.R. Mickelson and C.B. Pena-Valdivia (eds.), Developing drought and low N tolerant maize. Proceedings of a Symposium, March 25–29, 1996, CIMMYT, EL Batan, Mexico. Mexico D.F., CIMMYT.

Badu-Apraku, B., A.F. Lum, M.A.B. Fakorede, A. Menkir, Y. Chabi, C. The, M. Abdulai, S. Jacob, and S. Agbaje. 2008a. Performance of early maize cultivars derived from recurrent selection for grain yield and *Striga* resistance. Crop Sci. 48:99–112.

Badu-Apraku, B., A. Menkir, S.O. Ajala, R.O. Akinwale, M. Oyekunle, and K. Obeng-Antwi. 2010. Performance of tropical early-maturing maize cultivars in multiple stress environments. Can. J. Plant Sci. 90:1 22.

Badu-Apraku, B., A. Menkir, M.A.B. Fakorede, A.F. Lum, and K. Obeng-Antwi. 2006a. Multivariate analyses of the genetic diversity of forty-seven *Striga* resistant tropical early maturing maize inbred lines. Maydica 51:551–559.

Badu-Apraku, B., S. Twumasi-Afriyie, P.Y.K. Sallah, E.A. Asiedu, W. Haag, K.A. Marfo, S. Ohemeng-Dapaah, and B.D. Dzah. 2006b. Registration of "Obatanpa GH" maize. Crop Sci. 46:1393–1395.

Bänziger, M., and H.R. Lafitte. 1997a. Breeding for N-stressed environments: how useful are N-stressed selection environments and secondary traits? p. 401–404. In: G.O. Edmeades, M. Banziger, H.R. Mickelson and C.B. Pefia Valdivia (eds.), Developing drought and low N tolerant maize. Proceedings of a Symposium, March 25–29, 1996, CUNNTT, El Batan, Mexico.DF.

Bänziger, M., and H.R. Lafitte. 1997b. Efficiency of secondary traits for improving maize for low-nitrogen target environments. Crop Sci. 37:1110–1117.

Bänziger, M., N. Damu, M. Chisenga, and F.T. Mugabe. 1999a. Evaluating the drought tolerance of some popular maize hybrids grown in sub-Saharan Africa. p. 61–63. In: Maize production technology for the future: Challenges and opportunities. Proceedings of Eastern and Southern Africa Regional Maize Conf., Sept. 21–25, 1998, CIMMYT - Kenya, Addis Ababa, Ethiopia.

Bänziger, M., G.O. Edmeades, and H.R. Lafitte. 1999b. Selection for drought tolerance increases maize yields across a range of nitrogen levels. Crop Sci. 39:1035–1040.

Bänziger, M., G.O. Edmeades, D. Beck, and M. Bellon. 2000. Breeding for drought and nitrogen stress tolerance in maize: from theory to practice. CIMMYT, Mexico, D.F.

Bennetzen, J.L., F. Gong, J. Xu, C. Newton, A.C.de Oliveira. 2000. The study and engineering of resistance to the parasitic weed *Striga* in rice, sorghum, and maize. p. 197–205. In: B.I.G. Haussmann, D.E. Hess, M.L. Koyama, L. Griver, H.F.W. Rattunde and H.H. Geiger (eds.), Breeding for *striga* resistance in cereals. Margraf Verlag, Ibadan, Nigeria.

Berner, D.K., J.G. Kling, and B.B. Singh. 1995. *Striga* research and control. A perspective from Africa. Plant Dis. 97:652–660.

Betrán, F.J., D. Beck, M. Bänziger, and G.O. Edmeades, 2003a. Secondary traits in parental inbreds and hybrids under stress and non-stress environments in tropical maize. Field Crops Res. 83:51–65.

Bolanos, J., and G.O. Edmeades. 1993. Eight cycles of selection for drought tolerance in lowland tropical maize. II. Responses in reproductive behavior. Field Crops Res. 31:253–268.

Bolanos, J., and G.O. Edmeades. 1996. The importance of the anthesis-silking interval in breeding for drought tolerance in tropical maize. Field Crops Res. 48:65–80.

Buah, S.S., M.S. Abdulai, S.S. Seini, K.O. Gyasi, A. Huudu, M.Y.B. Suglo, and M. Jatoe. 2009. Onfarm testing and demonstration of drought tolerant maize varieties and/or hybrids in the upper west region of Ghana. SARI, Wa, Ghana.

Cechin, I., and M.C. Press 1993. The influence of nitrogen on growth and photosynthesis of sorghum infected with *Striga hermonthica* from different provenances. Weed Res. 3:289–298.

Chapman S.C., and G.O. Edmeades. 1999. Selection improves drought tolerance in tropical maize populations. II Direct and correlated responses among secondary traits. Crop Sci. 39:1315–1324.

Comstock, R.E., and R.H. Moll, 1963. Genotype × Environment Interactions. Symposium on Statistical Genetics and Plant Breeding. National Academy Science National Research Council, Washington, DC, p. 164–196.

Comstock, R.E., and H.F. Robison, 1948. The components of genetic variance in population of biparental progenies and their use in estimating the average degree of dominance. Biometrics 4:254–266.

Denmead O.T., and R.H. Shaw. 1960. The effects of soil moisture stress at different stages of growth on the development and the yield of corn. Agron. J. 52:272–274.

Dankyi, A.A., P.Y.K. Sallah, A. Adu-Appiah, and Gyamera-Antwi. 2005. Determinants of the adoption of quality protein maize, Obatanpa, in southern Ghana.- Logistic regression analysis. Paper presented at the Fifth West and Central Africa Regional Maize Workshop, IITA-Cotonou, Benin Republic, May 2–7, 2005. WECAMAN/IITA.

DeVries, J. 2000. The inheritance of *Striga* reactions in maize. p. 73–84. In: Haussmann, et al. (eds.), Breeding for *Striga* resistance in cereals. Proceedings of a Workshop. IITA, Ibadan. Margraf Verlag, Weikersheim.

DTMA (Drought Tolerant Maize for Africa)2008. Summary of the results of multilocation and farmer participatory trials executed by DTMA partners in West and Central Africa in 2007. IITA, Ibadan, Nigeria.

Edmeades, G.O., M. Bänziger, S.C. Chapman, J.M. Ribaut, and J. Bolanos. 1995. Recent advances in breeding for drought tolerance in maize. p. 24–41. In: B. Badu-Apraku, M.O. Akoroda, M. Ouedraogo and F.M. Quin (eds.), Contributing to food self-sufficiency:

Maize research and development in west and central Africa. Proceedings of Regional maize workshop 1995 May 28 to June 02. IITA, Cotonou, Benin Republic.

Edmeades, G.O., J. Bolaños, M. Bänziger, J.-M. Ribaut, J.W. White, M.P. Reynolds, and R. Lafitte. 1998. Improving crop yields under water deficits in the tropics. p. 437–451. In: V.L. Chopra et al. (eds.), Crop productivity and sustainability: shaping the future. Proceedings of the 2nd International Crop Science Congress, New Delhi, Nov. 10–17, 1996. New Delhi: Oxford and IBH.

Efron, Y. 1985. Using temperate and tropical germplasm for maize breeding in the tropical area of Africa. p. 105–131. In: A. Brandolini and E. Salami (eds.), Istituto Agronomico per l'Oltremare, Florence, Italy.

Efron, Y., S.K. Kim, J.M. Fajemisin, J.H. Mareck, C.Y. Tang, Z.T. Dabrowski, H. Rossel, G. Thottappilly, and I.W. Buddenhagen. 1988. Breeding for resistance to maize streak virus: a multidisciplinary team approach. Plant Breed. 103:1–36.

Ejeta, G., L.G. Butler, and A.G. Babiker. 1992. New approaches to the control of *Striga*. *Striga* research at Purdue University, Agricultural Experimental Station, Purdue University, West Lafayette, IN.

Fajemisin, J.M., Y. Efron, S.K. Kim, F.H. Khadr, Z.T. Dabrowski, J.H. Mareck, M. Bjarnason, V. Parkinson, L.A. Evrett, and A. Diallo. 1985. Population and varietal development in maize for tropical Africa through resistance breeding approach. p. 385–407. In: A. Branddollini and F. Salami (eds.), Breeding strategies for maize production improvement in the tropics. FAO and Istituto Agronomico per l'Oltremare, Florence, Italy.

Fakorede, M.A.B. 1979. Interrelationships among grain yield and agronomic traits in a synthetic population of maize. Maydica 24:181–192.

Fakorede, M.A.B., and O.A. Akinyemiju. 2003. Climatic change: effects on maize production in a tropical rainforest location. p. 230–254. In: B. Badu-Apraku, M.A.B. Fakorede, M. Ouédraogo, R.J. Carsky and A. Menkir (eds.), Maize revolution in West and Central Africa. Proceedings of Regional Maize Workshop, May 2001, IITA-Cotonou, Benin Republic.

Fakorede, M.A.B., B. Badu-Apraku, A. Menkir, S.O. Ajala, and A.F. Lum. 2007. A review of NARES IARC Donor collaboration to develop demand driven technologies for improved maize production in West and Central Africa. p. 3–24. In: B. Badu-Apraku, M.A.B. Fakorede, A.F. Lum, A. Menkir and M. Ouedraogo (eds.), Demand-driven technologies for sustainable maize production in West and Central Africa. Proceedings of Fifth Biennial Regional Maize Workshop, IITA-Cotonou, Benin, May 3–6, 2005. WECAMAN/IITA, Ibadan, Nigeria.

Fakorede, M.A.B., J.M. Fajemisin, J.L. Ladipo, S.O. Ajala, and S.K. Kim. 2003. Development and regional deployment of streak virus resistant maize germplasm: an overview. Plant virology in sub-Saharan Africa: Proceedings organized by IITA, Ibadan, Nigeria.

Falconer, D.S. 1989. Introduction to quantitative genetics. Longman, London. 438.

Fukai, S., and M. Cooper. 1995. Development of drought-resistant cultivars using physiomorphological traits in rice. Field Crops Res. 40:67–86.

Gethi, J.G., and M.E. Smith. 2004. Genetic responses of single crosses of maize to *Striga hermonthica* (Del.) Benth. and *Striga asiatica* (L.) Kuntze. Crop Sci. 44:2068–2077.

Gurney, A.L., J. Slate, M.C. Press, and J.D. Scholes. 2006. A novel form of resistance in rice to the angiosperm parasite *Striga hermonthica*. New Phytol. 169:199–208.

Hallauer, A.R., and J.B. Miranda. 1988. Quantitative genetics in maize breeding. Iowa State University Press, Ames.

Haussmann, B.I.G., D.E. Hess, G.O. Omanya, R.T. Folkertsma, B.V.S. Reddy, M. Kayentao, H.G. Weiz, and H.H. Geiger. 2004. Genomic region influencing resistance to parasitic weed *Striga* hermonthca in two recombinant inbred populations of sorghum. Theor. Appl. Genet. 109:1005–1016.

Haussmann, B.I.G., D.E. Hess, B.V.S. Reddy, H.G. Welz, and H.H. Geiger, 2000. Improved methodologies for breeding *Striga*-resistant sorghum. Field Crops Res. 66, 195–211.

Hess, D.E., Ejeta, G., and Butler, L.G., 1992. Selecting sorghum genotypes expressing a quantitative biosynthetic traits that confers resistance to *Striga*. Phytochemistry 31:493–497.

Holland, J.B. 2006. Estimating genotypic correlations and their standard errors using multivariate restricted maximum likelihood estimation with SAS Proc MIXED. Crop Sci. 46:642–654.

IITA (International Institute of Tropical Agriculture)1992. Sustainable food production in sub-Saharan Africa. 1. IITA's contribution. IITA, Ibadan, Nigeria.

Kim, S.K., Y. Efron, F. Kadhr, J. Mareck, and J. Fajemisin, 1984. Combining ability estimates for tropical and converted temperate maize inbred lines. Agronomy Abstracts. ASA Annual Meetings, Las Vegas, Nov. 25–30 p. 74.

Kim, S.K. 1991. Breeding maize for *Striga* tolerance and the development of a field infestation technique. p. 96–108. In: S.K. Kim (ed.), Combatting *Striga* in Africa. Proceedings of International Workshop. IITA, ICRISAT, and IDRC, Aug. 22–24, 1988. IITA, Ibadan, Nigeria.

Kim, S.K. 1994. Genetics of maize tolerance of *Striga hermonthica*. Crop Sci. 34:900–907.

Kim, S.K., and V.O. Adetimirin. 1995. Overview of tolerance and resistance maize hybrids to *Striga hermonthica* and *Striga asiatica*. p. 255–262. In: D. C. Jewell, S.R. Waddington, J.K. Ransom and K.V. Pixley (eds.), Maize research for stress environments. Proceedings of Fourth Eastern and Southern Africa Regional Maize Conf. CYMMIT, Harare, Zimbabwe.

Kim, S.K., and V.O. Adetimirin. 1997. Responses of tolerant and susceptible maize varieties to timing and rate of nitrogen under *Striga hermonthica* infestation. Agron. J. 89:38–44.

Kim, S.K., and M.D. Winslow. 1991. Progress in breeding maize for *Striga*-tolerance/resistance at IITA. p. 494–499. In: J.K. Ransom (ed.), Proceedings of 5th International Symposium Parasitic Weeds, Nairobi. June 24–30, 1991. IITA, Ibadan, Nigeria.

Kling, J.G., J.M. Fajemisin, B. Badu-Apraku, A. Diallo, A. Menkir, and B.A. Melake. 2000. *Striga* resistance breeding. p. 103–118. In: B.I.G. Haussmann, D.E. Hess, M.I. Koyama, L. Grivet, H.F.W. Ratunde and H.H. Geiger (eds.), Breeding for *striga* resistance in cereals. Proceedings of Workshop. IITA, Ibadan, Nigeria, Aug. 18–20, 1999. Margraf Verlag, Weikersheim, Germany.

Lafitte, H.R., and G.O. Edmeades. 1994. Improvement for tolerance to low soil nitrogen in tropical maize. I. Selection criteria. Field Crops Res. 39:1–14.

Lagoke, S.T.O., V. Parkinson, and R.M. Agunbiade. 1991. Parasitic weeds and control methods in Africa. p. 3–15. In: S.K. Kim (ed.), Combating *Striga* in Africa. Proceedings of International Workshop organized by IITA, ICRISAT, and IDRC. Aug. 22–24, 1998. IITA, Ibadan, Nigeria.

Lane, J.A., D.V. Child, T.H.M. Moore, G.M. Arnold, and J. A. Bailey. 1997. Phenotypic characterization of resistance in *Zea diploperennis* to *Striga hermonthica*. Maydica 42:45–51.

Lee, S.J., W. Yan, K.A. Joung, M.C. Ill. 2003. Effects of year, site, genotype, and their interaction on the concentration of various isoflavones in soybean. Field Crops Res. 81:181–192.

M'Boob, S.S. 1986. A regional program for West and Central Africa. p. 190–194. In Proceedings of FAO/OAU All African Government Consultation on *Striga* Control, Maroua, Cameroon, Oct. 20–24, 1986.

Maiti, R.K., K.V. Ramaiah, S.S. Biesen, and V.L. Chidley. 1984. A comparative study of the haustorial development of *Striga asiatica* (L.) Kuntze on sorghum cultivars. Ann. Bot. 54:447–457.

McCown, R. L., B.A. Keating, M.E. Probert, and R.K. Jones. 1992. Strategy for sustainable production in semi-arid Africa. Outlook Agr. 21:21–31.

Menkir, A. 2003. The role of GIS in the development and targetting of maize germplasm to farmers' needs in West and Central Africa. p. 16–30. In: B. Badu-Apraku, M.A.B. Fakorede, M. Ouédraogo, R.J. Carsky and A. Menkir (eds.), Maize revolution in West and Central Africa. Proceedings of Regional Maize Workshop, May 2001, IITA-Cotonou, Benin Republic.

Menkir A., and A.O. Akintunde. 2001. Evaluation of the performance of maize hybrids, improved open-pollinated and farmers' local varieties under well watered and drought stress conditions. Maydica 46:227–238.

Menkir, A., B. Badu-Apraku, C. Thé, and A. Adepoju. 2003. Evaluation of heterotic patterns of IITA lowland white maize inbred lines. Maydica 48:161–170.

Menkir, A., B. Badu-Apraku, C.G. Yallou, A.Y. Kamara, and G. Ejeta. 2007. Breeding maize for broad-based resistance to *Striga hermonthica*. p. 99–114. In: G. Ejeta and J. Gressel (eds.), Integrating new technologies for *Striga* control: towards ending the witch-hunt. World Scientific Publishing Co., Pte., Ltd., Singapore.

Meseka, S.A., A. Menkir, and A.S. Ibrahim. 2006. Genetic analysis of drought tolerance in maize inbred lines: preliminary results. p. 515. In: B. Badu-Apraku, M.A.B. Fakorede, A. F. Lum, A. Menkir and M. Ouedraogo (eds.), Demand-driven technologies for sustainable maize production in West and Central Africa. Proceedings of Fifth biennial regional maize workshop, IITA-Cotonou, Benin, May 3–6, 2005. WECAMAN/IITA, Ibadan, Nigeria.

MIP-1996. Maize improvement program archival report, 1989–1992. Part 1: Maize population improvement. CID, IITA, Ibadan, Nigeria.

Monneveux, P., C. Sánchez, D. Beck, and G.O. Edmeades. 2006. Drought tolerance improvement in tropical maize source populations: evidence of progress. Crop Sci. 46:180–191.

Morris, C.F., K.G. Campbell, and G.E. King. 2004. Characterization of the end-use quality of soft wheat cultivars from the eastern and western US germplasm 'pools'. Plant Genet. Res. 2:59–69.

Mumera, L.M., and F.E. Below. 1993. Crop ecology, production and management. Crop Sci. 33:758–763.

National Research Council. 1988. Report of an Ad Hoc Panel of the Advisory Committee on Technology Innovation, Board on Science and Technology for International Development, National Research Council in Cooperation with the Board on Agriculture, National Research Council. National Academy Press, Washington, DC.

NeSmith D.S., and J.T. Ritchie. 1992. Effects of water-deficits during tassel emergence on development and yield components of maize (*Zea mays* L.) Field Crops Res. 28:251–256.

Ober, E.S., M.L. Bloa, C.J.A. Clark, A. Royal, K.W. Jaggard, and J.D. Pidgeon. 2005. Evaluation of physiological traits as indirect selection criteria for drought tolerance in sugar beet. Field Crops Res. 91:231–249.

Odhiambo, G.D., and J.K. Ransom 1994. Long-term strategies for *Striga* control. p. 263–266. In: D.C. Jewell, S.R. Waddington, J.K. Ransom and K.V. Pixley (eds.), Maize research for stress environments. Proceedings of 4th Eastern and Southern Africa Regional Maize Conf., Harare, Zimbabwe, 28 March–April 1994. CIMMYT, Harare, Zimbabwe.

Oikeh, S.O., and W.J. Horst. 2001. Agro-physiological responses of tropical maize cultivars to nitrogen fertilization in the moist savanna of West Africa. p. 804–805. In: W.J. Horst, M. Kamh, J.M. Jibrin and V.O. Chude (eds.), Plant-nutrition, food security and sustainability of agro-ecosystems. Kluwer Academic Publisher, Dordrecht, The Netherlands.

Onyibe, J.E., C.K. Daudu, J.G. Akpoko, R.A. Gbadegesin, and E.N.O. Iwuafor. 2003. Pattern of spread of extra-early maize varieties in the Sudan savanna ecology of Nigeria. p. 382–394. In B. Badu-Apraku, M.A.B. Fakorede, M. Ouédraogo, R.J. Carsky and A. Menkir (eds.), Maize revolution in West and Central Africa. Proceedings of Regional Maize Workshop, May 2001, IITA-Cotonou, Benin Republic.

Okai, D.B., S.A. Osei, A.K. Tua, S. Twumasi-Afriyie, W. Haag, B.D. Dzah, K. Ahenkora, and E.I.K. Osafo. 1994. Quality protein maize as a broiler feed ingredient. Proc. Ghana Anim. Sci. Symp. 22:45–49.

Osei, S.A., A. Donkor, C.C. Atuahene, D.B. Okai, A.K. Tua, W. Haag, B.D. Dzah, K. Ahenkora, and S. Twumasi-Afriyie. 1994. Quality protein maize as a broiler feed ingredient. Proc. Ghana Anim. Sci. Symp. 22:45–49.

Ransom, J.K., R.E. Eplee, and M.A. Langston. 1990. Genetic variability for resistance to *Striga asiatica* in maize. Cereal Res. Commun. 18:329–333.

Ransom, D.G., P. Haffter, J, Odenthal, A. Brownlie, and E. Vogelsang. 1996. Characterization of zebrafish mutants with defects in embryonic hematopoiesis. Development 123:311–319.

Rodenburg, J., L. Bastiaans, and M.J. Kropff. 2006. Characterization of host tolerance to *Striga hermonthica*. Euphytica 147:353–365.

Sallah, P.Y.K., S. Twumasi-Afriyie, and C.N. Kasei, 1997. Grain productivities of four maturity groups of maize varieties in the Guinea savanna. p. 173–178. In: B. Badu-Apraku, M.O. Akroda, M. Ouedraogo and F.M. Quin (eds.), Contributing to food self-sufficiency: Maize research and development in West and Central Africa. Proceedings of Regional Maize Workshop, IITA-Cotonou, Benin Republic, 29 May–2 June 1995. WECAMAN/IITA.

Setimela, P.S., B. Vivek, M. Bänziger, J. Crossa, and F. Maideni. 2007. Evaluation of early to medium maturing open-pollinated maize varieties in SADC region using GGE giplot based on the SREG model. Field Crops Res. 103:161–169.

Shaxson, L., and C. Riches. 1998. Where once there was grain to burn: a farming system in crisis in eastern Malawi. Outlook Agr. 27:101–105.

Sprague, G.F., and S.A. Eberhart. 1977. Corn Breeding. p. 305–362. In: G.F. Sprague (ed.), Corn and corn improvement. 2nd ed. Agron. Monogr. 18. ASA, CSSA, SSSA, Madison, WI.

Twumasi-Afriyie, S., P.Y.K. Sallah, M. Owusu-Achaw, K. Ahenkora, R.F. Soza, W. Haag, B.D. Dzah, B.D. Okai, and A. Akuamoah-Boateng. 1997. Development and promotion of quality protein maize in Ghana. p. 140–148. In: B. Badu-Apraku, M.O. Akroda, M. Ouedraogo and F.M. Quin (eds.), Contributing to food self-sufficiency: Maize research and development in West and Central Africa. Proceedings of a Regional Maize Workshop, IITA-Cotonou, Benin Republic, 29 May–2 June 1995. WECAMAN/IITA.

United Nations Statistics Division. 2008. National accounts estimates of main aggregates. Available at http://data.un.org/Data.aspxq_2008_&d [Accessed 30 Nov. 2009].

Vivek, B.S., A.F. Krivanek, N. Palacios-Rojas, S. Twumasi-Afriyie, and A.O. Diallo. 2008. Breeding quality protein maize (QPM): protocols for developing QPM cultivars. CIMMYT, Mexico, DF.

Vogt, W., J. Sauerborn, and M. Honisch. 1991. *Striga hermonthica* distribution and infestation in Ghana and Togo on grain crops. p. 372–377. In: J.K. Ransom, L.J. Musselman, A.D. Worsham and C. Parker (eds.), 5th International Symposium Parasitic Weeds, CIMMYT, Nairobi, Kenya, June 24–30, 1991.

Wolfe, D.W., D.W. Henderson, T.C. Hsiao, and A. Alvio. 1988. Interactive water and nitrogen effects on maize. II. Photosynthetic decline and longevity of individual leaves. Agron. J. 80:865–870.

Wright, S. 1921. Correlation and causation. J. Agr. Res. 20:557–585.

Yallou, C.G., A. Menkir, V.O. Adetimirin, and J.G. Kling. 2009. Combining ability of maize inbred lines containing genes from *Zea diploperennis* for resistance to *Striga hermonthica* in maize. Euphytica 115, 149–158.

Yan, W., and I. Rajcan. 2002. Biplot evaluation of test sites and trait relations of soybean in Ontario. Crop Sci. 42:11–20.

Yan, W., and M.S. Kang, 2003. GGE Biplot analysis: a graphical tool for breeders, geneticists, and agronomists. CRC Press, Boca Raton, FL.

Yan, W., and I. Rajcan. 2002. Biplot evaluation of test sites and trait relations of soybean in Ontario. Crop Sci. 42:11–20.

Yan, W., M.S. Kang, B. Ma, S. Woods, and P.L. Cornelius. 2007. GGE biplot vs. AMMI analysis of genotype-by-environment data. Crop Sci. 47:643–655.

4

Almond Breeding

Thomas M. Gradziel
Department of Plant Sciences
University of California at Davis
Davis, CA 95616, USA

Pedro Martínez-Gómez
Departamento de Mejora Vegetal
CEBAS-CSIC, P.O. Box 164
E-30100 Espinardo (Murcia), Spain

ABSTRACT

Almond, a productive and highly nutritious tree nut with easy transportability and long storability, originated in southeast Asia with subsequent global dissemination to regions having the required Mediterranean-type climate of moderate winters and hot, dry summers. Improvement from ancient times to present typically involves the asexual propagation of elite selections from intraspecific and occasionally interspecific hybridizations. This hybridization/cloning strategy effectively generates and captures highly complex though poorly understood genetic, epigenetic, and genomic interactions, making it among the most potent breeding techniques. The multifaceted synergistic interactions contributing to high productivity and regional adaptability thus represent significant challenges to continued cultivar improvement but also highlight opportunities to more fully characterize and harness these higher-order breeding manipulations. Basic breeding goals include genetic improvement and cultivar development. Genetic improvement typically has a well-defined, focused objective such as the transfer of self-compatibility to locally adapted genetic backgrounds. In contrast, success at cultivar development is indicated by sizable commercial plantings over the long production time required for commercially profitability, and so is rarely determined by superior performance in one or a few traits, but rather by the absence of deficiencies for the large number of characteristics required for commercial success. This need to simultaneously

---

Plant Breeding Reviews, Volume 37, First Edition. Edited by Jules Janick.
© 2013 Wiley-Blackwell. Published 2013 by John Wiley & Sons, Inc.

optimize a large number of essential traits remains the greatest challenge to breeding strategies including the use of molecular-based techniques. Single-sequence repeats (SSRs) are currently proving useful for characterizing genetic identity and relationships, and when combined with single nucleotide polymorphisms (SNPs), are being utilized to better characterize genotypes and genetic maps. Other genomic markers being utilized include those associated with agronomic traits based on either cDNA sequences (expressed sequences tags, ESTs) or databases (cloned gene analogs, CGAs) or single point mutations (SNPs). While genetic improvement strategies, including marker assisted selection (MAS), are becoming increasingly efficient at the partitioning and so manipulating the principal additive genetic interactions affecting the target trait, these resource intensive and inherently reductionist approaches may lead to reduced efficacy of commercial cultivar development if not fully complemented with the equally essential holistic cultivar development approaches.

KEYWORDS: *Amygdalus dulcis*; epigenetic; genetic improvement; hybridization; molecular marker assisted selection; nonadditive genetic variance; *Prunus communis*; *Prunus dulcis*

I. INTRODUCTION
II. BOTANY
    A. Taxonomy, Morphology, and Composition
    B. Origin and Dissemination
    C. Reproductive Biology
        1. Flowering and Pollination
        2. Self-incompatibility
        3. Fruit Development
III. GENETIC DIVERSITY
    A. Biodiversity and Ecosystems
    B. Conservation
    C. Cytogenetics
    D. Interspecific Hybridization in Breeding
IV. GENETIC IMPROVEMENT
    A. Breeding Objectives
    B. Early Genetic Studies
    C. History of Crop Improvement
    D. Selection
    E. Cultivar Development
V. MOLECULAR APPROACHES
    A. Almond Genomics
    B. Molecular Analysis
    C. Molecular Breeding
    D. Epigenetics
    E. Transgenics
VI. FUTURE PROGRESS
LITERATURE CITED

I. INTRODUCTION

Almond (*Prunus dulcis* (Miller) D. A. Webb, syn. *Prunus amygdalus* Batsch., *Amygdalus communis* L., *Amygdalus dulcis* Mill.) originated in the areas around the Fertile Crescent and represents one of the first tree crops to be domesticated. Cultivation spread throughout the Mediterranean regions that have the required temperate, dry climates, and eventually throughout the world. California accounted for 80% of the global commercial production of 921,246 tonnes (t) in 2010–2011 with a production of 738,448 t from plantings of over 300,000 ha (Table 4.1). Almond has become the largest agricultural export of the state of California and the largest specialty crop export in the United States (Almond Board of California 2011) with a farm value exceeding US$2.8 billion. Spain has the largest area under cultivation, estimated at over 436,500 ha. The remaining world production comes from about 20 countries particularly Australia, Turkey, Iran, and Tunisia. Limited almond production extends into the Balkan Peninsula including areas of Bulgaria, Romania, and Hungary. Additional areas exist in central and south-western Asia including, Iraq, Israel, Ukraine, Tajikistan, Uzbekistan, Afghanistan, and Pakistan, extending into western China. Many almond species are native to these regions where almond growing is often under dryland, low-input culture. Increasing almond production is also occurring in the southern hemisphere countries having similar climates, particularly regions of Australia, central Chile, Argentina, and South Africa. Commercial production is typically limited to Mediterranean climates with mild winters and hot, dry summers. A high susceptibility to fungal and bacterial diseases of the blossoms, leaves, branches, and fruit also reduce production in areas having rain and/or high humidity during the growing season (Kumar and Uppal

Table 4.1. Commercial production of almonds in major producing countries.

Location	Production (t)
USA (California)	738,448
Australia	38,555
Spain	34,926
Turkey	15,876
Iran	12,247
Tunisia	12,247
Others	68,947
Total	921,246

Source: Almond Board of California (2011).

1990; Ogawa and English 1991). Similarly, excessive moisture in the root zone can result in tree losses due to root rots or asphyxia. Almond is also among the earliest temperate tree crops to bloom, which further limits production to areas relatively free from spring frosts. Because almond is self-sterile, it requires cross-pollination, which acts to promote genetic variability and so adaptability to new microenvironments.

The natural range of the various almond species from north-western China to the northern Indus Valley in the east, to Mesopotamia and southern Europe in the west, overlapped areas important in the transition of humans from hunter/gatherers to more permanent settlements. These "cradles of civilization" were inherently "cradles" of plant domestication, which undoubtedly involved selection within the numerous wild almonds. The edible kernels of wild almonds and related species were important food staples from ancient times. Stone tools used for the apparent cracking of almond shells documents the harvesting of wild almonds in northern Israel by our human ancestors as early as 780,000 years ago (Martinoli 2005; Weiss et al. 2004). Around 11,000 BCE, almonds, pistachios, and lentils were being utilized at Franchthi cave in southern Greece, indicating that the farming of legumes and nuts preceded that of grain in Greece and possibly the rest of Asia Minor (Farrand 1999; Hansen and Renfrew 1978).

Almonds have been an important source of food for thousands of years in parts of Asia and Europe. Within each region, the best wild seedlings were routinely selected for propagation by local farmers while natural selection continued its unrelenting pressures toward greater adaptation to local environments, including regionally important disease and insect pests. The self-sterile nature of almond insured a continuous exchange and mixing among cultivated and wild germplasm including in many cases, related species (Grasselly 1972; Socias i Company 2002). Superior genotypes would be identified and vegetatively propagated. Most modern cultivars in Asia, the Mediterranean area, and more recently in California, originated as such time-tested clonal selections. The subsequent selection over hundreds of years and hundreds of thousands of clonal propagations has also identified improved bud sports for many of these well-established cultivars. Both genetic and cytogenetic changes such as deletions, point mutations, aneuploidy, and translocation, (Jáuregui et al. 2001; Martínez-Gómez 2003; Gradziel 2003), as well as epigenetic alterations such as gene activation and gene silencing would be selected, though because the subsequent selections are vegetatively propagated, the specific nature of control is rarely analyzed. The complex genetic interactions contributing to high productivity and regional adaptability thus

4. ALMOND BREEDING

represent significant challenges to continued cultivar improvement, but also represent valuable opportunities to characterize and harness these higher-order genetics.

II. BOTANY

A. Taxonomy, Morphology, and Composition

The almond fruit is classified as a drupe with a pubescent skin (exocarp), a fleshy but thin hull (mesocarp), and a distinct hardened shell (endocarp). The hull undergoes limited enlargement during development, later becoming dry and leathery, and dehiscing at maturity (Fig. 4.1). The mature endocarp ranges from hard to soft and papery depending upon genotype. Horticulturally, almonds are classified as a "nut" in which the edible seed (the kernel or "meat") is the commercial product. The kernel includes an embryo surrounded by the pellicle.

Fig. 4.1. Cultivated almond shoot showing leaf, fruit, and kernel morphology.

Almond, as in all *Prunus*, initiates flower buds laterally on current season growth, which then bloom and fruit the following year (Fig. 4.1). In general, there are three basic classes of bearing habits: most flower buds on 1-year old shoots as in 'Ai', most flowers on spurs as in 'Tuono'; and mixed as in 'Mission' (syn. 'Texas') and 'Nonpareil'. A mixture of both bearing habits is considered advantageous. Shoot bearing habits are associated with precocious production while spur habit greatly increases the bearing surface and harvest index. Foliage density is, in turn, determined by the branching habit and the size and distribution of leaves. Foliage density differences can be visually characterized among cultivars. Leaf size varies with position, however, with shoot leaves tending to be large and spur leaves small. A classification of growth habits based on variations in primary, secondary, and tertiary shoot development have been described by Gradziel et al. (2002) and Kester and Gradziel (1990).

Almond fruit of different cultivars vary in size, shape, pubescence, retention of the pistil remnants, and nature of the suture line (Monastra et al. 1982). In 'Drake,' the suture line shows a relatively deep depression while 'Nonpareil' has a relatively smooth line, and 'Mission' fruit show two prominent vertical ridges. The pattern by which 'splitting' occurs in the hull also differs and can be representative of cultivars. Four basic types have been described: ventral split opening on one side (Nonpareil), ventral and dorsal split (IXL), four-way split (California), and dorsal split (Jeffries). The thickness and weight of the mature hull may also differ significantly among cultivars. Some hulls, such as 'Mission', are thin and dry and contribute only a small portion of the entire fruit. Others, such as 'Nonpareil,' are thick and fleshy and provide a relatively large proportion of the weight. In California, hulls are used for livestock feed and the food value is better with larger hulls. Hull characteristics also affect the relative ease with which nuts are removed from the tree at harvest, the ability of nuts to field-dry rapidly during harvest and the ease of hull removal. These processes are more critical with soft-shelled cultivars used in California where worm infestations and concealed damage from wet field conditions can be serious problems.

Shell hardness is associated with the total amount of lignin deposited to the shell during nut development. Shelling proportion (dry weight of kernel/dry weight of in-shell nut) is used to obtain a quantitative measure of shell density and is utilized in commercial activities to calculate kernel yield of different cultivars. Markings on the outer shell are characteristic of individual cultivars as well as different almond species. Within *P. dulcis*, the markings or pores tend to be mostly circular, less frequently elongated and occasionally a mixture of both. Pores may be large or small, many or few. Other species have smooth and thin shells as with *P. orientalis, P. turcomenica*, and *P. bucharica* or are distinctly grooved or scribed as with *P. kuramica, P. tangutica*, and *P. persica*. The integrity of the shell, particularly at the suture, is important, since poorly sealed shells have kernels exposed and so susceptible to disease and worm damage. The shell consists of an outer and an inner layer separated by channels through which vascular fibers develop. As the hull dehisces and separates from the nut, the outer shell layer may remain attached to the hull and separate from the inner shell layer. The latter type is often associated with high shelling percentages and poor shell seal.

The almond is a large, nonendospermic seed and has two large cotyledons. Kernel size is often expressed by linear dimensions of length, width, and thickness. These parameters are established during the first growth phase of nut development in the spring and are completed by early summer. Crop density within cultivars, is inversely related to average kernel size. Among kernels of a given cultivar, a high correlation also exists between dry weight and linear dimensions of length and width. Weight increases continuously until maturity. Improper filling may be caused by adverse growing conditions, moisture stress, early ripening, or other environmental and cultural stresses. Shape is a function of differences in length, width, and thickness. The unique shapes of certain cultivars tend to establish specific marketing categories and uses. Irregularities in width and thickness may change the visual effect significantly. A high correlation was found to exist between width and length among kernels of the same cultivar even when compared in different years and from different locations (Kester and Gradziel 1996). The correlation between thickness and either width or length, however, was much lower. As size dimensions decrease, thickness is not necessarily related. Consequently, the relative width to length may appear different for different genotypes otherwise having a similar kernel mass. Shape is usually described from a side view of the kernel. Kernels may be round, oval, ovate, oblong with edges that are relatively straight when viewed on one edge, and rounded to various degrees on the other. Thickness (viewed from the edge) may vary from base to tip. Unequal thickness can result in unequal roasting during processing.

The almond kernel is consumed either in the natural state or processed. Because of its protein and fat content and good flavor, crunchy texture, and good visual appeal, it has many important food uses (Rosengarten 1984). As an ingredient in many manufactured food products, kernels may be roasted dry or in oil followed by salting with various seasonings (Schirra 1997; Woodroof 1979). The processed kernel is used either blanched or unblanched. Blanching removes the pellicle (skin) using hot water or steam. Large amounts of kernels are combined with chocolate in confectionery. Almond kernels can be sliced or diced to be used in pastry, ice cream, breakfast cereals, and vegetable mixtures. The kernels are also ground into paste to be used in bakery products and in the production of marzipan. The flavor and texture of almonds can be intensified or moderated through proper selection of cultivar combinations, origin, moisture content and processing, and handling procedures (Kester et al. 1993). Variation in amygdalin content account for some flavor differences among cultivars, particularly the distinct amaretto flavor common in certain Mediterranean almonds (Dicenta and García 1993b; Vargas et al. 2001). Californian cultivars had amygdalin contents ranging from 0.33% to 0.84% with only 'Peerless' outside this range at 1.75% (dry weight). In contrast, the Italian cultivars varied from 0.73% to 1.95% with only two cultivars below that range (Schirra 1997). Even higher amygdalin levels will result in bitter almond seeds that can be blended as paste with sweet almonds to achieve the desired cherry or amaretto flavor. For the bitter reaction to take place, the substrate amygdalin and a β-glucosidase enzyme must come into contact following damage to and lysis of the cells. Bitterness results from hydrolysis of the glucoside amygdalin by the β-glucosidase enzyme that produces benzylaldehyde (which confers the "cherry" or "amaretto" flavor) and cyanide (which can be poisonous) (Kester and Gradziel 1996). Benzaldehyde is also known in the chemical and flavoring industries as "oil of bitter almond" because of its preponderance in bitter rather than sweet almonds. This trait is also typical of the wild almond species where it protects the seed from herbivores.

Almonds are among the most nutrient dense of all tree nuts (Kendall et al. 2003). They are an excellent source of essential fatty acids, vitamins, and minerals (Table 4.2) (Saura-Calixto et al. 1981). Raw almonds are one of the best plant sources of protein. While certain nut storage proteins can pose an allergenic health threat to consumers, Sathe et al. (2001) found no significantly elevated allergy risk in a range of cultivated almond as well as breeding lines derived from interspecies hybrids.

Almonds are also one of the best natural sources of vitamin E (Sabate and Haddad 2001) that is believed to play a role in preventing heart disease, certain kinds of cancer, and cataract formation (Kodad et al. 2006). About 30 g of almonds (20–25 kernels) contains 37% of the recommended daily value of vitamin E, 21% of magnesium, and 15% of the recommended daily value of phosphorus. Almonds also represent a convenient source of folic acid and fiber (Vezvaei and Jackson 1996; Schirra 1997). Historical uses of sweet and/or bitter

Table 4.2. Nutrient composition of the raw almond kernel per 100 g fresh weight of edible portion.

Principle	Value
Energy	578 kcal
Protein	21.26 g
Carbohydrates	19.74 g
Glucose	4.54 g
Starch	0.73 g
Fats	
Saturated fatty acids	3.88 g
Monounsaturated fatty acids	32.16 g
Polyunsaturated fatty acid	12.21 g
Vitamins	
Folate, total	29 mcg
Vitamin E	25.87 mg
Nutrients	
Calcium	248 mg
Magnesium	275 mg
Phosphorus	474 mg
Potassium	728 mg
Sodium	1 mg

Source: Adapted from Socias i Company et al. (2007).

almond ointment included the treatment of asthma, pattern baldness, and as a soothing salve for skin burns.

Almond kernels are also a source of high-quality oil (Abdallah et al. 1998; Kodad et al. 2005). The oil, which can constitute over 50% of the kernel dry weight, is primarily composed of the more stable oleic acid, making it desirable from ancient times to the present for use as a base for various ointments and pharmaceuticals. The high levels of this monounsaturated fat may be partly responsible for the observed association between frequent nut consumption and reduced risk of coronary heart disease (Folgoni et al. 2002; Lovejoy et al. 2002). Recent evidence has suggested that the incidence of deaths due to coronary heart disease, hypertension, congestive heart failure, and stroke is decreased in people who eat a serving of nuts several times per week (Socias i Company et al. 2007). Because of their high lipid content of 50%–55%, almond kernels are a concentrated energy source (Fraser et al. 2002). The oil is primarily monounsaturated, being approximately 70% oleic and 15% linoleic acid that, in addition to the previously described health benefits, results in an agreeable supple, buttery flavor and high nutritional value as well as long-term stability in storage (Fulgon et al. 2002). The hull, which is analogous to the flesh of the closely related peach (*P. persica* L.), contains about 25% DW sugar that facilitates its use as a livestock feed. A thorough review of almond nutritional and food quality traits, including opportunities for their genetic manipulation, has recently been compiled by Socias i Company et al. (2007).

B. Origin and Dissemination

Early researchers proposed that cultivated almond resulted from selection from within a species listed originally as *A. communis* (syn. *P. communis* Archang.) based on studies of two natural populations containing large numbers of sweet seeded individuals rather than the bitter kernels typically found in the wild (Watkins 1979). One population is located in the Kobet Dag mountain range in central Asia between present day Iran and Turkmenistan and the second population occurs on the lower slopes of the Tian Shan Mountains between Kyrgyzstan and western China. The natural range of *A. communis* was proposed to have extended across Iran, the Transcaucausus, and eastern Turkey, and into present-day Syria, and thus overlapped with known sites of early almond cultivation (Denisov 1988; Kester et al. 1991). According to this view, the distinction between cultivated and wild forms gradually disappeared with direct and indirect human selection. However, because the purportedly natural sweet-kernel populations closely resembles the phenotypic range of present day cultivated almonds, it has more recently been suggested that the Kobet Dagh and Tian Shan populations are, in fact, more recent remnants or escapes from later domesticated or semi-domesticated orchards (Ladizinsky 1999). An emerging consensus proposes that cultivated almond represents a generalized, fungible kernel phenotype, probably derived from *P. fenzliana* but with contributions through natural interspecific cross-hybridizations with a range of related species occurring naturally within this range (Fig. 4.2), possibly including *P. bucharica*, *P. kuramica*, and *P. triloba* (Gradziel 2010; Zeinalabedini et al. 2010).

The subsequent and widespread dispersal of cultivated almonds occurred in three stages: Asiatic, Mediterranean, and Californian. The Asiatic stage included the initial domestication and the subsequent spread throughout central and south-western Asia often along major prehistoric trade routes. The range centers on present-day Iran extending east to western China, southeast to India and northern Pakistan, northwest through Turkey, and southwest into the uplands and deserts of central Israel and Syria. Almonds are reported in Hebrew literature as early as 2000 BCE. Their culture continues to the present time within the

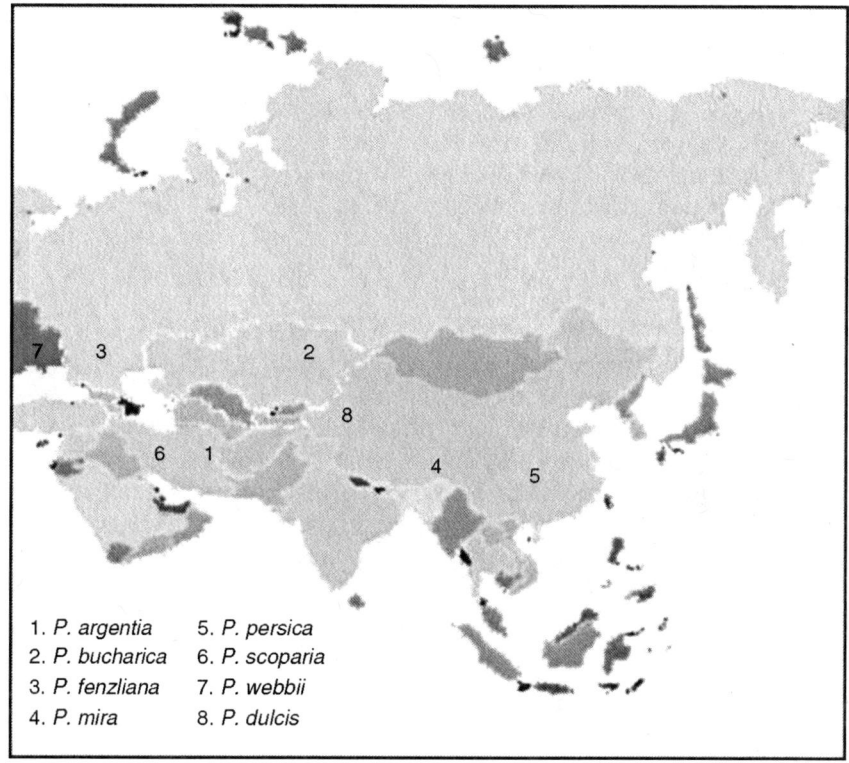

Fig. 4.2. Map of Asia showing centers of origin of selected almond species.

Asiatic region, where in many areas almonds are grown under dry land, subsistence agricultural practices similar to those used thousands of years ago.

In the Mediterranean stage, almonds appear to have been brought into Greece prior to 300 BCE, eventually becoming introduced to all compatible areas of the Mediterranean. Initial introductions may have come from the early ocean trading Phoenicians and Greeks during establishment of colonies in Sicily and other Mediterranean sites (Bacarella et al. 1991). Cultivation typically occurred within 80 km of the Mediterranean coast extending onto the slopes of river valleys as well as the interior areas in Spain. Subsequent introductions occurred around 500–600 CE with the conquest of North Africa by Arabs who also brought almonds into southern Spain and Portugal. Two thousand years of continuous cultivation in the Mediterranean basin has concentrated almond plantings into specific regions where well-defined seedling ecotypes have evolved. A tolerance to drought and susceptibility to high soil moisture placed almonds in a mixed culture system with olives, carob, and other desert adapted crops. Almonds were typically found at higher elevations on well-drained slopes to avoid spring frosts and saturated soils. In these more marginal environments, cultural practices evolved that minimized inputs of labor, fertilizers, and use of supplementary water. Locally adapted seedling populations eventually led to a number of local selections adapted to very specific climatic and culture conditions. Selection toward greater local adaptation appears to have been augmented by a more recent introgression of genes from nearby wild almond species. Godini (2000) and Socias i Company (1990) provide evidence for the introgression of self-compatibility and morphological features from *P. webbii* in the development of commercially important cultivars along the northern shore of the Mediterranean Sea. Both natural and artificial selection appears to have occurred. For example, the presence of the bitter gene would be desirable in the wild as it confers resistance to kernel feeding (homozygote kernels are bitter) and foliage feeding (heterozygote and homozygote leaves are bitter) but would be undesirable for human consumption. Most European (and New World) cultivated almonds are heterozygous for bitterness owing to its discouragement of foliar feeding, so that many open pollinated seed-derived local land races typically segregate for bitter kernels (Grasselly 1972). The need to graft-over bitter seedlings within these populations eventually led to selection of local vegetative clones that subsequently became characteristic of these regions (Bacarella et al. 1991).

The Californian stage initially began as an extension of Mediterranean culture, utilizing a hard-shelled germplasm originally brought from Spain. Later, soft-shell types more compatible for California were introduced from France. High-input orchard practices, however, soon differentiated Californian production from that of Europe and Asia. Important California cultural changes included the movement of almond production from more marginal coastal sites to the very productive Central Valley, the development of new rootstocks and orchard management practices for these highly productive sites, the selection of consistently high-yielding cultivars, and the standardization of markets based upon a relatively few cultivar types. The combination of highly adapted cultivars and rootstocks, favorable soil and climate, abundant water and effective management has given California growers the highest productivity in the world. Production per hectare continues to show upward trends with yields surpassing $4\,\text{t}\,\text{ha}^{-1}$ presently possible for some cultivar/site combinations. Cultivars vary in size, shape, vigor, branching pattern, growth, and bearing habit (Gulcan 1985; Brooks and Olmo 1997). Tree size is a relative term that depends not only on the individual genotype but also on orchard age, site (climate, soil), and management (irrigation, planting density, fertilization, rootstock, and pruning). Tree size is related to precociousness and productivity. Some cultivars such as 'Carmel' show reduced growth with age, partly as a consequence of precocious production, while others such as 'Nonpareil' tend to maintain vigor, resulting in larger trees. Size of an individual tree is directly correlated to yield and must be balanced against tree spacing and density to optimize production per hectare. Size of a tree also directly affects management efficiency, depending on the type of cultural system utilized. For mechanized harvest, fewer trees per unit are desirable. If trees are too large, however, they become difficult to shake, prune, and spray. Most cultivated almonds fall within the tree size range of medium to large, depending upon age and site.

C. Reproductive Biology

1. Flowering and Pollination. The almond produces a typical peryginous self-incompatible *Prunus* flower. Honeybees, foraging for pollen or for nectar secreted at the base of the flower are important pollinators (Thorp and Roper 1994). Flowers of different cultivars and related species may differ in petal size, shape, color, number of stamens, and arrangement and length of stamens relative to the stigma. The number of the anthers within each flower varied among almond species and genotypes and a typical value for cultivated almond was 29 anthers per flower. The average number of anthers per flower for *A. orientalis* almond species was 21 while the number of anther per flower for *A. turcomanica* was 14 (Bayazit et al. 2011). The number of stamens may vary from 12 to about 40 with the usual number being 26–34. The distributions of stamen number within seedling populations from parents of different stamen number indicate quantitative inheritance with a tendency toward dominance of larger numbers. Although the usual pistil number is one, some genotypes such as "Eureka" tend to produce two, which may result in a double fruit. Two flowers can sometimes be produced within the same flower bud with similar results. The structure of the pistil and style also varies. Some styles are straight and elongated, extending above the anthers by petal-fall. In other individuals, the stigmas and the anthers are approximately at the same level, a condition associated with increased chances for self-pollination. Floral bud differentiation takes place during summer, primarily in August, and floral development continues into the fall and winter (Polito and Micke 1994). Time of flowering is one of the most important adaptive traits of almond as it determines vulnerability to spring frosts and rains. Flowering time is determined by chilling requirements to overcome dormancy and heat requirements for subsequent bud growth and development. Actual timing of bloom can vary from year to year depending upon the temperature patterns before and during bloom (DiGrandi-Hoffman et al. 1994). In general, the sequence of bloom among different cultivars tends to be fairly constant but relative bloom time between specific cultivars can sometimes be reversed because of differing requirements for initial chilling or subsequent heat. This relationship is important commercially since it is desirable to have a pollenizer flower just before the high-value cultivar in order to maximize its cross-pollination and subsequent yield. For this reason bloom time is often given relative to a high-value cultivar. In California, 'Nonpareil' is commonly used whereas in the Mediterranean area, 'Marcona' is frequently the standard.

2. Self-incompatibility. The cultivated almond, as well as most related almond species, expresses gametophytic self-incompatibity, which discourages self-fertilization, favors cross-pollination, and thus maintains high genetic variability within seedling populations. Genetic control of pollen-pistil self-incompatibility is through a single gene (S) that exists as a series of alleles including alleles for self-compatibility. Each diploid genotype carries two alleles of the series. Pollen grains, which have a haploid genome, are unable to fertilize a self-incompatible pistil possessing the same allele. Pollenizers also show cross-incompatibility with other cultivars with the same S-alleles. Cross-incompatibility groups (CIGs) have been identified and incompatibility alleles have been assigned to many of them (Tamura et al. 2000; Boškovic et al. 2003; Channuntapipat et al. 2003; López et al. 2004). These groups are important because they guide the selection of cultivar combinations used in orchard planting and provide important gene markers for pedigree studies. CIG groups have now been identified for all major California cultivars (Barckley et al. 2006). Self-fruitfulness refers to the ability of a plant to be consistently fertilized from self-pollen. This requires a combination of both self-compatibility and successful self-pollination. Different degrees of self-compatibility and self-fruitfulness exist (Gradziel et al. 2002).

3. Fruit Development. Following fertilization, the growth of the fruit, seed, and embryo follow the typical three stages of development in which the pericarp, seed, and nucellus develop during stage I, the endosperm and embryo enlarge during stage II, and the dry weight of the embryo increases during stage III. Time of nut maturity is an important commercial trait. Physiological processes, which accompany almond fruit ripening, include dehiscence of the hull or mesocarp, hull-split, fruit abscission, and dehydration or the loss of moisture in the hull and nut. The entire process of hull and nut maturation and drying may require 2–6 weeks to complete. Usually, maturity is most reliably characterized by the initiation and progress of hull splitting. The dates for the initiation of 5%–10% splitting and the completion of splitting are useful criteria when comparison is made to standard cultivars. Moisture stress can accelerate hull splitting but adequate water is required for the hull to ripen properly. If splitting begins prematurely and the nut dries too rapidly, the hull may close tightly on the shell, becoming difficult to remove. In California, the pattern of maturation across the range of almond genotypes extends from early August to late October. Following hull-split, the almond hull and kernel rapidly desiccate to below 7% moisture where mold contamination is effectively suppressed.

III. GENETIC DIVERSITY

A. Biodiversity and Ecosystems

Wild populations of almond species representing a wide range of morphological and geographical forms have evolved throughout central and south-western Asia (Fig. 4.2 and Table 4.3). Some of the more than 30 species described by botanists may represent subspecies or ecotypes within a broad collection of genotypes adapted to the range of ecological niches in the deserts, steppes, and mountains of central Asia (Grasselly 1972). Browicz (1969) separated almond species into two subgroups: Amygdalus (leaves conduplicate in bud and 20–30 or more stamens) and Dodecandra (leaves convolvulate in bud and fewer than 17 stamens). The most north-easterly germplasm is located in western China and Mongolia and includes *P. mongolica*, *P. pedunulata*, and *P. tangutica* (*P. dehiscens*). The remainder occupy a more or less contiguous area in west central Asia. Almonds in the most northern area include species in Section Chamaeamygdalus and extend from the Balkan Peninsula to the Altai Mountains. The most southern and xerophytic germplasm includes species in the Spartiodes Section, which can have leafless slender shoots at maturity, and the Lyciodes (Dodecandra) Section, which are very dwarfed and thorny. Species in the Section

Euamygdalus resembles cultivated almonds and include many species extending from central Asia to southern Europe (Fig. 4.2) as well as the peaches *P. persica*, *P. mira*, and *P. davidiana*.

Rehder (1940) has placed all wild and domesticated almond species in the genus *Prunus*. *P. dulcis* (cultivated almond), *P. persica* (cultivated peach), *P. mira*, *P. argentea* (*A. orientalis*), *P. bucharica*, and *P. fenzliana*, are placed in the subgenus *Amygdalus*, section Euamygdalus, *P. scoparia* is placed in the section Spartiodes, and *P. webbii* placed in the section Lycioides (Table 4.3). However, many Mediterranean and Central Asian researchers prefer the classification of Browicz and Zohary

Table 4.3. Botanical relationship of *Prunus* species in subgenus *Amygdalus*.

Almond group
Section Euamygdalus Spach
    Prunus dulcis (Miller) D.A. Webb
    P. bucharica Korshinsky
    P. communis (L) Archangeli
    P. fenzliana Fritsch
    P. kuramica Korchinsky
    P. orientalis (Mill.), syn. *P. argentea* (Lam)
    P. kotschyi [Boissier and Hohenm (Nab.) and Rehd.]
    P. korschinskii Hand-Mazz.
    P. webbii (Spach) Vieh.
    P. zabulica Serifimov
    P. tangutica Batal.(syn. *P. dehiscens*) Koehne
Section Spartioides Spach
    P. scoparia Spach
    P. spartioides Spach
    P. arabica Olivier
    P. glauca Browicz
Section Lycioides Spach
    P. spinosissima Franchet
    P. turcomanica Lincz.
    P. lycioides Spach
    P. eburnean Spach
    P. brahuica Boiss
    P. erioclada Bornm.
Section Chameamygdalus Spach
    P. nana (Stock)
    P. ledebouriana Schle.
    P. petunnikowi Lits.
    P. georgica Desf.

Peach group
    P. persica (L.) Batsch
    P. mira Koehne
    P. davidiana (Carriere) Fransch

(1996) where *P. persica*, *P. mira*, and *P. davidiana* are in the genus *Prunus* while almond and the other almond-like species discussed here are placed in the genus *Amygdalus*. While acknowledging the easy hybridization between almonds and peaches and also the high level of synteny between these genomes, these researchers argue that the divergent evolution of almonds in the harsh climates of central Asia vs. peach in the temperate to subtropical climates of south-eastern China have led to dramatically different growth and development patterns and that from an ecological and even taxonomic perspective, these wide divergences suggest their placement in separate genera. Almond and peach thus represent a unique situation in crop plant genetics, where very similar genomes are expressed as very different plant forms. This genome–phenome disjunction may prove useful for the elucidation and eventually manipulation of epigenetic mechanisms that are now recognized to have profound effects on fundamental plant developmental pathways thus affecting final form and function.

B. Conservation

Cultivated almonds show high levels of genetic variability due to their interspecies-origin and because their self-sterility makes them obligate out-crossers. Commercial cultivars within individual production areas, however, often show a limited genetic base due to their derivation from only a few founder genotypes selected for their desirable regional value (Felipe et al. 1992). For example, most commercially important California cultivars originated from crosses between only two parents: 'Nonpareil' and 'Mission' (Bartolozzi et al. 1998; Hauagge et al. 1987; Kester and Gradziel 1996). In addition, only a limited number of almond accessions are currently maintained in the main germplasm collections worldwide (FAO 2009) (Table 4.4).

Several early reports have documented transfer of commercially useful genes from related wild almond species through either natural or controlled crosses (Denisov 1988; Gradziel and Kester 1998; Socias i Company and Felipe 1988, 1992; Ritcher 1969; Grasselly 1972; Felipe 2000). More recently Denisov (1988), Kester et al. (1991), and Socias i Company (1990) have pioneered the use of wild species germplasm to create improved almond cultivars. The historical use of these species and their hybrids as almond rootstocks facilitated subsequent introgressions. The use of wild species directly as a rootstock for dryland almond has been widely reported. For example, Atlı (2008) reported that *A. orientalis* (*P. argentea*) could be utilized as a rootstock for cultivated almond without any incompatibility problem.

Table 4.4. Main almond germplasm banks in the world.

Country	Research center	City	Accessions
Europe			
France	INRA-Avignon	Avignon	180
Italy	University of Udine	Udine	65
Spain	CEBAS-CSIC	Murcia	70
	CITA	Zaragoza	80
	IRTA-Mas Bove	Reus	83
Asia			
Iran	National Plant Gene Bank	Karaj	67
Syria	Centre for Studies of Arid Zones	Sednaya	130
Turkey	Ege University	Izmir	51
America			
USA	USDA-UC-Davis, California	Davis	165
Africa			
Morocco	INRA-Rabat	Rabat	120
Oceania			
Australia	University of Adelaide	Adelaide	45

Source: FAO (2009).

The researchers also reported that while 'Nonpareil' grafted on 'Texas' gave an accumulative yield of $83.8\,\text{kg}\,\text{da}^{-1}$ at the end of the 4-year period, 'Nonpareil' grafted on *A. orientalis* gave $351.1\,\text{kg}\,\text{da}^{-1}$ for the same time interval. The researchers also reported evidence for a high level of spontaneous interpecific hybridization in the wild between species with overlapping ranges, including *P. spartioides* in Iran, *P. bucharica*, and *P. fenzliana* in Russia, *P. webbii* in Turkey, and *P. fenzliana*, *P. bucharica*, *P. kuramica*, *P. argentea*, *P dehiscens*, and *P. kotschyi* at lower incidence in these (Fig. 4.2) and nearby areas (Gradziel et al. 2001; Denisov 1988; Grasselly 1972; Rickter 1969). Interspecific crosses between related species (mainly *P. persica* × *P. dulcis* but also *P. webbii* × *P. dulcis*) have been used for almond rootstock breeding in France, the United States, Spain, and Yugoslavia (Gradziel et al. 2001; Denisov 1988; Grasselly 1972; Rickter 1969).

C. Cytogenetics

The chromosome number of *P. dulcis* as well as *P. fenzliana*, *P. nana* (*P. tenella*), *P. bucharica*, *P. kotschyi*, and *P. scoparia* is $2n = 16$, which is the same as peach *P. persica* (Kester et al. 1991). Martínez-Gómez et al. (2005) presented an improved technique for counting chromosomes in almond by slide preparation and staining. In this protocol, they used treated almond root tips with 2% colchicine for 3 h at 5°C. For the fixation and hydrolyzation steps, they used mixture of methanol, propionic acid and chloroform (6:3:2), stored in 70% ethanol and 1 N HCl respectively, followed by staining with acetic acid orcein (45%). They confirmed the number of chromosomes of $2n = 2x = 16$ with most chromosomes being symmetrical and metacentric. Martínez-Gómez and Gradziel (2003) used a similar protocol to show an aneuploidy chromosome complement $(2n - 1 = 15)$ for stunted seedlings compared to normal growth with diploid chromosome complement $(2n = 16)$. Yousefzadeh et al. (2010) presented a cytological study of different wild almond species and confirmed the number of chromosomes, chromosome type, number of satellites. The number of chromosomes for all species was $2n = 2x = 16$. Chromosomes were mostly metacentric, one to two were submetacentric, one telocentric, and one with a satellite.

D. Interspecific Hybridization in Breeding

Greater genetic variability and so increased breeding options for desired traits such as self-fertility and disease resistance are currently being pursued through the incorporation of breeding material from other regions (Kester and Gradziel 1996; Socias i Company 1998; Martínez-Gómez et al. 2003). Because of the possible interspecies origin of many of these cultivars (Kester et al. 1991; Ladizinsky 1999; Socias i Company 2002), improvement of certain traits may have also benefitted from the introduction of genes indirectly from related species. Hybridization between *P. dulcis* and other almond species has often taken place naturally wherever different species come into contact. *P. webbii* grows throughout the Mediterranean region and its range intersects with cultivated almond in Italy (Sicily), Spain, and Greece. Natural hybridization has occurred and introgression evidently results since in the Apulia region of Italia, *P. webbii* has been shown to be the source of self-fertility (Godini 2002).

Controlled crosses of *P. dulcis* with other almond species in Sections Euamygdalus and Spartiodes are readily achieved (Gradziel et al. 2001b; Gradziel 2001). Hybridization with Section Lycioides is possible though somewhat more difficult and even more difficult with Chameamydalus. Despite their physical and developmental differences, crosses with cultivated peach (*P. persica*) and wild relatives *P. mira* and *P. davidiana* can be easily accomplished and have proven to be particularly valuable as rootstocks as well as sources of commercially useful traits (Gradziel et al. 2001; Gradziel 2001).

IV. GENETIC IMPROVEMENT

Breeding goals can be divided into two basic categories: genetic improvement and cultivar development. Genetic improvement typically has a well-defined, focused objective such as improved disease resistance within locally adapted genetic background. In contrast, success at cultivar development is indicated by sizable commercial plantings over the long production time required for commercial profitability. For example, a successful California almond cultivar is expected to have an average annual kernel production of over $7\,t\,ha^{-1}$ and an orchard-life expectancy of at least 20 years in order to be commercially viable. Cultivar success, then, is rarely determined by superior performance in one or a few traits, but rather is determined by the absence of deficiencies for the large number of fruit and tree characteristics required for commercial viability (Gradziel 2012). The need in almond crops to simultaneously optimize a large number of essential traits remains the greatest challenge to breeding strategies including the use of MAS and other molecular-based techniques.

A. Breeding Objectives

Breeding objectives typically fall in three general areas: increase yield, improve market quality, and decrease production costs. Because almond is self-sterile, insufficient cross-pollination is frequently a major determinant of commercial yield (Asai et al. 1996; Micke 1994). Self-fruitfulness results from the combination of self-compatibility (i.e., self-pollen shows compatible growth to fertilization on pistils of its own flower) and autogamy (i.e., a flower structure promoting consistent self-pollination). Autogamy appears to be controlled by a number of genes (Kester and Gradziel 1996) affecting flower structure as well as the more dynamic aspects of the flowering process including timing of anther dehiscence (Gradziel and Weinbaum 1999) and pattern of stigma growth relative to maturing anthers (Godini 2002). Although highly autogamous selections have been identified, the genetic manipulation of this trait remains uncertain. Self-compatibility, as with self-incompatibility, is controlled by a major gene (Dicenta and García 1993b), though modifier genes also play important roles (Gradziel et al. 2002; Socias i Company et al. 2005). While many almond species demonstrate some level of self-compatibility, in a cultivated almond background only the self-compatible genes from *P. mira*, *P. persica*, and *P. webbii* have resulted in fruit set above the 30% considered desirable for commercial production (Gradziel et al. 2001b). Breeding populations developed from interspecies crosses segregate for self-compatibility in the expected Mendelian ratios for a single gene (Dicenta and García 1993a; Socias i Company and Felipe 1988, 1992; Gradziel et al. 2001). *P. mira*, the species that, when crossed to almond shows the highest selfing percentages following introgression of the self-compatibility gene, also shows high levels of self-pollination (Gradziel et al. 2001). Long-term efforts to breed self-compatible almonds have been reviewed by Socias i Company (1990).

Almond quality has also become an important goal for breeding (Socias i Company et al. 2008), despite the difficulties in defining a kernel quality ideotype because of the sizable differences in consumer preferences (Janick 2005). Kernel quality must consider not only the chemical composition conferring a specific organoleptic quality, but also physical traits affecting use. Thus, a different type of shell is preferred depending on the customs of individual regions, being hard in most Mediterranean countries and soft in California and regions with a similar growing system. The chemical composition of almond kernels also represents an evolving goal for breeding because of the beneficial aspects of almond on human health. Such benefits appear related to the antioxidant compounds of almond kernels, the high content of oleic acid among the fatty acids and the desirable nutrient and fiber content. Although these aspects have not yet been routinely incorporated into the new releases, they are receiving increasing attention not only among almond breeders, but also among growers, processors, and consumers. A comprehensive review of almond quality components as well as their heritability, when known, has recently been summarized by Socias i Company et al. (2009).

Decreased production costs are being pursued though improved genetic resistance to economically important diseases and pests in both scions and rootstocks. The most serious foliage diseases of almond include shothole caused by *Stigmina carpophila* (syn. *Coryneum beijerinkii*), travelure (*Fusicladium amygdali*), polystigma (*P. occhraceum*) fusiccocum (*Fusicocum amygdali*), and anthracnose (*Gloeosporium amygdalinum* and *Colletotrichum acutatum*). Relative susceptibilities of important cultivars in different countries have been determined and potential sources of resistance have been identified (Kester et al. 1991). Blossom and twig blight, the major crop limiting fungal disease worldwide, is caused by *Monilinia laxa* and *Monilinia cinerea*. These fungi attack the flowers and are most serious in years when rain occurs during bloom. Other fungi, including *Botrytis cinerea* can also be a serious problem under these conditions. Aflatoxin producing *Aspergillus flavus* infections of the kernel is a major problem, particularly where insect damage is common (Dicenta et al. 2003; Gradziel and Kester 1994;

Gradziel et al. 2000; Gradziel and Wang 1994). Although disease control has been possible through fungicides, the need to consider natural resistance becomes more important with the continued loss of agrochemicals. Almonds can also be infected by the same range of viruses as other *Prunus* including the ALAR viruses (ringspot, prune dwarf, line pattern, calico, and apple mosaic) and NEPO viruses (tomato black ring, tomato ring spot, and yellow bud mosaic). Leaf and flower mosaic phenotypes can result from the combination of several viruses. Several complexes of virus-like disorders that produce "stem pitting" and "graft union brown-line" are known but not well understood (Uyemoto and Scott 1992). Many cultivars of almond, however, appear to be immune to the Dideron isolate of plum poxvirus that remains a serious problem for most other stone fruits (Martínez-Gómez et al. 2004).

In California, navel orangeworm (*Paramyelois transitella*) and peach tree borer (*Anarsia lineata*) can cause serious damage to nuts at harvest (Rice et al. 1996). This problem is related to the vulnerability of soft, paper-shell, and poorly sealed sutures common to California cultivars, including 'Nonpareil,' 'Ne Plus Ultra,' 'Winters,' and 'Merced' (Gradziel and Martínez-Gómez 2002). Partial control is achieved through integrated pest management, particularly orchard sanitation (IPM Manual Group of University of California at Davis 1985). Resistance through better-sealed shells has been observed in some cultivars including 'Carmel', 'Mission,' and 'Butte' (Hamby et al. 2011). This problem is not serious in the Mediterranean area because of the characteristic well-sealed, very hard, and thick shells of the major cultivars.

Mite species, including Pacific spider mite (*Tetranychus pacificus*), two-spotted spider mite (*T. urticae*), European red mite (*Panonycus ulmi*), and brown almond mite (*Bryobia rubriculus*) can adversely affect production and may be locally important, particularly in conditions of moisture stress. Variation in susceptibility exists among different cultivars. The almond wasp (*Eurytoma amygdali*) is an important pest from the Middle East extending into Greece. It attacks the young developing nut. Other significant Mediterranean pests that attack the trunk and branches of trees include *Scolytus amygdali* and *Capnodis tenebrionis*. *Capnodis* is a species of borer that attacks the trunk of trees in the Mediterranean basin, particularly trees that are under stress.

B. Early Genetic Studies

Heritability estimates for almond have been reviewed by Kester and Gradziel (1996), Dicenta et al. (1993a,b), and Socias i Company et al. (2007) and summarized in Table 4.5. Traditional breeding, by necessity,

Table 4.5. Estimates of broad sense heritabilities for important almond traits.

Trait	Heritability
Physiological traits	
Blooming time	0.67
Blooming duration	0.20
Blooming intensity	0.54
Leafing time	0.83
Ripening season	0.69
Production intensity	0.45
Morphological traits	
Bud density	0.30
Branching habit	0.19
Nut and shell traits	
Weight	0.81
Length	0.5
Width	0.37
Thickness	0.28
Width/length ratio	0.46
Thickness/length ratio	0.53
Thickness/width ratio	0.3
Hull dehiscence	0.02
Hull pubescence	0.28
Retention of outer shell	0.34
In-shell weight	0.81
Shell color	0.05
Shell hardness	0.55
Shell thickness	0.51
Shell seal	0.14
Shell type	0.55
Width of shell opening	0.21
Kernel traits	
Weight	0.64
Length	0.77
Width	0.62
Thickness	0.71
Width/length ratio	0.46
Thickness/length ratio	0.43
Thickness/width ratio	0.21
Double kernels	0.51
Kernel color	0.42
Kernel crease	0.79
Worm damage	0.3

Source: Modified from Kester et al. (1977) and Socias i Company et al. (2011).

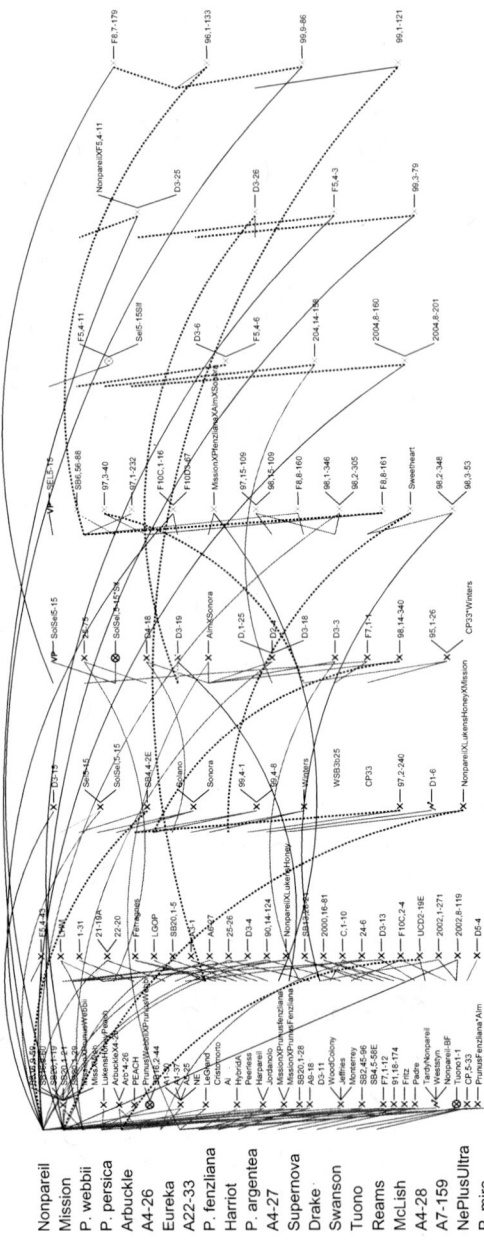

Fig. 4.3. Flowchart showing the lineages of advanced selections and parents currently used in breeding for self-compatibility and improved productivity, including disease and pest resistance. Sources of self-compatibility have been independently transferred from the almond variety 'LeGrand', the Italian variety 'Tuono', the induced-mutation "Supernova" and related species including *Prunus webbii*, *P. mira*, and *P. persica* (peach). Additional sources of disease resistance and improved kernel and tree quality have been transferred from heirloom almond varieties, *P. fenzliana*, *P. persica*, and *P. argentea*. Several advanced selections have incorporated traits such as self-compatibility from multiple sources as breeding experience has shown improved performance and improved stability over years and locations when multiple, diverse sources were combined. Similar results have been found for disease and pest resistance.

230

targeted those alleles whose heritability is large enough to be differentiated from background environmental variance. As new germplasm is incorporated into the breeding program, however, new genes and genetic relationships are introduced that can change final heritability values. With the extensive new germplasm incorporated into almond breeding programs over the past two decades (Figs. 4.3 and 4.4) novel and often exotic genes have been introgressed to improve important

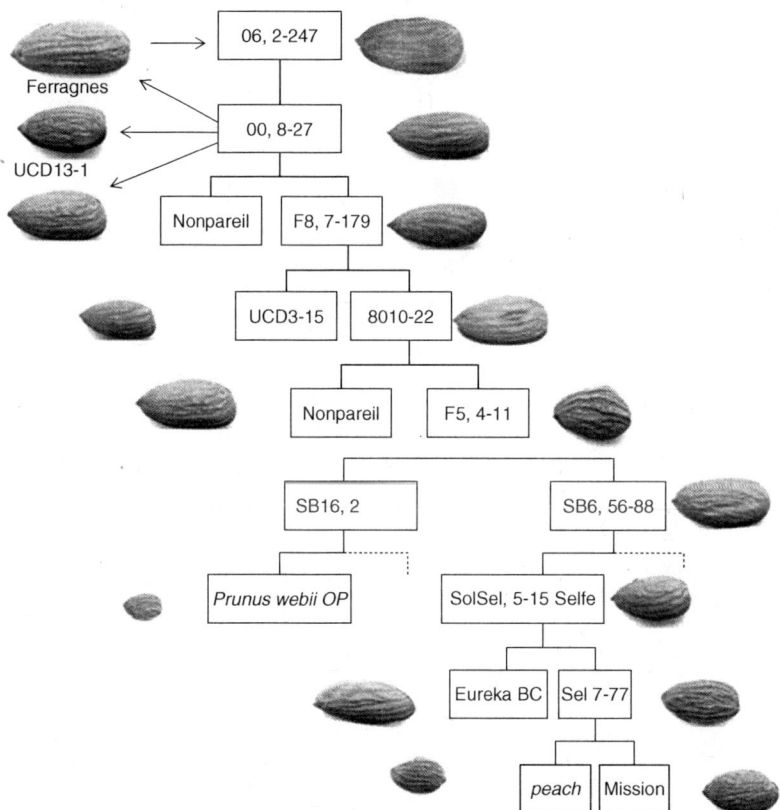

Fig. 4.4. Lineage (pollen parent to right; seed parent to left) showing transfer of self-compatibility from peach and the wild almond *Prunus webbii* to advanced Californian breeding selections. Early breeding efforts involved a complex series of crosses to transfer traits (self-compatibility and hull rot resistance) from wild relatives to cultivated almonds. A series of recurrent backcrosses to 'Nonpareil'-type almonds then transferred desired traits to a California adapted background. More recent crosses target high commercial quality with maximum productivity and high genetic variability.

traits and previously established heritability values may no longer be accurate and need to be re-established on a case-by-case basis.

Genetic studies have demonstrated a positive response to selection for late flowering (Kester et al. 1977; Dicenta et al. 1993a; Socias i Company et al. 1999). Flowering density and productivity are also two important traits, which have been studied by Kester and Asay (1975), Grasselly and Crossa-Raynaud (1980), Vargas and Romero (1984) and Dicenta et al. (1993a). A few studies have been performed regarding the time of maturity (Kester and Asay 1975; Dicenta et al. 1993b). Dicenta et al. (2005) and García-Gusano et al. (2010) correlated the chilling requirements for flower breaking dormancy with the chilling requirements for seed breaking dormancy. These authors indicated the possibilities of early selection of late-flowering almonds as a function of seed germination or leaf emergence time of seedlings. There are many studies regarding the transmission of kernel traits (Kester et al. 1977; Vargas and Romero 1984; Dicenta et al. 1993b). In addition, kernel bitterness has been characterized as a monogenic trait, the bitter genotype being recessive (Dicenta and García 1993b; Vargas et al. 2001) though heritability appears more complex in almond by peach breeding lines (Kester and Gradziel 1996). Finally, self-compatibility was studied by different authors who determined its monogenic nature with a multi-allelic S series, and identified the S_f allele as responsible for self-compatibility (Socias i Company and Felipe 1988; Dicenta and García 1993a; Ortega and Dicenta 2003). Self-compatibility is expressed within the styles of flowers and results in the successful growth to fertilization of self-pollen tubes (Bošković et al. 1997, 2003). Multiple modifier genes also exist that under certain environments can confer relatively high self-compatibility levels in the absence of the S_f allele (Gradziel 2009).

Many traits are not always amenable to genetic analysis as they are affected by changing environmental conditions and/or they may only be visible in adults and so requiring a long time for their analysis. Molecular markers provide a possible solution to some of these problems and have already allowed fast, accurate, highly discriminative, and environmentally stable tests useful for variability analysis, pedigree determinations, or cultivar identification (Dangl et al. 2009). Moreover, some markers, such as isozymes, restriction fragment length polymorphisms (RFLPs), single-sequence repeats (SSRs) derived of the knowledge of specific genome sequences, allow the comparison of the variability among homologous regions of the same or different species (Wünsch and Hormaza 2002; Martínez-Gómez et al. 2003). Appropriate molecular markers combined with advanced statistical analysis techniques also offer the opportunity for more accurate discrimination between exotic and more traditional genes resulting in the opportunity for more efficient gene introgression. MAS has been particularly successful in the genetic improvement of selfpollinating crops where genes act primarily in an additive manner since most advanced selections have been inbred to near homozygosity. In out-crossed crops such as almond, however, high levels of heterozygosity exist (Sorkheh et al. 2009a,b). Additional and often exploitable genetic contributions result from interactions within individual loci (dominance), among different loci (epistasis and other genetic interactions), and even between genomes (as in the interspecies hybrid vigour of hybrid rootstocks) (Gradziel 2009; Socias i Company 2008). These interactions, however, are less amenable to molecular marker analysis and manipulation. A more detailed description of the use of these molecular markers is presented in the Section V.

C. History of Crop Improvement

In the early 1900s, formal plant breeding programs were established in most major production areas to accelerate the selective process through controlled crosses and related genetic manipulations. While many goals, such as improved yield and production efficiency were similar among programs, regional breeding goals often varied due to different environments and local disease and pest problems. At the same time, the globalization of the almond market imposed more stringent limits on acceptable kernel and shell characteristics. Despite inherent obstacles to rapid genetic improvement, including large plant size and the long seed-to-seed generation period of four years or more, many commercially successful cultivars have resulted from such controlled crossing programs in the last decades. Examples include the cultivars 'Ferragnes', 'Ferraduel', and 'Ferrastar' from France; 'Butte', 'Ruby', 'Sonora', 'Padre', and 'Winters' from California; and 'Guara' from Spain. Regional almond breeding programs and their primary objectives have been reviewed by Kester and Gradziel (1996). In addition to the goal of improved adaptation to local environments, cultivar improvement programs must also address changing market demands. Since both market requirements and regional adaptation placed considerable limits on the final genetic make-up, most breeding programs pursue the incremental improvement of locally established varieties, typically by the sequential addition of new genetic value (disease resistance, nut quality, maturity time, and productivity) (see Fig. 4.4). Traits such as increased yields, improved quality, and those contributing to decreased production costs have been found to be largely inherited in a quantitative manner (Kester et al. 1977; Spiegel-Roy and Kochba 1981) with a few exceptions (such as self-compatibility and kernel bitterness). Consequently, the efficiency of breeding programs depends on the information available on the transmission of traits targeted for improvement. In addition, developing new almond cultivars is a long and tedious process involving the generation of large population of seedlings from which the best genotypes are selected.

D. Selection

While recurrent mass selection and synthetics have been utilized in European and Asian breeding programs in the early to mid-1900s for low-input, low-output almond production (Gradziel 2011), virtually all modern almond breeding programs employ hybridizations followed by clonal fixation.

Cloning can capture even highly complex and poorly understood genetic interactions making it arguably the most efficient breeding technique for combining, in true-breeding cultivars, the fullest range of desirable genetic, epigenetic, and genomic interactions. This capacity makes cloning particularly promising for the characterization and eventual manipulation of these largely underutilized interactions. For example, clone analysis offers unique opportunities for the study of epigenetic interactions since different and often heritable phenotypes (juvenility, imprinting, and gene-silencing) of the same clone (genotype) in the same environment would be the expression of epigenetic rather than genetic factors. For example, noninfectious bud-failure in almond appears to be an epigenetic-like clonal aging condition where the genetic (DNA) composition of affected cultivars appears unchanged but where gene activity is altered in a heritable manner (Kester et al. 2004). Although it is a major production problem in almond, it appears a poor candidate for MAS since the DNA appears identical in both affected and unaffected genotypes (Dangl et al. 2009). Similarly, genome–genome interactions that appear to play important roles in enhancing vegetative vigor as characterized by interspecies hybrid rootstocks (Armstrong 1957; Rom and Carlson 1987) appear be the result of both genetic and genomic differences between the parents, possibly including differences in chromosome orientations, scaffold structure, histone composition, methylation patterns, and synteny differences (Grant-Downton and Dickenson 2006). Thus, the level of genetic gain from the hybridization/cloning approach is limited only by the quality and diversity of the breeding parents and the size of the progeny population. Cloning of interspecies hybrids has been shown to be very effective for the breeding of vigorous and often disease resistant rootstocks for almond such as the 'Hansen' and 'Nickels' peach by almond hybrids (Gradziel 2009). Vegetative growth vigor in interspecies hybrids, which is sometimes termed "luxuriance" to distinguish it from intraspecies hybrid vigor or heterosis, can often transgress well beyond that of even a highly-vigorous parent, and appears to involve both gene–gene and genome–genome interaction. Although providing valuable tools for a more thorough dissection/characterization of these crop improvement opportunities, molecular–genetic analysis, as currently employed, may ultimately hinder breeder utilization of these germplasm resources because of its very specialized and inherently reductionistic focus.

An additional advantage of clone-based breeding methods is the ability to accumulate desirable mutations. Naturally occurring mutations are often identified as bud sports (novel phenotypes originating from a single bud) that, while typically rare, become increasingly likely with larger planting size and time periods. Desirable mutations in an established cultivar have the advantage of providing a discrete improvement in an otherwise well-established genotype (i.e., a cultivar whose cultural management and marketing has already been well worked out), making them very desirable. An example of a beneficial bud sport is 'Tardy-Nonpareil', which flowers approximately 10 days after standard 'Nonpareil' and so has greater frost/disease avoidance. For several reported almond bud sports, however, molecular analysis has shown the actual origin was from sexual crosses (Dangl et al. 2009). An example is 'Carmel', which has been reported to be a bud sport of 'Nonpareil'. 'Carmel' DNA fingerprint, however, does not match that of 'Nonpareil' but is consistent with it being a progeny between 'Nonpareil' and 'Mission'. In the early to mid-1900s in California, almond scion cultivars were often grafted onto the lower market value and greater disease resistant 'Mission' seedling rootstocks (which were usually pollinated by 'Nonpareil' as it initiated flowering approximately a week earlier). In a small number of propagations, the scion bud failed and a rootstock bud grew instead. Where the rootstock phenotype was clearly different, it was recognized as an 'escape' and rebudded. But where it is similar enough, it was frequently mistaken as a bud sport, which, if of sufficiently good quality, was propagated as a new cultivar.

E. Cultivar Development

The definitive aim of plant breeding is the development of successful cultivars. A successful cultivar can be conveniently defined as providing a net improvement over the cultivar to be replaced. That is, it must be at least as good as the cultivar it is to replace in the areas of horticulture, quality, disease/pest resistance, market, yet possess improvements valuable enough to result in sizable commercial plantings. While powerful molecular-based strategies are becoming available for genetic improvement, the major barrier to successful cultivar development is not the process of genetic improvement but rather the process of simultaneously maintaining commercial quality for the wide range of other essential traits. This is the reason bud sport mutations have been a valuable source of new cultivars since they can confer a distinct improvement to an otherwise genetically unreshuffled, commercially proven cultivar. A well-established dogma of tree fruit breeding is that the success of a new cultivar is determined not by its exceptional performance in specific areas but rather a uniformly superior performance across a broad range of characteristics or traits. Consequently, it is the absence of serious deficiencies that will ultimately determine commercial success of a new variety. This is particularly relevant in tree crops were orchards are expected to be productive for 20 years or more in order to be commercially viable, and where failed cultivars cannot be readily ploughed under and replanted as with cereal and vegetable crops. The ecologist and author Jared Diamond (1998) has termed this decisive vulnerability to a broad spectrum of potential deficiencies the "*Anna Karenina*" effect based on Leo Tolstoy's classic opening sentence in his novel of that name: "*All happy families are alike; each unhappy family is unhappy in its own way.*" In addition to kernel quality, good performance is required for numerous traits in a broad range of essential categories, including tree structure, productivity and longevity, disease and insect resistances, harvest time, uniformity and ease-of-harvest, precocity, freedom from alternate bearing, postharvest performance, rootstock compatibility, market type, and consumer preference. Thus, while genetic improvement may benefit from a focused, reductionist approach to trait improvement, successful cultivar development requires the simultaneous, holistic manipulation of a large number of essential traits. A traditional additive-gene-based MAS approach would quickly become overwhelmed by the number of required markers. This incongruity, while complicating cultivar development may also be undermining future breeding progress. Genetic improvement strategies, including MAS, are becoming increasingly efficient at the partitioning and so manipulating the principal additive genetic interactions affecting the target trait, but because they are resource intensive, these inherently reductionist approaches may lead to reduced effectiveness of more traditional tree cultivar development if not fully complemented with the equally essential holistic cultivar development approaches.

Almond and other tree crops are also unique in that successful cultivars need to demonstrate this superior fitness over a much broader range of environmental variation (position on tree, planting site, age of tree, age of clone, varying disease and insect pressures, changes in climate and weather pattern). Superior fitness over a range of environments is ultimately more important to final cultivar productivity than exceptional performance within a narrow environmental niche (Gradziel 2012). Tree crop cultivars thus resemble the clonal colony or "genet" of ecology such as the "Pando" clone of Quaking Aspen (*Populus tremuloides*) in the Wasatch Mountains of Utah, USA, which cover 43 ha and so is often considered the world's largest organism by mass. Since the estimated 47,000 individual clonal trees ("ramets" in ecological terms) that constitute the *Pando* clone have developed over a wide range of differing ecological niches, its competitive advantage appears to result from a broad adaptability rather than being highly adapted to a specific niche. In comparison, the 'Nonpareil' almond clonal cultivar is planted on over 116,000 ha (at over 48 trees ha^{-1}) in California alone, with additional plantings in Europe, Asia, North Africa, South America, and Australia. Although there are other cultivars that out-yield or have higher market value than 'Nonpareil' in certain production areas and periods (Connell et al. 2010), 'Nonpareil' continues to dominate this crop because of a superior overall fitness (i.e., economic returns over the typical orchard life of 20+ years) with over 5.6 million trees currently planted over an ecologically diverse 700 km stretch of central California ranging from Redding at latitude 40°34′36″N to Bakersfield at 35°22′24″N.

Improved environmental buffering has been shown to be associated with the higher genetic heterozygosity typical of almond cultivars (Socias i Company and Felipe 1988; Dicenta and García 1993a; Ortega and Dicenta 2003). Extended periods of selection for broad environmental adaptability would occur (for many European and Asian almond cultivars, selection has been occurring for hundreds to thousands of years) and would thus identify rare, elite selections where the maximum potential of additive, dominance, epistatic, genomic, and epigenetic interactions was combined. Clonal propagation allows the capture of these rare elite genotypes for future plantings as well as future genetic improvements through bud sport mutation or further, albeit rare, favorable recombinations.

In addition, bearing in almond is primarily on perennial spurs (Sorkheh et al. 2010). Spur production, with its highly efficient fruit-to-vegetative ratio (harvest index) often dominates crop production in mature, highly productive orchards. In almond, these spur-based production-units have also been shown to be fairly autonomous and competitive, that is carbohydrate flow is primarily from nearby leaves to the more competitive local sink, whether individual developing nut or vegetative apices (Grossman and DeJong 1998). While the clone (genet) or individual tree (ramet) may be the target of natural selection of wild clonal colonies, the fruiting-unit (compet) may prove as important or more important in the more synthetic and often more intensive crop breeding selection. Recent research in almond has shown that the most productive orchards are those where the quantity (number), quality (fecundity), and longevity of spurs are optimized during the multiyear, peak production phase of the planting (Gradziel and Lampinen 2011).

Productivity remains the most important attribute in new almond cultivars but because of its complexity and all-inclusive nature is often managed as a nebulous quantitative trait, which frustrates a more thorough analysis and manipulation by both traditional and molecular approaches. Molecular approaches such as association mapping, offer unprecedented opportunities to more fully characterize important components of yield as a basis for future genetic and cultural manipulation but require a more detailed understanding of the biological basis (Sorkheh et al. 2010). It is informative how biotechnology progress over the last two decades has advanced to the point where sequencing individual almond breeding lines can now be readily achieved, yet our understanding of the physiological and developmental components of a trait as critical as yield has made only rudimentary progress over the same time period. This precarious biological knowledge-base, along with the traditionally insular nature of molecular genetic analysis remains a major impediment to more efficient cultivar breeding in tree nut crops.

V. MOLECULAR APPROACHES

Estimates of the genome size of *Prunus* ranges from 290 million to 309 million base pairs (being ~294 million base pairs in almond) (Arumuganathan and Earle 1991). The recent sequencing of the peach genome by Sosinski et al. (2010) indicated a size of 227 million base pairs for this species. This genome size is relatively small in the plant kingdom, which ranges from the 125 million base pairs for *Arabidopsis thaliana* to 5 billion base pairs for maize and 50 billion base pairs for *Lilium* spp. (lilies).

The most important molecular markers used in almond studies are isozymes, restriction fragment length polymorphisms (RFLPs), randomly amplified polymorphic DNA (RAPDs), simple the sequence repeats (SSRs), and markers based on unique DNA sequences. Isozyme analysis was one of the first molecular marker evaluations available to almond studies and offered codominant expression and good reproducibility but were limited by the small number of loci, which could be analyzed by conventional staining methods, as well as a low genetic variation at most loci. Nonetheless, it was isozyme studies that first documented extensive genetic variability in almonds overall, as well as the limited genetic base of many almond breeding programs (Hauagge et al. 1987; Arulsekar et al. 1989; Vezvaei et al. 1995). RFLPs are also codominant but can detect a virtually unlimited number of markers. RFLPs have been used for discovering linkages between markers, for constructing genetic maps, for cultivar identification and for the characterization of genetic variability. RAPDs based on PCR amplification of arbitrary primers have been useful for characterizing germplasm variability (Bartolozzi et al. 1998; Martins et al. 2003) but had limited application for cultivar identification and map construction since they are dominant markers with occasional difficulties with repeatability.

Presently, SSR or microsatellite markers, which are also based on PCR amplification, have proven more useful for characterizing genetic relationships (Martínez-Gómez et al. 2003a), cultivar identification (Martínez-Gómez et al. 2003b; Martins et al. 2003), and map construction (Dirlewanger et al. 2004) due to their high polymorphism, codominant inheritance, abundance, and the frequent successful amplification of SSR markers developed in related species (Martínez-Gómez et al. 2007). Currently, the use of single nucleotide polymorphisms (SNPs) combined with SSR in the characterization of almond genotypes and the elaboration of genetic maps is also being evaluated (Tavassolian et al. 2010).

A. Almond Genomics

In almond, genomic resources are limited and subordinate to the development in other more important *Prunus* species such as peach. In almond, as in peach, the more recent markers being used are those associated with agronomic traits and based on either cDNA sequences (expressed sequences tags, ESTs) or databases (cloned gene analogs, CGAs) or single point mutations (SNPs).

EST analysis has provided a preliminary picture of the numerous almond genes involved in fruit and tree development and it also has provided an extensive resource for gene cloning and genetic mapping. A recent collection of ESTs from different *Prunus* (mainly peach but also almond) based on cDNA libraries has been released to public databases, and more than 83,751 putative unigenes have been detected in the Genome Database for Rosaceae (GDR) (http://www.bioinfo.wsu.edu/gdr/). In this database, a comparison of *Prunus* maps is performed together with the development of ESTs. In addition, current studies on mapping EST, physical mapping and transcriptomics are being developed (Zhebentyayeva et al. 2008). This database has also been a valuable resource in genomic and genetic research in peach. SNPs are the most frequent molecular marker in the genome having a codominant inheritance. Unigene contigs have been pursued for SNPs in the GDR database, with a total of 5,284 SNPs and a frequency of 0.65 SNPs/100 bp (Meneses et al. 2007). This study is complementary to the other works regarding EST development in *Prunus* performed by different research groups in Italy as part of the work of the Italian National Consortium for Peach Genomics (http://www.itb.cnr.it/estree/) (Pozzi et al. 2007). In this database a collection of 75.404 sequences analyzed from different peach cDNA libraries is also presented. In addition, approximately 200 ESTs were selected for mapping on a physical framework map and a total of 33,189 SNPs was identified.

The analysis of different ESTs using the National Center for Biotechnology Information (NCBI) databases (http://www.ncbi.nlm.nih.gov) indicated significant similarity to protein coding sequences in the database and opened interesting new possibilities for peach and almond genomics. As part of the effort to increase and enrich the genomics resources in different *Prunus* species, researchers are fabricating different peach microarrays using unigene sets as probes (Pozzi et al. 2007). The development of microarrays could also be a good approach for the study of nut quality in almond with good potential for the development of markers associated with important horticultural characteristics. Microarray analysis is a powerful way to utilize sequence information as it allows the simultaneous monitoring of many genes in a single experiment. Recently, transcriptomic studies have been carried out in the study of apricot fruit development and ripening using the available peach microarray that contains around 5,000 oligonucleotides, each correspond to a single unigene (Manganaris et al. 2008). Similarly, the physical mapping of rDNA genes by Corredor et al. (2004) has allowed the establishment of a more precise karyotype for almond. In addition, cloning of genes expressed during seed development has been reported by García-Mas et al. (1996). Suelves and Puigdomenech (1998) have described the cloning of the mandelonitrile lyase gene responsible for the creation of both cyanide and the amaretto flavor of bitter almonds. Finally, using high-information content fingerprinting (HICF), the first physical map has been generated for peach (Zhebentyayeva et al. 2008). This map is composed of 2,138 contigs containing 15,655 BAC (bacterial artificial chromosome) clones and can be considered the first step to the subsequent complete genome sequencing of peach, opening the door to the complete sequencing of other *Prunus* species such as almond. This physical map also integrates 2,633 markers including peach ESTs, cDNAs and RFLPs and is anchored to the *Prunus* T × E reference map using 152 core genetic probes.

The International Peach Genome Initiative has recently released the complete peach genome sequence (http://www.rosaceae.org/peach/genome). This peach v1.0 consists of eight pseudomolecules (the first eight scaffolds) representing the eight chromosomes of peach and they are numbered according to their corresponding linkage groups (Sosinski et al. 2010). This complete peach genome sequence will be of great interest for future molecular studies in *Prunus* species particularly for closely related species such as almond.

B. Molecular Analysis

SSR analysis confirmed previous isozymes studies that showed almond to be the most polymorphic species within the major *Prunus* tree crop species (Martínez-Gómez et al. 2007) making it an ideal candidate for map construction. Extensive research, particularly in Europe (Ballester et al. 1998, 2001; Corredor et al. 2004; Dirlewanger et al. 2004; Martínez-Gómez et al. 2007), led to the development of a high-density almond map, which includes 562 markers (361 RFLPs, 185 SSRs, 11 isozymes, and 5 STSs) covering a total distance of 519 cM with an average density of 0.92 cM marker^{-1} and largest gap of 7 cM (Dirlewanger et al. 2004).

The order of molecular markers observed in the almond map was similar to maps developed with other *Prunus* species suggesting a high level of synteny within the genus (Dirlewanger et al. 2004; Martínez-Gómez et al. 2007). These homologies among *Prunus* genomes support the opportunity for successful interspecific gene introgression as demonstrated by the successful transfer of traits from closely related species to almond (Gradziel et al. 2001a; Martinez Gomez et al. 2003b). The high level of synteny within the genus also supports the transferability of genetic information developed from linkage maps of other *Prunus* species. The availability of high-density linkage maps has allowed

Table 4.6. Markers associated to main agronomic traits in almond.

Trait	Symbol	Linkage group	Marker	References
Flower color	B	G1	RFLP	Jáuregui et al. (2001)
Shell hardness	D	G2	RFLP	Arús et al. (1999)
	D	G2	SSR	Sánchez-Pérez et al. (2007)
Nematode resistance	Mi	G2	RFLP	Bliss et al. (2002)
Anther color	Ag	G3	RFLP	Joobeur et al. (1998)
Blooming time	Lb	G4	RAPD	Ballester et al. (2001)
	Lb	G4	SSR	Sánchez-Pérez et al. (2007)
	Lb	G1, G4, G6, G7	SSR	Sánchez-Pérez et al. (2009)
Chill requirements	CR	G1, G3, G4, G7	SSR	Sánchez-Pérez et al. (2009)
Heat requirements	HR	G1, G3, G4, G7	SSR	Sánchez-Pérez et al. (2009)
Blooming density	Bd	G4	SSR	Sánchez-Pérez et al. (2007)
Leafing time	Lf	G1, G4	SSR	Sánchez-Pérez et al. (2007)
Reaping time	Rd	G4, G5	SSR	Sánchez-Pérez et al. (2007)
In-shell weight	Shw	G1, G2	SSR	Sánchez-Pérez et al. (2007)
Kernel weight	Kw	G1, G4	SSR	Sánchez-Pérez et al. (2007)
Kernel taste	Sk	G5	RFLP	Bliss et al. (2002)
	Sk	G5	RFLP	Joobeur et al. (1998)
	Sk	G5	SSR	Sánchez-Pérez et al. (2007)
	Sk	G5	SSR	Sánchez-Pérez et al. (2010)
Double kernel	Dk	G4	SSR	Sánchez-Pérez et al. (2007)
Self-compatibility	S	G6	RFLP	Ballester et al. (1998)
	S	G6	RFLP	Arús et al. (1999)
	S	G6	SSR	Sánchez-Pérez et al. (2007)
	S	G6	RFLP	Bliss et al. (2002)

recent successes in establishing the approximate map position of major genes in almond (Table 4.6). Other important achievements include the use of bulk segregant analysis (BSA) to map the self-incompatibility gene (Ballester et al. 1998), as well as a major gene controlling delayed flowering time (Ballester et al. 2001; Grasselly 1978; Socias i Company et al. 1999). Root-knot nematode resistance in an almond–peach hybrid has also recently been mapped by Dirlewanger et al. (2004).

Sánchez-Pérez et al. (2007) identified the map positions of two major genes for kernel taste (Sk) in linkage group five (G5) and for self-incompatibility (S) in G6. In addition, QTLs mapped include two for leafing date (Lf-Q1 and Lf-Q2) in G1 and G4, one for shell hardness (D-Q) in G2, one each for double kernel (Dk-Q) and productivity (P-Q) in G4, one for blooming date (Lb-Q) in G4, two for kernel weight (Kw-Q1 and Kw-Q2) in G1 and G4, and two for in-shell weight (Shw-Q1 and Shw-Q2) in G1 and

G2. Finally, four SSR loci BPPCT011, UDP96-013, UDP96-003, and PceGA025) were linked to the important agronomic traits of leafing date, shell hardness, blooming date, and kernel taste (Table 4.6, Fig. 4.3). More recently, Acta Sánchez-Pérez et al. (2009) identified a major QTL for flowering time in G4 (*Lb*) together with other minor QTLs in G1, G6, and G7. In agreement with these, a major QTL for chilling requirements was also located in G4 together with other minor QTLs in G1, G3, G6, and G7. Finally five minor QTLs were located controlling heat requirement in G1, G3, G4, G6, and G7.

A recent strategy for the location of new markers in an established genetic linkage map is the "selective" or "bin" mapping approach as developed in the *Prunus* reference linkage map. This technique allows mapping with the use of a subset of plants of a population from which a map is already available. The advantage of this strategy is that it allows mapping with less time and cost and is adequate for simplifying the construction of high-density maps. The use of this set of six individuals promises to be a very useful resource for peach genetic linkage studies in the future. The reference map has been divided in 67 "bins" or regions (from 8 to 25 cM) to locate the future markers (Howad et al. 2005).

Major efforts have also been directed toward cloning and characterizing the economically important self-incompatibility gene in almond (Bacarella et al. 1991). The cDNA encoding almond S-RNase was first cloned by Ushijima et al. (1998). To better understand the nature of the self-incompatibility gene, Ushijima et al. (2001) later cloned and characterized the cDNA-encoding mutated S-RNase from the almond cultivar 'Jeffries,' which appeared to have dysfunctional *S*-allele haplotype in both pistil and pollen.

C. Molecular Breeding

Despite these recent advances in the application of the newer biotechnologies, almond, as well as other tree crops, lag behind the progress typically observed for annual crops. This is, in large part, the consequence of the inherent difficulties in doing genetic studies on such large-sized and long generation-time plants (Martínez-Gómez et al. 2003). These inherent obstacles to traditional breeding make the opportunities with the new technologies much more revolutionary, however, when applied to tree crops. When fully integrated with the array of breeding methods developed to capitalize on the inherent advantages of tree crops, such as the capability to capture desirable genetic/epigenetic arrangements through vegetative propagation, breeding potential could be expected to surpass that for seed propagated annual crops. Almond is currently well positioned to be a leader in this effort.

Molecular markers, because they offer the opportunity for fast, accurate, and environment-independent evaluation at the seedling stage, promise to dramatically increase selection efficiency. In addition, specific markers offer the advantage of codominant expression, good reproducibility, and allow the ability to compare genetic variation among homologous regions of the same or different species (Martínez-Gómez et al. 2003). A detailed review of biotechnology research with almond has recently been provided by Martínez-Gómez et al. (2007). A classical breeding approach toward these goals would involve an initial hybridization between selected parents, followed by introgression of the traits of interest, typically by backcrossing to the parent with the most promising commercial potential. Other new biotechnology approaches, particularly gene mapping and gene tagging, offer the promise of greater efficiencies in the areas of gene discovery and gene introgression (Martínez-Gómez et al. 2007).

Selection by molecular markers is particularly useful in tree crops such as almond with a long juvenile period, and when the expression of the gene is recessive or the evaluation of the character is otherwise difficult, as with resistance to biotic or abiotic stress (Luby and Shaw 2001). If sufficient mapping information is known, marker assisted selection (MAS) can dramatically shorten the number of generations required to "eliminate" the undesired genes of the donor in backcrossing programs. Marker loci linked to major genes can be used for selection, which is sometimes more efficient than direct selection for the target gene (Knapp 1998). As previously discussed, the use of mapping populations segregating for the characters of interest has been the principal strategy for the analysis of marker-trait association in almond (Table 4.6). The analysis of cosegregation among markers and characters allows establishing the map position of major genes and QTLs responsible for their expression (Fig. 4.5).

PCR-based markers of almond self-incompatibility S-alleles have been successfully used to identify different self-incompatibility genotypes (Barckley et al. 2006; Channuntapipat et al. 2003; Tamura et al. 2000). Similar results were obtained by Boškovic et al. (2003) who identified major almond cultivar stylar S-RNase by electrophoresis in vertical polyacrylamide gels. PCR-based markers of almond self-incompatibility S-alleles have been employed to facilitate the integration of self-compatible S-alleles from related species (Gradziel et al. 2001b). Screening efficiency and flexibility has been greatly increased with the development of successful multiplex PCR techniques by Sánchez-Pérez et al. (2004).

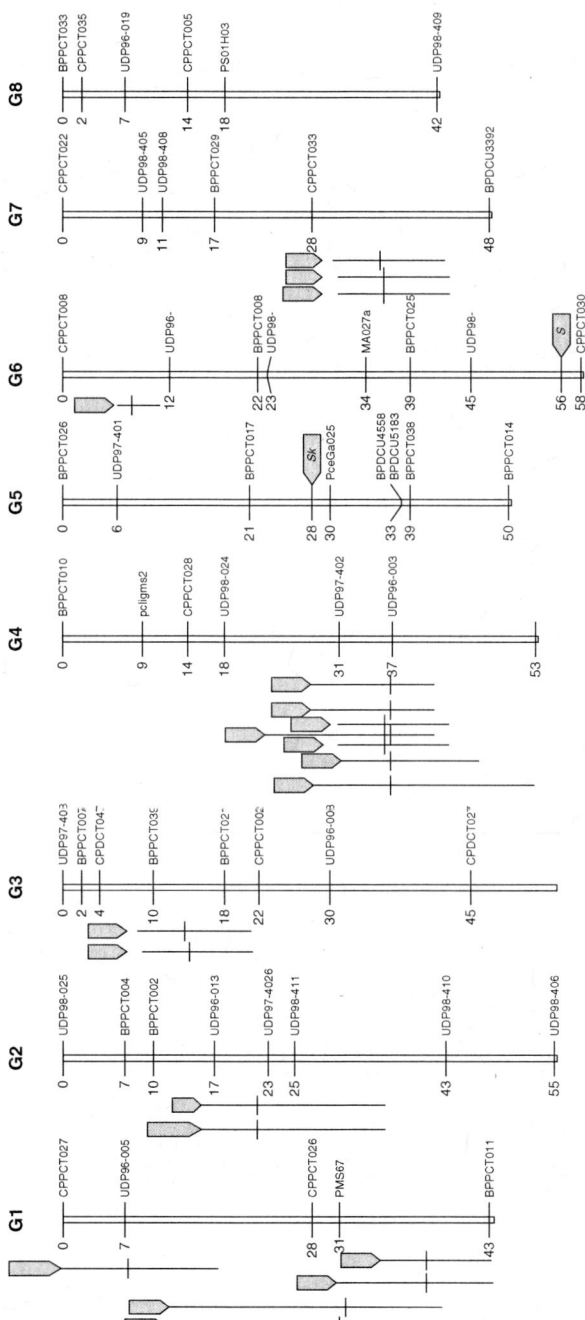

Fig. 4.5. Molecular linkage map constructed with the JOINMAP software of the 'R1000' × 'Desmayo Largueta' almond progeny, with the approximate location of major almond genes and QTLs indicated in Table 4.6.

Using advanced cloning strategies, Ushijima et al. (2003) have described the structural and transcriptional analysis of a pollen-expressed F-box gene with haplotype-specific polymorphism strongly associated with self-incompatibility.

Molecular markers are currently being employed to elucidate the genetic basis of plant processes controlled by multiple genes. For example, Campalans et al. (2001) have described a differential expression technique based on cDNA-AFLP (amplified restriction fragment polymorphism derived technique for RNA fingerprinting) to characterize genes involved in drought tolerance in almond. Results identified increased drought tolerance in specific genes associated with leaf function.

D. Epigenetics

The probable interspecies and clonal origin of many modern almond cultivars suggest promising opportunities for the manipulation not only of the traditional genetic (i.e., Mendelian) determinants but also the epigenetic controls that are only recently becoming characterized. Epigenetic modification may have particular value for almond breeding because epigenetic variability appears to be greatly enhanced with interspecies hybrids (Grant-Downton and Dickinson 2006) and because commercially valuable epigenetic variants can be effectively captured in cultivars by the vegetative propagation common in tree crop cultivar dissemination (Kester et al. 2004). Epigenetic-like changes have been documented in clonal differences within cultivars and in a more fully characterized epigenetic disorder known as noninfectious bud-failure, which threatens over 50% of California production. This disorder is expressed as a deterioration of the clone vitality with increasing age, leading to bud failure in individual trees and branches. Initial symptoms include the necrosis of the growing point of vegetative buds during the fall. The resulting shoot phenotype, as expressed the following spring, is a failure of terminal and/or sub-terminal vegetative buds to grow. When the terminal bud fails, "dieback" results. However, lower and later developing buds may survive providing a "flush" of new growth at basal and subterminal sites of the shoot. "Rough-bark" areas sometimes also develop in narrow bands on the shoots. New shoots from surviving buds grow vigorously and then again fail at the terminal buds. Repetition of this sequence in consecutive years results in an erratic growth pattern often referred to as "crazy-top." Kester et al. (2004) have shown that control of this type of epigenetic disorder can be achieved through well-designed certification programs similar to those used to control vegetatively propagated viruses. Such programs have three basic steps: identification of single tree sources that test negatively for the disorder in clonal-source screening trials (Kester et al. 2004), maintenance and registration of a limited number of trees of the selected clone-source in a foundation orchard, and limited multiplication of registered material to provide certified trees for commercial nurseries (Uyemoto and Scott 1992).

Because epigenetic changes do not respond to traditional breeding methods designed to manipulate classic Mendelian genes, they are generally perceived as undesirable and routinely rogued out. However, because both genetic and epigenetic composition can be captured through clonal propagation, the same methods used to rogue out epigenetic changes can also be utilized to capture desirable epigenetic arrangements. An example would be the widespread practice among nurseries in selecting superior clonal sources of important vegetatively propagated cultivars (Hartman et al. 2002). Epigenetic capture offers unique advantages to breeding programs utilizing wide crosses, since the interspecific hybridization process has been shown to increase the levels of epigenetic variability resulting in novel and sometimes transgressive phenotypes (where the trait is expressed at levels beyond the sum of the parents). The value of epigenetic selection in almond is documented by commercial California plantings where approximately 25% of current plantings represent new cultivars developed in the last 50 years but where two cultivars, 'Nonpareil' and 'Carmel', which had been declining in importance owing to increasing incidence of non-infectious bud-failure, currently make over 50% (or approximately seven million trees) of current plantings (Almond Board of California 2011) utilizing rejuvenated clonal sources developed through epigenetic selection (Kester et al. 2004).

E. Transgenics

The recent development of powerful new biotechnologies has also advanced plant-breeding efforts through the direct incorporation of foreign genes using genetic engineering strategies. Genetic engineering can reduce the time and space required to improve fruit and tree characteristics compared with traditional plant breeding methodologies. A specific genetic change, with no addition of associated detrimental genes, can be accomplished rapidly, whereas traditional breeding programs require more generations and extensive acreage. Genetic engineering can also increase the diversity of genes and germplasm available by allowing the stable integration of foreign DNA into the plant genome. In the case of fruit trees such as almond, genetic engineering represents an alternative to overcome handicaps of traditional breeding programs (i.e., long juvenility period, self-incompatibility, evaluation of agronomic traits in the field) (Singh and Sansavini 1998). However, the presence of genetic mosaics in regenerated transgenic plantlets could also lead to later problems with losses of cultivar trueness-to-type in vegetatively propagated crops (Marcotrigiano and Gradziel 1997).

Most transgenic *Prunus* plants, including almonds, have been obtained via *Agrobacterium*-mediated transformation because of its efficacy (Archilletti et al. 1995; Singh and Sansavini 1998; Miguel and Oliveira 1999; Petri and Burgos 2005; Costa et al. 2007). Much of this research involved the introduction of reporter genes in seedling tissues. The *gus* gene, which codes for β-glucuronidase, has been successfully utilized as a marker gene in genetic transformation of almond (Archilletti et al. 1995; Miguel and Oliveira 1999). In addition, *npt*II (neomycin phosphotranspherase) gene, which confers kanamycin resistance, has been used in this species (Archilletti et al. 1995). The recovery of transgenic forms of established cultivars is most desirable owing to their proven commercial value. While almond seedlings are readily transformed using *Agrobacterium* mediated approaches, the regeneration of plantlets from established cultivar cells has proven very difficult (Archilletti et al. 1995; Miguel and Oliveira 1999). This difficulty is believed to be due to the recalcitrance of cultivar cells to initiate the required organogenesis presumably because they have lost their juvenility with their advanced clonal age (Costa et al. 2007; Santos et al. 2009). In addition, while new genetic engineering techniques offer significant advantages for the discrete addition of new genes to commercially established cultivars, the current dearth of transgenes useful to tree crop breeding, limits its present application. A final barrier is the reluctance of the economically important European and Asian markets to accept transgenic fruits and nuts.

VI. FUTURE PROGRESS

The need to simultaneously optimize the large number of traits essential for commercial success remains the greatest challenge to cultivar breeding. For many almond cultivars, selection has been occurring for hundreds to thousands of years for such rare but elite individuals where the maximum potential of additive, dominance, epistatic, genomic, and epigenetic interactions is combined. Clonal propagation allows the capture of these rare elite genotypes for future plantings as well as future genetic improvements through bud sport mutation or favorable recombinations. Molecular markers, because they offer the opportunity for fast, accurate, and environment-independent evaluation at the seedling stage, offer valuable opportunities to increase selection efficiency. Selection by molecular markers is particularly useful in tree crops such as almond with a long juvenile period, and when the expression of the gene is recessive or the evaluation of the character is otherwise difficult, as with resistance to biotic or abiotic stress. When sufficient mapping information is known, marker assisted selection will also dramatically shorten the number of generations required to rogue-out undesired genes of the donor in backcrossing programs. Markers linked to major genes can be used for selection, which may sometimes be more efficient than direct selection for the target gene. Because molecular marker assisted selection primarily target genes with additive effects, they allow increased efficiencies for certain types of genetic improvements but may hinder actual cultivar development because of their parochial nature. The inherent capacity of hybrid/clone-based cultivars to capture the fullest range of beneficial genetic, epigenetic, and genomic interactions for applied crop improvement provides both a prerequisite and unique opportunity for researchers to evolve beyond the current reductionistic additive-gene approach, but would require (perhaps stimulate) significant parallel progress in our understanding of the basic underlying developmental and inheritance mechanisms at the epigenetic, genomic as well as genetic level. An even greater challenge/opportunity would be the progression from the present focus on single trait genetic improvement to an emphasis on the concurrent management/advancement of the multitude of traits required for commercial success. When fully integrated with the traditional array of breeding methods developed to capitalize on the inherent advantages of tree crops, such as the capability to capture desirable genetic/epigenetic arrangements through vegetative propagation, breeding potential could be expected to surpass that for seed propagated annual crops.

LITERATURE CITED

Abdallah, A., M.H. Ahumada, and T.M. Gradziel. 1998. Oil content and fatty acid composition of almond kernels from different genotypes and California production regions. J. Am. Soc. Hort. Sci. 123:1029–1033.

Almond Board of California. 2011. Almond Almanac. Modesto.

Archilletti, T., P. Lauri, and C. Damiano. 1995. *Agrobacterium*-mediated transformation of almond leaf pieces. Plant Cell Rep. 14:267–272.

Armstrong, D.L. 1957. Cytogenetic study of some derivatives of the F1 hybrid *Prunus amygdalus* × *P. persica*. Ph.D. thesis, Univ. California, Davis.

Arulsekar, S., D.E. Parfitt, and D.E. Kester. 1989. Comparison of isozyme variability in peach and almond cultivars. J. Hered. 77:272–274.

Arumuganathan, K., and E.D. Earle. 1991. Nuclear DNA content of some important plant species. Plant Mol. Biol. Rep. 9:208–218.

Arús, P., J. Ballester, B. Jáuregui, T. Joobeur, M.J. Truco, and M.C.deVicente. 1999. The European *Prunus* mapping project: update of marker development in almond. Acta Hort. 484:331–336.

Asai, W.K., W.C. Micke, D.E. Kester, and D. Rough. 1996. The evaluation and selection of current varieties. p. 52–60. In: Almond production manual. University of California 3364.

Atli, H.S. 2008. Determination of rootstocks characterizations of *Amygdalus orientalis* Mill, types. Publication 37. Republic of Turkey, Ministry of Agriculture and Rural Affairs, General Directorate of Agricultural Research.

Bacarella, A., G. Chironi, and G. Barbera. 1991. Aspetti tecnici, economici e di mercato del mandorlo in Sicilia. Quarderni di Ricerca di Sperimentazione (Palermo, Sicily) 40: 1–191.

Ballester, J., R. Boškovic, I. Batlle, P. Arús, F. Vargas, and M.C.deVicente. 1998. Location of the self-incompatibility gene on the almond linkage map. Plant Breed. 117:69–72.

Ballester J., R. Socias i Company, P. Arús, and M.C.deVicente. 2001. Genetic mapping of a major gene delaying blooming time in almond. Plant Breed. 120:268–270.

Barckley, K., S.L. Uratsu, T.M. Gradziel, and A.M. Dandekar. 2006. Multidimensional analysis of S-alleles from cross-incompatible groups of California almond cultivars. J. Am. Soc. Hort. Sci. 131:632–636.

Bartolozzi, F., M.L. Warburton, S. Arulsekar, and T.M. Gradziel. 1998. Genetic characterization and relatedness among California almond cultivars and breeding lines detected by randomly amplified polymorphic DNA (RAPD) analysis. J. Am. Soc. Hort. Sci. 123:381–387.

Bayazit, S., O. Caliskan, and B. Imrak. 2011. Comparison of polen production and quality characteristics of cultivated and wild almond species. Chilian J. Agr. Res. 71:536–541.

Bliss, F.A., S. Arulsekar, M.R. Foolad, A.M. Becerra, A. Gillen, M.L. Warburton, A.M. Dandekar, G.M. Kocsisne, and K.K. Mydin. 2002. An expanded genetic linkage map of *Prunus* based on an interspecific cross between almond and peach. Genome 45:520–529.

Bošković, R., K.R. Tobutt, I. Batlle, and H. Duval. 1997. Correlation of ribonuclease zymograms and incompatibility genotypes in almond. Euphytica 97:167–176.

Boškovic R., K.R. Tobutt, I. Batlle, H. Duval, P. Martínez-Góme, and T.M. Gradziel. 2003. Stylar ribonucleases in almond: correlation with and prediction of self-incompatibility genotypes. Plant Breed. 122:70–76.

Brooks, R.M., and H.P. Olmo. 1997. The Brooks and Olmo register of fruit and nut varieties. 3rd ed. ASHS Press, Alexandria, VA.

Browicz, K. 1969. Amygdalus. In: K.H. Rechinger (eds), Flora Iranica. 66:166–168.

Browicz, K., and D. Zohary. 1996. The genus *Amygdalus* L. (Rosaceae): species relationships, distribution and evolution under domestication. Gen. Res. Crop Evol. 43:229–247.

Campalans, A., M. Pages, and R. Messeguer. 2001. Identification of differentially expressed genes by the cDNA-AFLP technique during dehydration of almond (*Prunus amygdalus*). Tree Physiol. 21:633–643.

Channuntapipat, C., M. Wirthensohn, S.A. Ramesh, I. Batlle, P. Arús, M. Sedgley, and G. Collins. 2003. Identification of incompatibility genotypes in almond using specific primers based on the introns of the S-alleles. Plant Breed. 122:164–168.
Connell, J. H., T.M. Gradziel, B.D. Lampinen, W.C. Micke, and J. Floyd. 2010. Harvest maturity of almond cultivars in California's Sacramento Valley. Options Méditerranéennes. Serie A, Seminaires Méditerranéennes. 94:19–23.
Corredor, E., M. Román, E. García, E. Perera, P. Arús, and T. Naranjo. 2004. Physical mapping of rDNA genes to establish the karyotype of almond. Ann. Appl. Biol. 144:219–222.
Costa, M., C. Miguel, and M.M. Oliveira. 2007. Improved conditions for *Agrobacterium*-mediated transformation of almond. Acta Hort. 738:575–581.
Dangl, G.S., J. Yang, T. Gradziel, and D.A. Golino. 2009. A practical method for almond cultivar identification and parental analysis using simple sequence repeat markers. Euphytica 168:41–48.
Denisov, V.P. 1988. Almond genetic resources in the USSR and their use in production and breeding. Acta Hort. 224:299–306.
Diamond, J. 1998. Guns, germs and steel: the fates of human societies. W.W. Norton, New York.
Dicenta, F., and J.E. García. 1993a. Inheritance of self-compatibility in almond. Heredity 70:313–317.
Dicenta, F., and J.E. García. 1993b. Inheritance of kernel flavour in almond. Heredity 70:318–321.
Dicenta, F., J.E. García, and E. Carbonell. 1993a. Heritability of flowering, productivity and maturity in almond. J. Hort. Sci. 68:113–120.
Dicenta, F., J.E. García, and E. Carbonell. 1993b. Heritability of fruit characters in almond. J. Hort. Sci. 68:121–126.
Dicenta, F., M. García-Gusano, E. Ortega, and P. Martínez-Gómez. 2005. The possibilities of early selection of late-flowering almonds as a function of seed germination or leafing time of seedlings. Plant Breed. 124:305–309.
Dicenta, F., P. Martínez-Gómez, E. Martinez-Pato, and T. Gradziel. 2003. Screening for *Aspergillus flavus* resistance in almond. HortScience 38:266–268.
DiGrandi-Hoffman, G., R. Thorp, G. Lopez, and D. Eisikowitch. 1994. Describing the progression of almond bloom using accumulated heat units. J. Appl. Ecol. 82:1–17.
Dirlewanger, E., P. Cosson, W. Howad, G. Capdeville, N. Bosselut, M. Claverie, R. Voisin, C. Poizat, B. Lafargue, O. Baron, F. Laigret, M. Kleinhentz, P. Arús, and D. Esmenjaud. 2004. Microsatellite genetic linkage maps of myrobalan plum and an almond–peach hybrid—location of root-knot nematode resistance genes. Theor. Appl. Genet. 109:827–832.
FAO. 2009. World Information and Early Warning System on PGRFA (Plant Genetic Resources for Food and Agriculture). http://apps3.fao.org/wiews/wiews.jsp.
Farrand, W.R. 1999. Depositional History of Franchthi Cave, Fascicle 12: Sediments, Stratigraphy and Chronology. Indiana University, Bloomington.
Felipe, A.J., and R. Socias i Company. 1992. Almond germplasm. HortScience 27:863–866.
Felipe, A.J. 2000. El Almendro, I. El material vegetal. University of Zaragoza, Spain.
Fraser, G.E., H.W. Bennett, K.B. Jaceldo, and J. Sabate. 2002. Effect on body weight of a free 76 kilojoule (320 calorie) daily supplement of almonds for six months. J. Am. Coll. Nutr. 21:275–283.
Fulgoni, V.L., M. Abbey, P. Davis, D. Jenkins, J. Lovejoy, M. Most, J. Sabate, and G. Spiller. 2002. Almonds lower blood cholesterol and LDL-cholesterol but not HDL-cholesterol in human subjects: results of a meta-analysis. FASEB J. 16:A981–A982.

García-Gusano, M, P. Martínez-García, and F. Dicenta. 2010. Seed germination time as a criterion for the early selection of late-flowering almonds. Plant Breed. 129:578–580.

García-Mas, R., R. Messeguer, P. Arús, and P. Puigdomènech. 1996. Accumulation of specific mRNAs during almond fruit development. Plant Sci. 113:185–192.

Godini, A. 2000. About the possible relationship between *Amygdalus webbii* Spach. and *Amygdalus communis* L. Nucis 9:17–19.

Godini, A. 2002. Almond fruitfulness and role of self-fertility. Acta Hort. 591:191–203.

Gradziel, T.M. 2001. Almond species as sources of new genes for peach improvement. Acta Hort. 592:81–88.

Gradziel, T.M. 2003. Interspecific hybridizations and subsequent gene introgression within *Prunus* subgenus. Acta Hort. 622:249–255.

Gradziel, T.M. 2009. Almond (*Prunus dulcis*) breeding. p. 1–31. In: S.M. Jain and M. Priyadarshan (eds.), Breeding of plantation tree crops. Springer Science, New York.

Gradziel, T.M. 2011. Almond origin and domestication. Hort. Rev. 38:23–82.

Gradziel, T.M. 2012. Traditional genetics and breeding. In: A.G. Abbott and C. Kole. (eds.), Genetics, genomics and breeding of stone fruits. Science Publications, Plymouth, UK.

Gradziel, T.M., and D.E. Kester. 1994. Breeding for resistance to *Aspergillus flavus* in almond. Acta Hort. 373:111–117.

Gradziel, T.M., and D.E. Kester. 1998. Breeding for self-fertility in California almond cultivars. Acta Hort. 470:109–117.

Gradziel, T.M., and B. Lampinen. 2011. Constraints to almond productivity. Proceedings of International Congress on Almond and Pistaschio. San Liurfa, Turkey, 2010.

Gradziel, T.M., and P. Martínez-Gómez. 2002. Shell seal breakdown in almond is associated with the site of secondary ovule abortion. J. Am. Soc. Hort. Sci. 127:69–74.

Gradziel, T.M., and D. Wang. 1994. Susceptibility of California almond cultivars to aflatoxigenic *Aspergillus flavus*. HortScience 29:33–35.

Gradziel, T.M., and S.A. Weinbaum. 1999. High relative humidity reduces anther dehiscence in apricot, peach and almond. HortScience 34:322–325.

Gradziel, T.M., D.E. Kester, and P. Martínez-Gómez. 2002. A development based classification for shoot form in almond. J. Am. Pom. Soc. 56:106–112.

Gradziel, T.M., P. Martínez-Gómez, and A.M. Dandekar. 2001a. The use of S-allele specific PCR analysis to improve breeding efficiency for self-fertility in almond. HortScience 36:440–440.

Gradziel, T.M., P. Martínez-Gómez, F. Dicenta, and D.E. Kester. 2001b. The utilization of related almond species for almond variety improvement. J. Am. Pomol. Soc. 55:100–109.

Gradziel, T.M., N. Mahoney, and A. Abdallah. 2000. Aflatoxin production among almond genotypes is not related to either kernel composition or *Aspergillus flavus* growth rate. HortScience 34:937–939.

Gradziel, T.M., P. Martínez-Gómez, A. Dandekar, S. Uratsu, and E. Ortega. 2002. Multiple genetic factors control self-fertility in almond. Acta Hort. 591:221–227.

Grant-Downton, R.T., and H.G. Dickinson. 2006. Epigenetics and its implications for plant biology 2. The epigenetic epiphany': epigenetics, evolution and beyond. Ann. Bot. 97:11–27.

Grasselly, C. 1972. L'Amandier; caracteres morphologiques et physiologiques des varietes, modalite de leurs transmissions chez les hybrides de premiere generation. University of Bordeaux, France.

Grasselly, C. 1978. Observations sur l'utilization d'un mutant l'Amandier a' floraison tardize dans un programme d'hybridizaiton. Ann. Amelior. Plantes 28:685–695.

Grasselly, C., and P. Crossa-Raynaud. 1980. L'amandier. G.P. Maisonneuve et Larose. Paris, XII.

Grossman, Y.L., and T.M. DeJong. 1998. Training and pruning system effects on vegetative growth potential, light interception and cropping efficiency in peach trees. J. Am. Soc. Hort. Sci. 123:1058–1064.

Gülcan, R. 1985. Almond descriptors (rev.). IBPGR, Rome.

Hamby, K. L.W. Gao, B. Lampinen, T. Gradziel, and F. Zalom. 2011. Hull split date and shell seal in relation to navel orangeworm (Lepidoptera: Pyralidae) infestation of almonds. Hort. Entom. 104:965–969.

Hansen, J., and J. Renfrew. 1978. Palaeolithic–neolithic seed remains at Franchthi Cave, Greece. Nature 271, 349–352.

Hartmann, H.T., D.E. Kester, R.L. Geneve, and F.T. Davies, Jr., 2002. Hartmann and Kester's plant propagation: principles and practices. Prentice Hall, NJ.

Hauagge, R., D.E. Kester, and R.A. Asay. 1987. Isozyme variation among California almond cultivars: inheritance. J. Am. Soc. Hort. Sci. 112:687–693.

Howad, W., T. Yamamoto, E. Dirlewanger, P. Cosson, G. Cipriani, A.J. Monforte, L. Georgi, A. G Abbott, and P. Arús. 2005. Mapping with a few plants: using selective mapping for microsatellite saturation of the *Prunus* reference map. Genetics 171:1305–1309.

IPM Manual Group of U.C. Davis. 1985. Integrated pest management for almonds. Pub. 3308. University of California—Division of Agriculture and Natural Resources, Berkeley, CA.

Jáuregui, B., M.C.deVicente, R. Messeguer, A. Felipe, A. Bonnet, G. Salesses, and P. Arús. 2001. A reciprocal translocation between 'Garfi' almond and 'Nemared' peach. Theor. Appl. Genet. 102:1169–1176.

Joobeur, T, M.A., Virue, l M.C. deVicente, B. Jáuregui, J. Ballester, M.T. Dettori, I. Verde, M. J. Truco, R. Messeguer, I. Battle, R. Quarta, E. Dirlewanger, and P. Arús. 1998. Construction of a saturated linkage map for *Prunus* using an almond × peach F_2 progeny. Theor. Appl. Genet. 97:1034–1041.

Kendall, C.W., D.J. Jenkins, A. Marchie, Y. Ren, P.R. Ellis, and K.G. Lapsley. 2003. Energy availability from almonds: implications for weight loss and cardiovascular health. A randomized controlled dose-response trial FASEB J. 17:A339.

Kester, D.E., and T.M. Gradziel. 1996. Almonds (*Prunus*). p. 1–97. In: J. Janick and J.N. Moore (eds.), Fruit breeding. Wiley, New York.

Kester, D.E., T.M. Gradziel, and C. Grasselly. 1991. Almonds (*Prunus*). p. 701–758. In: J.N. Moore and H.J. Ballington (eds.), Genetic resources of temperate fruit and nut crops. International Society of Horticulture and Science, The Netherlands.

Kester, D.E., and R. Asay. 1975. Almonds. p. 367–384. In: J. Janick and J.N. Moore, (eds.), Advances in fruit breeding. Purdue University Press, West Lafayette, Indiana.

Kester, D.E., and T.M. Gradziel. 1990. Growth habit trait nomenclature in almond and peach phenotypes. HortScience 25:72 (Abstr.).

Kester, D.E., P. Raddi, and R. Asay. 1977. Correlations among chilling requirements for germination, blooming and leafing in almond. Proc. Am. Soc. Hort. Sci. 102:145–148.

Kester, D.E., P.E. Hansche, W. Beres, and R.N. Asay. 1977. Variance components and heritability of nut and kernel traits in almond. J. Am. Soc. Hort. Sci. 102:264–266.

Kester, D.E., A. Kader, and S. Cunningham. 1993. Almonds. p. 44–55. Encyclopedia of food science. Academic Press, London.

Kester, D.E., K.A. Shackel, W.C. Micke, M. Viveros, and T.M. Gradziel. 2004. Non-infectious bud failure in 'Carmel' almond: I. Pattern of development in vegetative progeny trees. J. Am. Soc. Hort. Sci. 127:244–249.

Knapp, S.J. 1998. Marker-assisted selection as a strategy for increasing the probability of selecting superior genotypes. Crop Sci. 38:1164–1174.

Kodad, O., M.S. Gracia-Gómez, and R. Socias i Company. 2005. Fatty acid composition as evaluation criterion for kernel quality in almond breeding. Acta Hort. 663: 301–304.

Kodad, O., R. Socias i Company, M.S. Prats and M.C. López Ortiz. 2006. Variability in tocopherol concentrations in almond oil and its use as a selection criterion in almond breeding. J. Hort. Sci. Biotechnol. 81:501–507.

Kumar, K., and D.K. Uppal. 1990. Performance of almond (*Prunus amygdalus* Batsch) selections in the subtropics. Acta Hort. 279:199–207.

Ladizinsky, G. 1999. On the origin of almond. Gen. Res. Crop Evol. 46:143–147.

López, M, M. Mnejja, M. Rovira, G. Colins, F.J. Vargas, P. Arús, and I. Batlle. 2004. Self-incompatibility genotypes in almond re-evaluated by PCR, stylar ribonucleases, sequencing analysis and controlled pollinations. Theor. Appl. Genet. 109:954–964.

Lovejoy, J.C., M.M. Most, M. Lefevre, F.L. Greenway, and J.C. Rood. 2002. Effect of diets enriched in almonds on insulin action and serum lipids in adults with normal glucose tolerance or type 2 diabetes. Am. J. Clin. Nutr. 76:1000–1006.

Luby, J.J., and D.V. Shaw. 2001. Does marker-assisted selection make dollars and sense in a fruit breeding program? HortScience 36:872–879.

Manganaris, G.A., C. Bonghi, P. Tonutti, and A. Ramina. 2008. A comparative transcriptomic approach to elucidate common and divergent mechanism involved in apricot and nectarine fruit development and ripening. p. 316. XIV Intl. Symp. Apricot Breeding and Culture. Matera (Italy).

Marcotrigiano, M., and T.M. Gradziel. 1997. Genetic mosaics and plant improvement. Plant Breed. Rev. 15:43–84.

Martínez-Gómez, P., and T.M. Gradziel. 2003. Sexual polyembryony in almond. Sex. Plant Repro. 16:135–139.

Martínez-Gómez, P., G.O. Sozzi, R. Sánchez-Pérez, M. Rubio, and T.M. Gradziel. 2003. New approaches to *Prunus* tree crop breeding. J. Food Agr. Environ 1:52–63.

Martínez-Gómez, P., S. Arulsekar, D. Potter, and T.M. Gradziel. 2003a. An extended interspecific gene pool available to peach and almond breeding as characterized using simple sequence repeat (SSR) markers. Euphytica 131:313–322.

Martínez-Gómez, P., S. Arulsekar, D. Potter, and T.M. Gradziel. 2003b. Relationships among peach and almond and related species as detected by SSR markers. J. Am. Soc. Hort. Sci. 128:667–671.

Martínez-Gómez, P., M. Rubio, F. Dicenta, T.M. Gradziel. 2004. Resistance to Plum Pox Virus (Dideron isolate RB3.30) in a group of California almonds and transfer of resistance to peach. J. Am. Soc. Hort. Sci. 129:544–548.

Martínez-Gómez, P., R. Sánchez-Pérez, F. Dicenta, and T.M. Gradziel. 2005. Improved technique for counting chromosomes in almond. Sci. Hort. 105:139–143.

Martínez-Gómez, P., K. Majourhat, M. Zeinalabedini, D. Erogul, M. Khayam-Nekoui, V. Grigorian, A. Hafidi, A. Piqueras, and T.M. Gradziel. 2007. Use of biotechnology for preserving rare fruit germplasm. Biorem. Biodiv. Bioavail. 1:31–40.

Martínez-Gómez, P., R. Sánchez-Pérez, F. Dicenta, W. Howard, P. Arús and T.M., Gradziel. 2007. Almond. p. 229–242. In: C. Kole (ed.), Genome mapping and molecular breeding. Springer, Heidelberg.

Martinoli, D.C. 2005. Plant food economy and environment during the Epipalaeolithic in southwest Anatolia: An investigation of the botanical macroremains from Öküzini and Karain. Ph.D. diss. Univ. Basel. http://edoc.unibas.ch/272/1/DissB_7223.pdf April, 2012.

Martins, M., R. Tenreiro, and M.M. Oliveira. 2003. Genetic relatedness of Portuguese almond cultivars assessed by RAPD and ISSR markers. Plant Cell Rep. 22:71–78.

Meneses, C., S. Jung, P. Arús, and W. Howad. 2007. In silico analysis and first applications of SNPs from the GDR database in peach. p. 160–166. XII EUCARPIA Fruit Section Sympos. Zaragoza, Spain.

Micke, W.C. 1994. Almond orchard management. Division of Agricultural Science Publications 3364. University of California, Berkeley.

Miguel, C.M., and M.M. Oliveira. 1999. Transgenic almond (*Prunus dulcis* Mill.) plants obtained by *Agrobacterium*-mediated transformation of leaf explants. Plant Cell Rep. 18:387–393.

Monastra, F.A. Crisafulli, F. Marchese, G. Ondradu, R. Pavia and L. Rivalta. 1982. Monografia di cultivar di mandorlo. Istituto Sperimentale Frutticoltura, Roma.

Ogawa, J., and H. English. 1991. Diseases of temperate zone tree fruit and nut crops. Division of Agriculture and Natural Resources—The University of California, p. 3345.

Ortega, E., and F. Dicenta. 2003. Inheritance of self-compatibility in almond: breeding strategies to assure self-compatibility in the progeny. Theor. Appl. Genet. 106:904–911.

Petri, C., and L. Burgos. 2005. Transformation of fruit trees. Useful breeding tool or continued future prospect? Trans. Res. 14:15–26.

Polito, V., and W.C. Micke. 1994. Bud development, pollination and fertilization. In: W.C. Micke, (ed.) Almond orchard management. Division of Agriculture and Natural Resources—The University of California, Berkeley.

Pozzi, C., A. Vecchietti, B. Lazzari, C. Ortugno, F. Barale, A. Severgnini, and F. Salamini. 2007. The ongoing peach genomics and funtcional genomics effort in Italy. p. 88–93. XII EUCARPIA Fruit Section Symposium. Zaragoza (Spain).

Rehder, A. 1940. Manual of cultivated trees and shrubs. MacMillan, New York.

Rice, R.E., W.W. Barnett, and R.A. Van Steenwyk. 1996. Insect and mite pests. p. 202–213. In: Almond Production Manual. University of California, Publ. 3364.

Rickter, A.A. 1969. Ways and methods of almond breeding (in Russian). Tr. Gos. Nikit. Bot. Sad. 43:81–94.

Rom, R.C., and R.F. Carlson. 1987. Rootstocks for fruit crops. Wiley, New York.

Rosengarten, F.Jr., 1984. The book of edible nuts. Walker and Co., New York.

Sabate, J., and E. Haddad. 2001. Almond-rich diets simultaneously improve plasma lipoproteins and alpha-tocopherol levels in men and women. Ann. Nutr. Metab. 45:596.

Sánchez-Pérez, R., F. Dicenta and P. Martínez-Gómez. 2004. Identification of S-alleles in almond using multiplex-PCR. Euphytica 138:263–269.

Sánchez-Pérez, R, D. Howad, F. Dicenta, P. Arús, and P. Martínez-Gómez. 2007. Mapping major genes and quantitative trait loci controlling agronomic traits in almond. Plant Breed. 126:310–318.

Sánchez-Pérez, R, F. Dicenta, and P. Martínez-Gómez. 2009. Transmission of chilling and heat requirements for flowering in almond and development of QTLs. p. 111–113. VI Intl. Pistacios and Almond Congr. Hort.

Sánchez-Pérez, R, D. Howad, J. García-Mas, P. Arús, P. Martínez-Gómez, and F. Dicenta. 2010. Molecular markers for kernel bitterness in almond. Tree Gen. Gen. 6:237–247.

Santos, A.M., M.J. Oliver, A.M. Sanchez, and M. Oliveira. 2009. An integrated strategy to identify key genes in almond adventitious shoot regeneration. J. Exp. Bot. 60: 4159–4173.

Sathe, S.K., S.S. Teuber, T.M. Gradziel, and K.H. Roux. 2001. Electrophoretic and immunological analyses of almond genotypes and hybrids. J. Agr. Food Chem. 49:2043–2052.

Saura-Calixto, F., M. Bauzá, F. Martínez de Toda, and A. Argamentería. 1981. Amino acids, sugars and inorganic elements in the sweet almond. J. Agr. Food Chem. 29:509–511.

Schirra, M. 1997. Postharvest technology and utilization of almonds. Hort. Rev. 20: 267–292.

Singh, Z., and S. Sansavini. 1998. Genetic transformation and fruit crop improvement. Plant Breed. Rev. 16:87–134.

Socias i Company R., O. Kodad, J.M. Alonso, and T.M. Gradziel. 2008. Almond quality: a breeding perspective. Hort. Rev. 34:197–238.

Socias i Company, R. 1990. Breeding self-compatible almonds. Plant Breed. Rev. 8: 313–338.

Socias i Company, R. 1998. Fruit tree genetics at a turning point: the almond example. Theor. Appl. Genet. 96:588–601.

Socias i Company, R. 2002. The relationship of *Prunus webbii* and almond revisited. Nucis-Newslett. 11:17–19.

Socias i Company, R., and A.J. Felipe. 1988. Self-compatibility in almond: transmission and recent advances. Acta Hort. 224:307–317.

Socias i Company, R., and A.J. Felipe. 1992. Almond: a diverse germplasm. HortScience. 27:863–869.

Socias i Company, R., A.J. Felipe, and J. Gomez Aparisi. 1999. A major gene for flowering time in almond. Plant Breed. 118:443–448.

Socias i Company, R., J.M. Alonso, and J. Gómez Aparisi. 2005. Evaluation of almond selections for fruit set under field conditions. Options Méditerranéennes, Serie A. 63:133–139.

Socias i Company, R., J.M. Alonso, O. Kodad and T.M. Gradziel. 2011. Almonds. p. 697–728. In: M.L. Badenes, and D.H. Byrne (eds.), Fruit breeding. Handbook of plant breeding 8. Springer, NewYork.

Sorkheh, K., B. Shiran, E. Asadi, H. Jahanbazi, H. Moradi, T.M. Gradziel, and P. Martínez-Gómez. 2009a. Phenotypic diversity within native Iranian almond (*Prunus* spp.) species and their breeding potential. Genetic Res. Crop Evol. 56:947–961.

Sorkheh, K, B. Shiran, S. Kiani, N. Amirbakhtiar, S. Mousavi, V. Rouhi, S. Mohammady, T. M. Gradziel, L.V. Malysheva-Otto, and P. Martínez-Gómez. 2009b. Discriminating ability of molecular markers and morphological characterization in the establishment of genetic relationships in cultivated genotypes of almond and related wild species. J. For. Res. 20:183–194.

Sorkheh, K., B. Shiran, M. Khodambashi, H. Moradi, T.M. Gradziel, and P. Martinez-Gomez. 2010. Correlations between quantitative tree and fruit almond traits and their implications for breeding. Scientia Hort. 125:323–331.

Sosinski, B.I. Verde, M. Morgante, and D. Rokhsar. 2010. The International peach genome initiative. A first draft of the peach genome sequence and its use for genetic diversity analysis in peach. 5th Intl. Rosaceae Genomics Conf. Nov. 2010. Stellenbosch, South Africa.

Spiegel-Roy, P., and J. Kochba. 1981. Inheritance of nut and kernel traits in almond (*Prunus amygdalus* Batsch). Euphytica 30:167–174.

Suelves, M., and P. Puigdomenech. 1998. Molecular cloning of the cDNA coding for the (R)-(+)-mandelonitrile lyase of *Prunus amygdalus*: temporal and spatial expression patterns in flowers and mature seeds. Planta 206:388–393.

Tamura, M., K. Ushijima, H. Sassa, H. Hirano, R. Tao, T.M. Gradziel, and A.M. Dandekar. 2000. Identification of self-incompatibility genotypes of almond by allele-specific PCR analysis. Theor. Appl. Genet. 101:344–349.

Tavassolian, I., G. Rabiei, D. Gregoy, M. Mnejja, M.G. Winthensohn, P.W. Hunt, J.P. Gibson, C.M. Ford, M. Sedgely, and S. Wu. 2010. Construction o fan almond linkage map in an Australian population Nonpareil x Lauranne. BMC Gen. 11:551.

Thorp, R., and G.M. Roper. 1994. Bee management for almond pollination. In: W.C. Micke, (ed.). Almond orchard management. Division of Agricultural Science Publications— University of California, Berkeley.

Ushijima, K, H. Sassa, R. Tao, H. Yamane, A.M. Dandekar, T.M. Gradziel, and H. Hirano. 1998. Cloning and characterization of cDNAs encoding S-RNases from almond (*Prunus dulcis*): Primary structural features and sequence diversity of the S-RNases in Rosaceae. Mol. Gen. Genet. 260:261–268.

Ushijima, K., H. Sassa, M. Kusaba, R. Tao, M. Tamura, T.M. Gradziel, A.M. Dandekar, and H. Hirano. 2001. Characterization of the S-locus region of almond (*Prunus dulcis*): analysis of a somaclonal mutant and a cosmid conting for an S haplotype. Genetics 158:379–386.

Ushijima, K., H. Sassa, A.M. Dandekar, T.M. Gradziel, R. Tao, and H. Hirano. 2003. Structural and transcriptional analysis of self-incompatibility (S) locus of almond (*Prunus dulcis*): identification of a pollen-expressed F-box gene with haplotype-specific polymorphism. Plant Cell 15:771–781.

Uyemoto, J.K., and S.A. Scott. 1992. Important disease of *Prunus* caused by viruses and other graft transmissible pathogens in California and South Carolina. Plant Dis. 76:5–11.

Vargas, F.J., and M.A. Romero. 1984. Considérations sur la sélection précoce dans des programmes d'amélioration de variétés d'amandier. Options Méditerranéennes. 2:143–145.

Vargas, F.J. M.A. Romero, and I. Batlle. 2001. Kernel taste inheritance in almond. Options Méditerranéennes. 56:129–134.

Vezvaei, A., T.W. Hancock, L.C. Giles, G.R. Clarke, and J.F. Jackson. 1995. Inheritance and linkage of isozyme loci in almond. Theor. Appl. Genet. 91:432–438.

Vezvaei, A. and J.F. Jackson. 1996. Almond nut analysis. p. 135-148. In: H.F. Linskens and J.F. Jackson (eds.), Modern methods of plant analysis. Vol. 18. Fruit analysis. Springer-Verlag, Berlin.

Vlasic, A. 1976. La cultivazione del mandorlo in Jugoslavia. In: L'amandier. Options Méditerranéennes. 32:75–77.

Watkins, R. 1979. Cherry, plum, peach, apricot and almond. *Prunus* spp. p. 242–247. In: N. W. Simmonds (ed.), Evolution of crop plants. Longman, London.

Weiss, E., W. Wetterstrom, D. Nadel, and O. Bar-Yosef. 2004. The broad spectrum revisited: evidence from plant remains. Proc. Natl. Acad. Sci. 26:9551–9555.

Woodrof, J.G. 1979. Tree nuts, production and processing products. Vol. III, 2nd ed. AVI Publications, Westport, CT.

Wünsch, A. and J.I. Hormaza. 2002. Cultivar identification and genetic fingerprinting of fruit tree species using DNA markers. Euphytica 125:56–67.

Yousefzadeh, A. S. Houshmand, B. Madani and P. Martínez-Gómez. 2010. Karyotypic studies in Iranian wild almond species. Caryologia 63:117–123.

Zeinalabedini, M., M. Khayam-Nekoui, V. Grigorian, T.M. Gradziel, and P. Martinez-Gomez. 2010. The origin and dissemination of the cultivated almond as determined by nuclear and chloroplast SSR marker analysis. Sci. Hort. 125:593–601.

Zhebentyayeva, T.N., G. Swire-Clark, L.L. Georgi, L. Garay, S. Jung, A. Forrest, B. Blackmon, R. Horn, W. Howad, P. Arús, D. Main, B. Sosinski, W.V. Baird, G.L. Reighard, and A.G. Abbott. 2008. A framework physical map for peach, a model Rosaceae species. Tree Gen. Gen. 4:745–756.

5

Breeding Loquat

Maria L. Badenes
Fruit Breeding Department
Instituto Valenciano de Investigaciones Agrarias
Apartado Oficial 46113
Moncada, Valencia, Spain

Jules Janick
Department of Horticulture and Landscape Architecture
Purdue University
West Lafayette, IN 47907-2010, USA

Shunquan Lin and Zhike Zhang
College of Horticulture
South China Agricultural University
Guangzhou 510642, China

Guolu L. Liang and Weixing Wang
College of Horticulture and Landscape Architecture
Southwest University
Chongqing 400716, China

ABSTRACT

Loquat (*Eriobotrya japonica* Lindl., Rosaceae, Maloideae) is a subtropical, evergreen fruit tree ($2n = 34$) indigenous to China, that blooms in the fall and early winter, and ripens in the spring. The association of fruit ripening with a traditional Spring Festival in South China makes it one of the most popular fruits of that area but loquat is now being grown in various other countries and is being exported from Spain. Most loquat cultivars are derived from clonal selection from open-pollination but at present various hybridization programs

---

Plant Breeding Reviews, Volume 37, First Edition. Edited by Jules Janick.
© 2013 Wiley-Blackwell. Published 2013 by John Wiley & Sons, Inc.

have been carried out for both scions and rootstocks. Seedless loquat has been achieved by application of gibberellic acid to triploids that occur infrequently by nonreduction in open-pollinated diploids. Advances in genomics offer possibilities for marker-assisted selection (MAS), and the transfer of useful genes into loquat from other species within the Rosaceae.

KEYWORDS: *Eriobotrya japonica*; Rosaceae; seedlessness; selection; synteny

I. INTRODUCTION
II. GERMPLASM
    A. Genus *Eriobotrya*
    B. Species *E. japonica*
    C. Rootstock
III. REPRODUCTIVE PHYSIOLOGY
    A. Flowering
    B. Blossoming
    C. Pollen Biology
        1. Morphology
        2. Germination
        3. Storage
    D. Pollination and Fertilization
        1. Environmental Factors
        2. Pollen Compatibility
    E. Embryology
        1. Micro- and Macrosporogenesis and Embryogenesis
        2. Seedlessness
        3. Seed Morphology and Physiology
    F. Fruit
IV. BREEDING OBJECTIVES
V. BREEDING METHODS
    A. Genetic Studies
    B. Selection
    C. Hybridization and Mass Selection
    D. Mutagenesis
    E. Breeding Seedless Loquat
        1. Induction of Seedlessness
        2. Production of Seedless Triploids in Loquat
        3. Morphology and Pomology of Seedless Triploid Loquat
    F. Molecular Breeding
        1. Molecular Markers
        2. Marker-Assisted Selection (MAS)
        3. Genetic Linkage Maps
        4. Transformation
VI. FUTURE PROGRESS
LITERATURE CITED

I. INTRODUCTION

Loquat (*Eriobotrya japonica*) is a subtropical evergreen fruit tree that blooms in the fall and early winter and ripens in the spring. The tree is cold hardy to $-10°C$ but fruits are damaged at $-3°C$. Production is particularly well suited for Mediterranean climates. Despite its Latin binomial loquat is indigenous to Southern China and its English name derived from the Chinese *luju* although it is better known there as *pipa*. Its various names include *bibassier* or *neflier du Japon* (French), *níspero japonés* (Spanish), *Japonische mispel* (German), *nepola Giaponese* (Italian), emeixa do Japao (Portugal), and Japanese plum or medlar (United States).

Loquat has been cultivated in China for over 2000 years and many species occur there in the wild state (Lin et al. 1999). Loquat was introduced to Japan in ancient times and described as early as 1189. Although long a favorite fruit in Japan, crop area has recently declined since cultivation is very labor-intensive. It was first described in the West by the German traveler and physician Englebert Kaempfer who observed it in Japan and named it *Amoenite Exotic* (1712), and more fully described by Thunberg (1784) who named it *Misfiles japonica* or Japanese medlar. In 1784, the loquat was introduced from Guangdong, China into the National Garden at Paris and reached the Royal Botanical Gardens at Kew, England in 1787. The genus was subsequently changed to *Eriobotrya* by John Lindley in 1812, in reference to the woolly clustered panicles. It was introduced to Florida from Europe and to California from Japan between 1867 and 1870. Chinese immigrants carried the loquat to Hawaii. Loquat was distributed around the Mediterranean to various countries including Algeria, Cyprus, Egypt, Greece, Israel, Italy, Spain, Tunisia, and Turkey. Cultivation spread to India, Nepal, Pakistan, and Southeast Asia including Korea, Laos, and Vietnam and reached Australia, New Zealand, Madagascar, and South Africa. It is now grown in Armenia, Azerbaijan, and Georgia and in the Americas including Argentina, Brazil, Chile, Ecuador, Guatemala, Mexico, and Venezuela.

Loquat fruits are round, ellipsoid, to obovate with diameter ranging from 2 to 5 cm and average weight from 30 to 70 g but larger fruit is possible. The thin peel and flesh is white or orange. Soluble solids content vary from 7% to 20%. The seeds, usually about three to five per fruit, are relatively large, each about 1.2–3.6 g and may be obtrusive when the fruit is consumed fresh. The fruit can be process in various forms including juice and wine. The flesh is about 70% of seeded fruit by weight and is aromatic, juicy, delicately flavored, and often considered delicious. However, loquat has some serious drawbacks including the high number of large seed, poor storage life, and susceptibility to bruising. Leaves and fruits have been traditionally considered to have high medical value and there is evidence of pharmaceutically active compounds (Lin et al. 2007).

Loquat is extremely popular in China because it is the first tree fruit of the season ripening around the popular Spring Festival. Production in China reached 512,000 tonnes (t) from 133,000 ha in 2010 and is increasing. Spain, the leading world exporter, produces 40,000 t from about 3,000 ha. Significant producers include Turkey, India, Japan, and Pakistan with small production in Brazil, Chile, Guatemala, Greece, Israel, Italy, Morocco, and Portugal. Although loquat is a minor crop in most of the world, there is considerable room for growth of the industry.

Breeding has long been identified as a priority for expansion of the loquat industry. Most major cultivars of loquat are derived from chance seedlings and there are many selections in the various provinces of China. Breeding programs based on hybridization have been initiated in several countries. The objective of this chapter is to review the current state of loquat genetics and breeding.

II. GERMPLASM

A. Genus *Eriobotrya*

The number of loquat species had been under dispute and the opinions of authors in different countries vary. As described in the literature and collected as specimens in herbariums in China, there are about 32 species or variants (Lin et al. 2004; Yang et al. 2005); most originated in Southern China with the rest from Southeastern Asia (Table 5.1). From 2005 to 2011, S. Lin and his team searched for these species or variants from China and Southeast Asia and found most of them. Only three species were not found such as *E. bengalensis* f. *contract* Vidal, *E. hookeriana* Decne, and *E. philippinensis* Vidal.

Species and variants in China could be divided into three groups (Fig. 5.1) based on stamen and style number and leaf size (Yang et al. 2007). The first group has about 15 stamens, two to four styles, and small leaves (width < 2 cm) and consists of three species: *E. seguinii* Card, *E. henryi* Nakai, and *E. angustissima* Hook. f. The second group has 20 stamens, five styles, and large leaves (width >5 cm) and consists of three species: *E. japonica* Lindl., *E. malipoensis* Kuan, and *E. elliptica* Lindl. and two additional variants of this species: var. *petelottii* Vidal and f. *petiolata* Hook. The third group has 20–30 stamens, 2–4 styles and medium sized leaves, and is divided into four subgroups based on their

5. BREEDING LOQUAT

Table 5.1. Loquat species and varieties as listed in diverse literature and/or indicated on specimens in Chinese herbaria.

Eriobotrya species	Representative area
E. angustissima Hook. f	Laos, Southern Vietnam
E. bengalensis Hook. f.	South-western China
f. *angustifolia* Vidal	Yunnan, China
f. *contract* Vidal	Vietnam
f. *intermedia* Vidal	Yunnan, China
E. cavaleriei Rehd.	Guangdong, China
E. deflexa Nakai	Taiwan, China
var. *buisanensis* Nakai	Taiwan, China
var. *koshunensis* Nakai	Taiwan, China
E. elliptica Lindl.	Tibet, China
var. *petelottii* Vidal	Vietnam
f. *petiolata* Hook.	Tibet, China
E. fragrans Champ.	Guangdong, China
var. *furfuracea*. Vidal	Vietnam
E. henryi Nakai	Yunnan, China
E. hookeriana Decne	Tibet, China
E. japonica Lindl.	Southern China
E. kwangsiensis Chun	Guangxi, China
E. latifolia Hook.	Moalmayne
E. macrocarpa Kurz	Burma
E. malipoensis Kuan	Yunnan, China
E. obovata W.W. Smith	Yunnan, China
E. philippinensis Vidal	Philippines
E. poilanei Vidal	Vietnam
E. prinoides Rehd. & Wils.	Yunnan, China
var. *laotica* Vidal	Laos
var. *dadunensis* H.Z. Zhang	Sichuan, China
E. salwinensis Hand.-Mazz.	Yunnan, China
E. seguinii Card	Guangxi, China
E. serrata Vidal	Yunnan, China
E. stipularis Craib	Cambodia
E. tengyuehensis W.W. Smith	Yunnan, China

distribution as shown in Fig. 5.1. Subgroup 1 is distributed around Southern China and South Asia and Southeast Asia. Subgroup 2 is distributed along valley of the Pearl River. Subgroup 3 is distributed along Salvin River, and Subgroup 4 is divided along the Mekong River.

Among these species, only *E. japonica* is cultivated for its fruits. *E. deflexa* and *E. prinoides* have been used as rootstock, but they are less widely used than *Photinia serrulata* Lindl. in China and *Cydonia, Malus, Pyrus,* and *Pyracantha* in Mediterranean regions. Although more than 95% of loquat are grafted on common loquat (*E. japonica*)

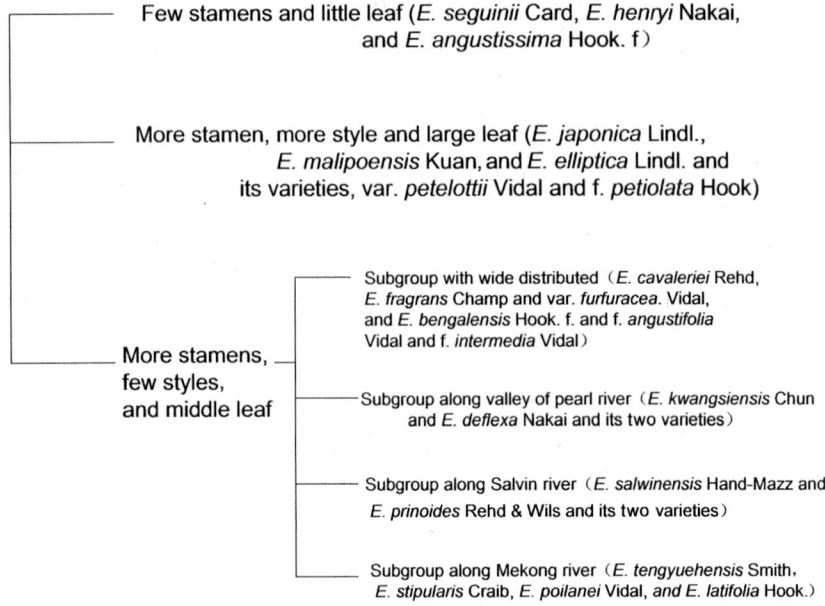

Fig. 5.1. Loquat species in China.

seedlings, other species of *Eriobotrya* could be used as rootstock and 10 species are under evaluation in China. *E. cavaleriei* Rehd. has been used to make wine and *E. prinoides* Rehd. & Wils. has been used in Chinese medicine with additional species under evaluation for this purpose.

B. Species *E. japonica*

There are more than 500 cultivars in China, most of which are seedling selections that are no longer in production although they are all conserved in the National Fruit Germplasm Repository in Fuzhou. There are 83 cultivars (or selections) in Zhejiang, more than 100 in Fujian, 57 in Jiangsu, 31 in Anhui, 18 in Guangdong, and 9 in Sichuan. However, less than 50 cultivars are widely cultivated around China. Germplasm banks are also located in Spain (ca. 100 cultivars), and Japan (<100 cultivars). Spanish commercial production depends on only four cultivars: 'Magdal,' 'Algerie,' 'Golden Nugget (from United States)', and 'Tanaka (from Japan),' with the majority of production from 'Algerie'. Three cultivars, 'Mogi', 'Tanaka', and 'Nakasakiwase', account for 95% of the total loquat commercial area in Japan. The characteristics of the major cultivars grown worldwide are presented in Table 5.2.

Table 5.2. Major loquat cultivars in the main producing areas of the world.

Country	Area	Name of cultivar	Origin	Outstanding characteristics
Brazil		Nectar de Cristal	Obtained by open pollination of Togoshi (Japan), 1970s	Yellowish-red flesh, high yield, fruit uniformity
		Parmogi	Obtained by open pollination of Mogi (Japan), 1970s	Yellowish-red flesh, high yield, pleasant taste
		Precoce de Itaquera	Selected from Japanese seedling	Yellowish-red flesh, very productive
China	Anhui	Guangrong	Selected from seedling of Dahongpao	Yellowish-red flesh, vigorous growth, stable yield, quite large fruit, good keeping quality
	Fujian	Changhong No. 3	Selected from a natural hybrid seedling of Changhong, 1990	Yellowish-red flesh, elongate-obovate fruits, 50 g, ripening in mid-April; high stable yield
		Guifei	Selected from seedling of whitish cultivar	Whitish flesh, fruit 52–68 g, Brix 13.8–15.2°
		Jiefangzhong	Dazhong seedling, selected 1950	Yellowish-red flesh, large fruits, average 70 g with some as large as 172 g; high yield
		Zaozhong No. 6	Jiefangzhong × Moriowase, 1992	Yellowish-red flesh, ripening in the beginning of April, average 53 g, attractive, good quality
	Jiangxi	Duhe	Introduced from unknown cultivar, 1958	Yellowish-white flesh, single seed, high yield, single seed, medium eating quality
	Sichuan	Dawuxing	Seedling	Yellowish-red flesh, large fruits with star-like calyx-lobe
	Zhejiang	Longquan No. 1	Seedling	Yellowish-red flesh, high yield
		Dahongpao	Old seedling cultivar	Whitish flesh, vigorish growth, stable yield
		Loyangqing	Selected from Dahongpao, 1980s	Yellowish-red flesh, strong disease resistance, high stable yield, good keeping quality

(continued)

Table 5.2. (*Continued*)

Country	Area	Name of cultivar	Origin	Outstanding characteristics
		Ninghaibai	Selected from whitish cultivar, 2004	Whitish flesh, fruit 52 g, Brix 13–16°, good eating quality
Egypt		Golden Ziad	Seedling of Premier	Yellowish-red flesh, high yield, early season
		Moamora Golden Yellow	Seedling of Premier	Yellowish-red flesh, high yield
India		Pale yellow		White flesh, fruit large
		Safeda		Flesh cream-colored, early to midseason
		Thames Pride		Yellowish-red flesh, bears heavily, early season, juicy, canned commercially
Israel		Akko 13	Japanese origin	Yellowish-red flesh, early season (March), juicy, agreeable flavor, good keeping quality, requires cross-pollination
		Zerifin	Seedling	Yellowish-red flesh, bears regularly and abundantly, excellent quality, stores well
Italy	Sicily	Marchetto (or Marturana)	Seedling from Ficarazzi, near Palermo	White–orange flesh, fruit 70–80 g, Brix 15°, easy peeling, does not ship will, adapted to local sales
		Nespolone di Trabia	Seedling from Trabia, near Palermo	Orange flesh, subacid, fruit 60–80 g, Brix 14.5°, ripening in May–June, very good eating quality, ships well
		Nespola Rossa	Seedling from Ficarazzi, near Palermo	Pink flesh, fruit 40–50 g, Brix 9.3°, harvested end of April, good pollinator for Nespolone di Trabia and Sanfilipparo, flowering Oct.–Dec.
		Sanfilipparo (or Gigante)	Derived from Nespolone di Trabia	Orange flesh, characteristics very similar to Nespolone di Trabia, but easily adaptable to different environments and climates; ships well
		Virticchiara	Seedling from Bocca di Falco, near Palermo	White-pink flesh, fruit 45–50 g, Brix 9.3°, sweet, very early ripening (April)

Country	Region	Cultivar	Origin/History	Description
Japan	Honshu; Shikoku Kyushu	Tanaka	Seed brought to Tokyo from Nagosaki 1888	Yellowish-red flesh, harvest in May, 60–76 g, good keeping quality
		Nakasaki-wase	Mogi × Hondawase, 1976	Yellowish-red flesh, prone to cold damage, very early ripening, excellent quality
		Suzukaze	Kusunoki × Mogi, 1974, Released in 1999	Yellowish-red flesh, fruit 55 g, Brix 12.7°, vigorous
		Yougyoku	Mogi × Morimoto, 1973, Released in 1999	Yellowish-red flesh, fruit 60 g, Brix 12.1°, sensitive to loquat canker
Spain	Alicante	Algerie	Seedling from Algeria	Yellowish-red flesh, 95% of Spanish production
	Andalusia	Golden Nugget	A clone of Tanaka obtained in California, 1888–1890	Yellowish-red flesh, 80% of total in Andalusia, good keeping, ships well
Turkey		Akko 13	Japanese origin	Yellowish-red flesh, juicy, sweet, agreeable flavor, dark orange good keeping
		Golden Nugget	A clone of Tanaka obtained in California, 1888–1890	Yellowish-red flesh, juicy, sweet flesh, apricot-like flavor, good keeping, ships well
		Hafif Cukurgobek	Seedling selection in Turkey	Yellowish-red flesh, sweet, juicy; pleasant flavor, good keeping
		Takaka	Japanese seedling	
United States	California	Advance	Seedling, selected 1897	Yellowish-red flesh, good pollinator
		Champagne	Seedling introduced, 1908	Yellowish-red flesh, juicy, excellent flavor, good for preserving
	Florida	Fletcher	Parentage unknown, 1957	Yellowish-red flesh, good keeping quality and flavor
		Wolfe	Seedling of advance, released 1965	Pale-yellow flesh, excellent flavor, stable yield, resistant to bruising

Cultivars in China can be divided into three groups: whitish flesh, large fruit with orange flesh, and medium-to-small fruit with orange flesh. Most cultivars in Japan belong to the medium-to-small fruit with orange flesh group but several belong to the other two groups. 'Shiro Mogi' is in the whitish-flesh group, and 'Tanaka' is in the group with large fruit size and orange flesh. In China, 70% of widely planted cultivars have orange flesh with the remainder having whitish flesh. Whitish-flesh cultivars, such as 'Zaozhong' and 'Baiyu', are the leading ones in Jiangsu Province. Among orange cultivars, there are different ecological types in various zones formed during the long course of their cultivation and acclimatization. Ecotypes in China have been divided into two cultivar groups: the north subtropical cultivar group (NSCG) and the south subtropical cultivar group (SSCG) (Ding et al. 1995). NSCG are primarily distributed in the mid- and north subtropical area, roughly in the provinces in the basin of the Yangtze River, located in the range of 27°–33°N where average annual temperature is 15°–18°C, with an absolute low temperature of −5° to −12°C, and 800–1500 mm of annual rainfall and snows and frost can occur. NSCG cultivars are characterized by strong cold-resistance. Fruits are mostly late ripening, with orange flesh, and small but with high quality. Representative cultivars are 'Luoyangqing' in Zhejiang, 'Zhaozhong' in Jiangsu, and 'Guangrong' in Anhui. These cultivars have been successfully introduced to the south subtropical zones and margins of the tropical zones and their fruits ripen as early as April. SSCG cultivars are grown in the south subtropical zone and margins of the tropical zone, approximately 19°–27°N, with only a few days of frost and snow or temperatures lower than 0°C, and with more than 1,500 mm of annual rainfall. The SSCG cultivars have poor cold-resistance but are high yielding and early, while their fruits are large but generally less flavorfull than NSCG fruits. Representative cultivars are 'Dawuxing' in Sichuan and 'Jiefangzhong' in Fujian. Flowers and fruits will be injured by cold if they are grown in the north subtropical zones. Attempts to introduce 'Jiefangzhong' to Zhejiang and Jiangsu have been carried out since the 1970s, but failed (Ding et al. 1995). Several Spanish cultivars with large fruit and orange flesh such as 'Marc' and 'Pelluches' are also included in the SSCG group.

C. Rootstock

Seedlings of *E. japonica* are the most widely used rootstocks and usually produce large trees (Hartmann and Kester 1975; Morton, 1987). In Japan, seedlings of domestic cultivars, such as 'Mogi' and 'Champagne', are usually used as rootstocks (Nesumi 2006) but clonal quince rootstocks are used in some modern orchards. In Mediterranean countries, loquat seedlings are also usually used as rootstocks since they are well adapted to calcareous soils, which are abundant in this region. In Spain, seedling loquats are the most common rootstocks.

There are reports of other rosaceous species being evaluated as rootstocks for loquat in various countries. These include hawthorn (*Cratagous scabrifolia* Rehd.), apple (*Malus* × *domestica* Borkh.), firethorn (*Pyracantha fortuneana* Roem.), medlar (*Mespilus vulgaris* Rchb.), pear (*Pyrus communis* L.), Chinese photinia (*Photinia serrulata* Lindl.), and quince (*Cydonia oblonga* Mill.).

Quince and pyracantha rootstocks may cause extreme scion dwarfing. Dwarfing on quince rootstocks has encouraged expansion of loquat cultivation in Israel since 1960. Quince rootstocks selections (A, C, BA29) produce smaller, more compact trees, a shorter juvenile period, and larger fruits with high sugar content and good color. However, quince rootstocks are very sensitive to calcareous soils and show graft union incompatibility with many loquat cultivars (Llácer et al. 2003; Polat and Kaska 1992; Polat 2007a). The growing of dwarf trees greatly reduces the labor of pruning, flower- and fruit-thinning, fruit bagging, and harvest. However, loquat on quince suffers from zinc deficiency, and some trees break off at the graft union (Blumenfield 1995). Quince rootstock, which tolerates heavier and wetter soils, is widely used in Egypt (Morton 1987). In Spain, under experimental conditions, quince conferred resistance to saline conditions due to their ability to reduce the transport of Na and Cl to the shoots (García-Legaz et al. 2005; Manuel et al. 2008). Chinese photinia was sometimes used as rootstocks for loquat cultivars in the Suzhou district of Jiangsu province, China, but this genus induces shallow-rooting and late-bearing.

Most of the 32 species of *Eriobotrya* have seemingly never been used as rootstocks for loquat in commercial plantings. Until recently, there had been only one report of another species of *Eriobotrya* beyond *E. japonica* used as rootstocks, that is, Taiwan loquat (*E. deflexa*) (Lin et al. 1999) but a number of candidate species could be used as rootstocks. The potential of six wild species of *Eriobotrya* as rootstock for domesticated loquat cultivars was evaluated in China (Zhang et al. 2010). When the domestic loquat, 'Zaozhong No. 6', was grafted onto different wild loquat species seedlings, differences in compatibility were observed and graft compatibility was lower for all combinations in comparison to 'Zaozhong No. 6' grafted onto domestic loquat seedlings. Delayed incompatibility was observed when wild Henry loquat (*E. henryi*) was used as a rootstock. Differences in soluble solid contents due to the use of different rootstocks were observed, and there was a significant negative correlation between graft compatibility and this difference. Rootstock affected anatomical changes in graft unions. Wild rootstocks influenced scion growth, fruit bearing, and fruit quality. The wild Fragrant loquat (*E. fragrans*) appeared to have value as a rootstock as it produces greater cold resistance and improved scion growth.

III. REPRODUCTIVE PHYSIOLOGY

A. Flowering

Loquat flower-buds differentiate from terminal shoot buds. Flowers initiated from May to August. Muramatsu et al. (1963) working in Japan found that flower-bud initiation can be affected by the time and duration of shading treatments with the crucial period for physiological differentiation 1 month after fruit harvest. Lin observed that the flower-bud physiological differentiation of 'Dahongpao' was initiated in late May, with two differentiated peaks, one is in mid-late June and the other is in mid-late July.

Flower development of loquat has been reviewed by Li (1982) and Lin (1992), and Rodriguez et al. (2007). In Ningbo City, Zhejiang province of China (29°N), the main shoot of inflorescence panicles differentiate in the beginning of August: the spring lateral shoot and the summer main shoot in the middle or the end of August, sepals and petals in the beginning of September, and stamens and pistils in the middle or end of September. Sperm and egg nuclei develop in October, and anthesis occurs in November. The summer lateral shoot begins to differentiate flower buds in September, 1 month later than the spring main shoot, but anthesis takes place in November. Until late October, the process of differentiation in these four kinds of shoots (spring main and lateral shoots and summer main and lateral shoots) is similar with pistil primorida reaching 1.5 mm (Chen 1958). The stigma of loquat is closed, with some internal transfer cells for nutrition transportation.

B. Blossoming

The main panicle typically usually bears 70–100 flowers on 5–10 branched, secondary axes. Depending on panicle size, flower number can range from 30 to 260. One month after the panicle appears, inflorescences begin to bloom and flowers remain from late November to early-mid January in China, and 1 month earlier in the Mediterranean basin countries. In most cultivars, the solitary floret at the top of main panicle axis blooms first, followed by the secondary axes in the middle of the panicle, and then the basal axes. If the panicle bends downward or droops, progression of flower opening is from the center of the bend to both sides. The opening sequence of the secondary-axes inflorescence is as follows: first, one floret at the base blooms, then the other at the apex or near the apex; and finally all the rest bloom in an acropetal direction. Solitary flower wither after 8–14 days.

Generally speaking, loquat blooms from late fall to mid-winter; thus low temperatures in early winter affects blossoming and fruit set. Loquat is floriferous between 11 and 14°C and flowering duration is prolonged below 10°C. Temperatures below −6°C causes severe damage to blossoms and result in seriously damage to trees (Wang 2003). Under ordinary climatic conditions in China, the best period of bloom and fruit set is from mid-November to mid-late December (Wu et al. 1991). Sustained low but nondamaging temperatures may induce cold resistance of the flower bud but slows the development.

C. Pollen Biology

1. Morphology. Yang et al. (2009c) observed pollen morphology of six species of *Eriobotrya* with scanning electron microscopy (SEM). Pollen is of the N3P4C5 model, prolate with three equally distributed colposates, triate-foveolate with many stripes, and a tectum punctured exine with distinct differences among the six species. Guo et al. (2010) found significantly different morphologies among 15 triploids derived from 'Longquan' loquat.

2. Germination. Loquat pollen begins to germinate when the temperature is above 10°C, and germination can reach 70% at the optimum temperature of 20°C. Germination is very low above 35°C, and no pollen germinates below 5°C. Exposure to suitable temperatures for a few hours is sufficient to induce good pollen germination. In the northern edge of the loquat cultivation zone in China, though the weather is cold during the blossom period; however, pollination, fertilization, and fruit-setting are not influenced because the noon temperature is above 15°C (Huang 2000). Both plant growth regulators (GA, IAA, and NAA) and plant nutrients (B, Mn, and Ca) promote pollen germination of loquat (Ding et al. 1991).

3. Storage. In cryopreservation studies of loquat pollen, 30% water content was adequate for survival (Wang et al. 2004). Thawing temperatures did not significantly affect pollen vitality after storage in liquid nitrogen (−196°C).

D. Pollination and Fertilization

1. Environmental Factors. Loquat is cultivated in the northern edge of the subtropical zone, so that low winter temperature is the main factor affecting pollination and fertilization (Zhang et al. 2005). Loquat cannot set fruit properly below 10°C. At blooming and fruit-setting stages, unsuitable weather such as continuous chilly temperature with rain, cold temperatures below $-3°C$, and lack of sunshine causes pollination and fertilization failure and low fruit set (Li and Tang 2006). Liang et al. (2011a) found that the mixed solution of boron, sucrose, calcium, and diethyl aminoethylhexanoate (DA-6) significantly promoted *in vitro* pollen tube elongation in 'Zaozhong No. 6' and 'Jiefangzhong'.

2. Pollen Compatibility. Pollen compatibility studies have been long neglected because the amount of flowers and fruits in loquat are usually sufficient to meet growers' need, and are usually excessive, requiring thinning for optimal commercial quality. It was long assumed that all loquat cultivars are cross- and selfcompatible. However, Xia (1993) discovered that cross-pollination between different loquat cultivars set better than selfing. It is now clear that selfcompatibility is not a common trait in loquat. Selfincompatibility is an evolutionary strategy used by flowering plants to prevent selffertilization and promote outcrossing (De Nettancourt 1977). However, the release of flower-visiting insects, such as bees (*Apis dorsata* and *A. mellifera*), during the flowering period can increase fruit set (Mann and Sagar 1987). Loquat specifically shares with the rest of the Rosaceae species a gametophytic selfincompatibility (GSI) system based on S-RNases (Igic and Kohn 2001). This system exhibits high polymorphism at the selfincompatibility locus. Several alleles have been identified in loquat and intercompatibility groups have been established (Carrera et al. 2009; Gisbert et al. 2009b). The high number of alleles found at the *S*-locus makes it particularly interesting for genetic diversity studies and it is essential for correct orchard planning and suitable design of breeding programs. As with other members of the Rosaceae, interplanting cultivars with the same *S*-alleles would be unsuitable for cross-pollination.

E. Embryology

1. Micro- and Macrosporogenesis and Embryogenesis. Microsporogenesis and the development of the male gametophyte was investigated by Li and Ding (1984) and Li et al. (1986). After one periclinal division, one or two lines of primary sporogenous cells are formed in the sporogoniums and then divide into many microspore mother cells. The meiosis of the microspore mother cell is simultaneous and produce androspore tetrads which developed into mature pollen grains each with two cells.

The macrospores appear when the flower buds are diamond-shaped. The layer cells under the placental intine epidermis increase rapidly and gradually expand to be ovular primordia, which change the diamond-shaped flower buds into round-ones. The sporogonium under the top nucellus epidermis has large nuclei and dense cytoplasm. After periclinal division, the sporogonium divide into a parietal cell and a sporogenous cell. The sporogenous cell develops into the embryo sac mother cell. After periclinal and anticlinal divisions, the parietal cell divides to form poly-layered periphery nucellar tissue and pushes the embryo sac mother cell to the deep nucellar tissue forming the crassinucellate ovule. The embryo sac mother cell produces four macrospores after meiosis; the macrospore in the chalazal end divides three times continuously and forms an 8-nucleate embryo sac. The other three macrospores degenerate and are eliminated. The megasporocyte divides into two nuclei that move to different poles to form a binucleate embryo sac that divide again to form a 4-nucleate embryo sac. After the third division, four nuclei are produced in both the micropylar and chalazal ends. The Polygonium-type embryo sac has eight nuclei immersed in the same cytoplasm and contains an egg apparatus (egg cell and two synergids), three antipodal, and a central cell containing two polar nuclei (Lin 1992).

The proembryo development of loquat is of the Cruciferae (onagrad) type (Zheng and Liang 1989). The zygote laterally splits to an embryoid with a small roof-cell at the chalazal end and a large basal cell at the micropylar end. The basal cell splits laterally again and forms an upside down T-shaped proembryo with four cells. After two or three longitudinal and lateral divisions, the basal cell develops into a suspensor with four to eight cells. Endosperm development of loquat belongs to the nuclear type. As the embryo develops to the globular stage, wall formation commences in the micropylar end of the embryo sac. At the heart stage, the endosperm around the proembryo in the micropylar end moves gradually to the chalazal end, followed by disintegration until elimination. When the typical dicotyledonous embryo evolves, the endosperm and nucellar cap is absorbed completely (Lin 1985).

2. Seedlessness. GA_3 applied before bloom can induce seedless fruits as a result of endosperm abortion or embryo degradation (Goubran and El-Zeftawai 1986). Endosperm abortion causes the lack of nutrition for embryo development, so the zygotic embryo divides abnormally and finally disintegrates. The ovule when deprived at the initial phase forms abortive seed forms (Deng et al. 2009).

3. Seed Morphology and Physiology. The mature loquat seed is oval-shaped and slightly flat, 2.3×1.4 cm in size. Each seed weighs 1.5–2.1 g and the weight ratio of cotyledons to embryo and hypocotyls is greater than 400. There is great diversity in seed characters within loquat germplasm. Number of seed per fruit may vary from one to eight, generally speaking three or four. Among 128 loquat germplasm accessions tested, most had two to four fully developed seeds and a few abortive seeds per fruit under favorable pollination conditions (Jiang et al. 2009). Seeds are triquetrous, beige, with a few spots and a small indehiscent sheath.

Loquat seed is recalcitrant and loses germinating ability rapidly. Seed moisture content and temperature are the two most important factors for seed storage (Ellis et al. 1991). Storage at 15°C with 51% relative humidity prolonged storage life (Chen et al. 1998a,b). At 10°–15°C, germination rate was 95% after 330 days.

Water imbibition in fresh loquat seed is restricted by the seed coat resulting in a low germination rate. Moisture content of loquat seed can reach 58.7% in the cotyledon and 73.2% in the embryonic axis (Chen et al. 1998a). Seed vigor decreases rapidly if too much water is lost and will be completely lost when the moisture content is reduced to 35.5% in cotyledons (or 28.0% in the embryonic axis. Germination is optimal at 25°C and is inhibited below 10°C or above 32°C. Germination can be increased by light desiccation (3%–5%) or GA_3 (0.1–0.5 mmol L^{-1}) but is inhibited by abscisic acid. Seed germination can be increased by stratification at 4°C for 30 days (Polat 1997).

F. Fruit

Loquat fruit, as other pomes, derives from the receptacle beneath the ovary and is considered a pseudocarp containing a fleshy edible receptacle in addition to the ripened ovary. There are two peaks and two troughs during fruit development. Fruit development can be divided into four stages: fruitlet development, seed development, fruit enlargement, and fruit ripening. During the whole process, the vertical diameter of the fruit grows faster than the transverse diameter, resulting in an elipsoid fruit shape in longitudinal section (Jia 2009). Loquat fruitlets freeze below −3°C (Fan et al., 2010).

Generally, the flesh of seeded loquat is thin and has many large seeds, with flesh recovery as low as 65%–70%, which adversely affects its commercial value. Therefore, seedless loquat would have great economic significance. Hu and Lin (2010) had suggested the induction of the seedless loquat by treatment with gibberellic acid (GA_3) and N-(2-chloro-4-pyridyl)-N-phenylureia (CPPU) treatment before fertilization at the flower-bud stage. The laboratory of G. Liang has used triploidy and GA application to develop seedless loquat (see Section V. E).

IV. BREEDING OBJECTIVES

Breeding objectives in loquat must include consumer requirements for appearance and quality as well as grower acceptability. Objectives for scion cultivars include low seed number or seedlessness, attractiveness, various colored skin and flesh, fruit size, increased quality especially high soluble solids, cold resistance, disease resistance, season of ripening, and productivity.

In the Southern subtropical areas of China, there is a demand for cultivars that ripen earlier than 'Zaozhong No. 6' which is harvested in February or March. In the Northern subtropical areas (Southern Shaanzi, Southern Gansu, and Central Jiangsu), late ripening (July to August) is desirable. Loquat is soft and susceptible to mechanical damage, which leads to difficulty in storage and transportation. However, it is possible to prolong the shelf life of loquat for up to 6 months. Desirable tree characters include short shoots to facilitate crop thinning and fruit bagging to reduce labor costs, which have doubled over the last decade in China.

Breeding objectives in the Mediterranean countries are focused on increasing the diversity of cultivars. In the case of Spain, 90% of the production relies on 'Algerie' and its bud mutations and 'Magdal'. New cultivars earlier that 'Algerie' with better fruit quality and tolerance to scab are sought. In other Mediterranean countries, the industry is based on clones that do not meet the international commercial standards of quality, and new cultivars with better fruit quality and yield are the main objectives.

Breeding of loquat rootstocks has been neglected for many years. Rootstocks influence scion growth, fruit bearing, and fruit quality. In addition to providing adaptability to soil conditions the major objectives for rootstocks include dwarfness and graft compatibility. Breeding objectives in Spain are focused on production of dwarfing rootstocks, tolerance to the calcareous soils, and tolerance to soilborne diseases such as *Phytophthora* and *Armillaria*.

The breeding of clonal rootstocks that would be vegetatively propagated has not been considered in China because of the problem of shallow rooting. Seedling rootstocks have tap roots that prevent orchards being blown down by typhoons and strong winds. However, the use of clonal rootstocks would reduce orchard variability. Thus, loquat rootstocks with strong root systems are an important breeding aim.

V. BREEDING METHODS

A. Genetic Studies

Genetic studies of loquat carried out in China have been summarized by Zheng (2007). Fruit characters included juvenility, size, pericarp and flesh color, maturity, and soluble solids and plant characters included resistance to leaf spot caused by *Fabraea maculata*. Early fruiting was shown to be quantitatively inherited. Bearing age ranges from 3 to 9 years averaging about 5. In general, selections from early parents bear fruit earlier than from late parents. There is evidence that early fruiting is maternally inherited. Heritability of fruit shape, weight, and quality was 96%, 80.8%, and 90.0%, respectively. Soluble solids inheritance is similar to that of fruit quality. No differences have been found in reciprocal crosses. Fruit maturity is quantitatively inherited.

Inheritance of pericarp and flesh color was complex suggesting incomplete dominance of one or several genes. Flesh color did not segregate if parents were white or yellow-white. Crosses involving gold flesh color showed segregation for white, gold, light salmon, and deep salmon. Salmon color was dominant to gold.

Seedlings from crosses involving only selections susceptible to leaf spot were all susceptible. Reciprocal crosses of resistant and susceptible cultivars showed evidence of maternal inheritance suggesting the use of resistant seed parents in crosses.

B. Selection

Till the 1980s, loquat planting in China was based mainly on seed propagation and the use of grafted selections was unpopular. Seed propagation from natural hybrids is associated with abundant genetic variation and provides abundant material for clonal selection. Both white-fleshed and orange-fleshed cultivars have been selected (Zheng 2007).

White flesh in loquat is appreciated for the tender, juicy, thick flesh, and sweet fine flavor. China has the most abundant white-fleshed germplasm. Cultivars selected from various parts of the country include 'Baili', 'Baiyu', 'Guanyu', 'Keyansuobaisha', 'Libai', 'Ninghaibai', 'Ruantiaobaisha', 'Wugongbai', 'Xinbai No. 1', 'Xinbai No. 3', and 'Xinbai

No. 8'. The orange-fleshed loquat selections include 'Anhuidahongpao', 'Bahong', 'Changhon No. 3', 'Dawuxing', 'Donghuzao', 'Hongdenglong', 'Jiefangzhong', 'Longmen No. 1', 'Luoyangqin', 'Luzhou No. 6', 'Puxinben', 'Puxuan1', 'Taicheng No. 4', and 'Yangmeizhou No. 4'.

Selection of rootstocks is underway in China. The Fruit Research Institute of the Fujian Academy of Agricultural Science has selected dwarfing rootstocks from selections of *E. japonica*, such as 'Daduhe' and 'Min'ai No. 1'.

C. Hybridization and Mass Selection

Hybridization between selected elite cultivars that carried the traits of interest followed by selection, characterization, and testing of seedlings is the prominent breeding method in loquat. Breeding programs are being carried out in China (Zheng 2007; Shih 2007), Japan (Terai 2002), and Spain (Gisbert et al. 2007). At present new cultivars still primarily come from selection made by growers, since dedicated breeding efforts are recently initiated and the loquat has a fairly long juvenility period. Most cultivars currently grown come from commercial fields or from evaluation of broader germplasm collections. Cultivar surveys have been carried out in all countries where the crop is present: China (Zheng 2007), Mediterranean countries (Llácer et al. 2003), Pakistán (Hussain et al. 2007), and Turkey (Karadeniz and Şenyurt 2007; Polat 2007b).

D. Mutagenesis

Jiang et al. (2007) demonstrated that loquat showed a strong sensitivity to gamma irradiation, with high mutation rate. The chemical mutagens, most useful, belong to the class of alkylating agents including ethyl methyl sulfate. Colchicine has been widely used to increase ploidy. 'Shiro Mogi' loquat was obtained from induced mutation of irradiated seed of 'Mogi' (Lin 1998, Prederi 2001). Compared to 'Mogi', 'Shiro Mogi' has erect stems and greater vigor with larger fruit. The blossom period is nearly 10 days later so that cold weather can be avoided. Harvest of 'Shiro Mogi' is later than 'Mogi' and fruit is juicy with a fine, tender texture, with 13%–14% total soluble solids. The fruit shape was rounder and the whitish flesh was thicker with less seeds than 'Mogi'.

E. Breeding Seedless Loquat

Seedlessness in fruits is a commercially valuable trait in a number of traditional fruit crops (Table 5.3). This trait is especially desirable in

Table 5.3. Seedless fruit crops and mechanisms.

Crop	Seedless mechanism	References
Apple (& pear)	Parthenocarpy, genetic sterility (apetelous), triploidy	Chan and Crain (1967)
Atemoya	Triploidy, GA application	George and Paull (2008)
Banana	Parthenocarpy, genetic sterility reinforced by triploidy	Simmonds (1976)
Citrus	Triploidy, irradiation induced sterility	Ollitrault et al. (2008)
Cucumber	Parthenocarpy, genetic sterility, growth regulator induced parthenocarpy	Paris and Maynard (2008)
Fig	Parthenocarpy, lack of pollinator	Condit (1947)
Grape	Parthenocarpy or stenospermocarpy (early embryo abortion)	Ledbetter and Ramming (1989)
Loquat	Triploidy, growth regulators (GA)	Lin et al. (1999); Liang et al. (2011a)
Pineapple	Parthenocarpy, genetic sterility, selfincompatibility, elimination of pollinators (humming birds)	Collins (1960)
Tomato	Growth-regulator induced parthenocarpy	Gorguet et al. (2008)
Watermelon	Triploidy	Kihara (1951)

loquat because it has large seeds as in *Prunus* and multiple seeds, usually between 4 and 6, as in *Malus* (Lin et al. 1999; Janick 2011) and seed weight is a significant portion (15%–20%) of total fruit weight.

1. Induction of Seedlessness. Seedlessness is often associated with parthenocarpic fruit development achieved without fertilization (vegetative) although in some cases stimulus by pollination is required (aitonomic). Apparent seedlessness (pseudoparthenocarpy) is associated with early embryo abortion (stenospermocarpy) and is common in grapes (Ledbetter and Ramming 1989) where traces of seed are observable.

There are a number of strategies to induce seedlessness. These include: (1) exploiting unbalanced gametes in triploidy (apple, atemoya, banana, citrus, loquat, and watermelon); (2) sterility either genetic or induced by irradiation (apple, cucumber, and citrus); (3) elimination of pollination in self-incompatible parthenocarpic plants (fig and pineapple); and (4) application of growth regulators, usually gibberellic acid (atemoya and loquat) or auxin (tomato).

The exploitation of triploidy to induce seedlessness is a promising breeding technique in those crops where the triploid condition is a suitable ploidy level as found in the Rosaceae. For example, 10% of

5. BREEDING LOQUAT

apple and pear cultivars are triploid although the frequency of occurrence is less that 1%. In many triploids such as watermelon, pollination by diploids is still required to stimulate fruit development.

2. Production of Seedless Triploids in Loquat. There are four potential routes to the induction of triploids.

Meiotic Nonreduction (NR). NR in mega- or microsporogenesis of diploids ($2n = 34$) produces diploid instead of haploid gametes either from a failure of division II in meiosis or fusion of sporocytes in division II (Fig. 5.2). NR gametes occur with a frequency of <1% in rosaceous species. In loquat, the frequency of nonreduced gametes was estimated as 0.28% from an analysis of meiosis in pollen mother cells (Yan et al. 2011).

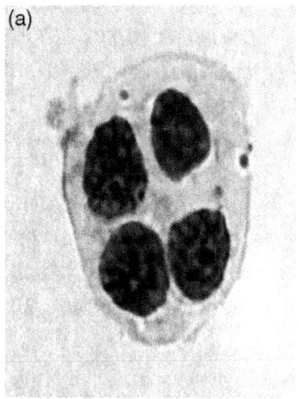

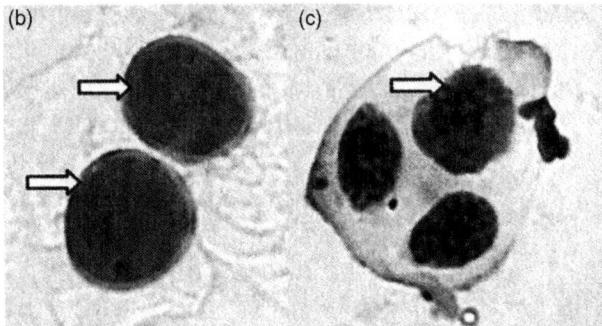

Fig. 5.2. "Tetrad stage" of diploid meiosis in microsporogenesis: (a) normal tetrad; (b) dyad due to failure of division II; (c) triad due to incomplete division II. Arrows show evidence of nonreduced microspores. *Source*: Yan et al. (2011).

Table 5.4. Frequency of ploidy levels in seedlings of diploid × diploid crosses in apple and loquat.

Crop	No. 2n × 2n seedlings	Distribution of seedlings % (no.)				
		Diploid	Triploid	Tetraploid	Pentaploid	Mixaploid
Apple	6,825	99.63	0.28 (19)	0.09 (6)		
Loquat	44,828	99.31	0.50 (225)	0.11 (50)	0.02 (10)	0.06 (26)

Sources: Einset (1959), Guo et al. (2007), Liang et al. (2011b).

The subsequent fusion of diploid and haploid gametes in diploid × diploid crosses leads to triploidy, often referred to as natural triploidy. In apple, triploid seedlings from 2n × 2n crosses occur with a frequency of 0.28% (Einset 1959) as compared to about 0.50% in loquat (Guo et al. 2007; Liang et al. 2011b) as shown in Table 5.4. In loquat, the frequency of triploidy varied with the seed parent from 0.18% to 1.62% (Guo et al. 2007). A total of 366 triploids in loquat have been identified by an analysis of 99,272 seed (G.L. Liang, unpublished).

Genomic *in situ* hybridization (GISH) of naturally occurring triploids using genomic DNA of the seed parent as the probe demonstrated that unreduced gametes resulted from NR in the seed parent with the haploid male gamete derived from either the seed parent or a different clone as a result of cross-pollination. (Liang et al. 2011c). However, karyotypic analysis of chromosomes of diploids and their related triploids (Fig. 5.3) indicate that some triploids may result from NR in microsporogenes based on an analysis of satellited chromosomes (Liang et al. 2011d).

Triploidy in loquat was determined by chromosome counts. Seeds extracted from fruits of individual cultivars, are disinfected and

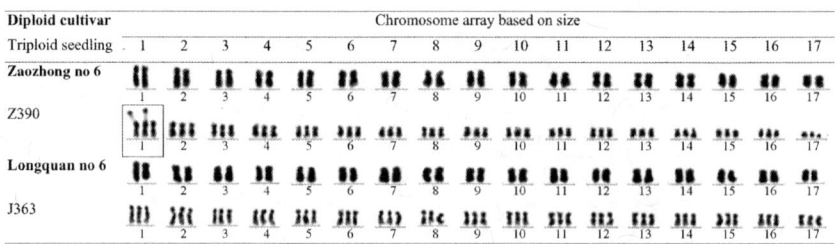

Fig. 5.3. Karyogram of two diploids and their related triploid seedlings. Satellite chromosomes are boxed. The two satellilted chromosome 1s in triploid Z390 must have been derived from nonreduction in a gamete of the diploid pollen parent containing one or two satellited chromosomes. *Source*: Liang et al. (2011d).

germinated in a peat mixture in the dark. Root tips were subjected to wall degradation by the hypotonic treatment method, pretreated with 0.002 M 8-hydroxyquinoline (a reagent to suppress spindle fiber formation); Carnoy's fixative (3 methanol:1 glacial acetic acid), enzymolysis (3% cellulose +3% pectinase), hypotonic treatment, fixation, smearing, flame drying, 5% Giemsa staining, microscopic examination, and photomicrography. Skilled technicians are able to evaluate 300 seedlings per day (Guo et al. 2007). Flow cytometry methods have been used in citrus for ploidy determination (Ollitrault et al. 2008) and may be useful in high-throughput assessment of ploidy in loquat seedlings.

Diploid × Tetraploid Crosses. If tetraploids are fertile, 100% of $4n \times 2n$ crosses are expected to be triploid, The production of natural tetraploids sometimes occurs from spontaneous somatic doubling but tetraploids can be readily induced by colchicine treatment in loquat (Kihara 1981). Triploids have also been achieved in loquat from $2n \times 4n$ crosses, where the pollen parent is tetraploid (Huang 1984, 1989). Blasco et al. (2011) applied different concentration and duration of colchicine to ungerminated seeds, apical buds, and whole seedlings, and obtained different rates of survival and ploidy level. Tetraploid plants were obtained in all treatments but the best treatment varied according to the plant material used. Since cytochimeras are a problem, in the Rosaceae, the tetraploid sectors must include LII of the meristem (Pratt 1983) to give rise to diploid gametes. The reliance on genetically prepotent tetraploids is a limitation of this system.

Culture of Triploid Nucellar Tissue. Since the endosperm of loquat is triploid, culture of endosperm tissue could lead to triploid clones (Chen and Lin 1991). Triploid plants from endosperm culture have been achieved in loquat (Lin et al. 1999).

Somatic Hybridization of $2n + n$ Cells. This technique involving protoplast fusion has been achieved in citrus. Haploid plants in citrus are produced, with very low efficiency, by pollination using irradiated pollen (Ollitrault et al. 2008) and can then be used to create haploid protoplasts. Protoplast isolation and culture for embryos has been achieved in loquat (Lin et al. 1989; Lin 1985, 1991, 1995) but protoplast fusion has not been reported. Development of haploids in loquat has been explored using microspore culture (Blasco et al. 2011; Padoan et al. 2011). Microspores from the late uninucleate to early binucleate pollen stages are most suitable for somatic embryogenesis. Callus production was 31.5% with 'San Filiparo' loquat in a medium supplemented with 1 mg L^{-1} of NAA and Zeatin.

3. **Morphology and Pomology of Seedless Triploid Loquat.** Triploid loquats set fruits in 2003 and 2004 but were aborted at a young stage without exogenous growth regulators. However the addition of GA during the flowering period resulted in fruit formation and it was later determined that GA at 100 ppm was the appropriate concentration (G. Liang, pers. commun.). By 2011, 86 triploids obtained from diploid cultivars had fruited and many selections had been propagated.

As compared to the diploid parents, triploid plants grow exuberantly producing large trees with few branches, dark green leaves with long dense villi, and notched leaf edges (Liang et al. 2011c).The circumference of trunks, diameter of annual branches, length, width and length × width of leaves of triploids were 1.7, 1.4, 1.6, 2.0, and 3.3 times greater than the diploid seed parent, respectively (Table 5.5). The number of branches and leaf index in triploids were 45% and 81% of diploids. Size of flowers and flower parts were significantly greater in triploids than diploids. Pollen germination of diploids averaged about 88% while triploid pollen germination ranged from 0% to 6.7%. Of 34 triploids examined, pollen germination was observed only in 19 plants, averaging 2.3%. (G. Liang, pers. commun.).

The seedless fruit show great variability is size and shape. Among 45 triploids analyzed, six had larger fruits and are considered elite or super strains (Table 5.6, Fig. 5.4). These triploids had thicker flesh with a flesh

Table 5.5. Comparison of plant morphology of diploid loquats and their related triploid seedlings.

Cultivar	Trunk circum. (cm)	No. of branches	Annual branch diam. (cm)	Leaf Length (cm)	Leaf Width (cm)	Leaf Length × width (cm²)	Leaf index
Dawuxing (2x)	25.5b[z]	7.2a	5.3b	27.5b	7.3b	201b	3.8a
3x seedling	50.0a	3.6b	7.1a	45.8a	15.6a	718a	2.9b
Longquan No.1 (2x)	29.7b	6.0a	6.5b	22.2b	7.0b	154b	3.2b
3x seedling	50.0a	2.4b	8.5a	36.3a	13.4a	488a	2.7a
Jinfeng (2x)	24.9b	7.0a	6.4	25.6b	7.2b	186b	3.5a
3x seedling	45.0a	3.0b	7.5	43.2a	15.6a	676a	2.8b
Zaohong No.3 (2x)	36.0b	7.0a	5.8b	25.7b	7.7b	197b	3.4a
3x seedling	49.0a	4.0b	7.4a	38.7a	14.1a	546a	2.8b

[z]Mean separation of 2x and related 3x means at 5% level.
Source: Liang et al. (2011e).

Table 5.6. Characteristics of fruit in diploid loquats and their related triploids.

Cultivars	Ploidy	Shape[z]	Flesh color[y]	No. of seeds	Fruit Weight (g)	Length (mm)	Width (mm)	Shape index	Edible portion (%)	TSS (%)	Composition (g 100 ml⁻¹) Sugar	Acid	Vit. C
Raotiaobaisha	2x	R	W	4–6	25.2 (34)[x]	37.8 (40)[x]	38.4 (40)[x]	0.97	62.0	14.3	8.97	0.47	1.79
H324	3x	LO	OY	0	50.3 (68)	56.2 (70)	40.1 (48)	1.40	86.0	12.5	8.16	0.36	1.82
Jinfeng	2x	O	OY	4–6	51.1 (133)	58.2 (86)	46.9 (68)	0.97	65.0	12.0	8.65	0.68	1.85
D425	3x	LO	OY	0	79.3 (103)	78.5 (94)	52.7 (61)	1.49	84.9	12.0	6.83	0.48	1.70
D327	3x	LO	OY	0	78.1 (85)	73.8 (90)	41.3 (50)	1.78	85.6	11.8	6.66	0.60	1.70
Dawuxing	2x	O	OY	4–6	58.7 (96)	62.5 (77)	45.1 (59)	0.97	65.0	12.8	8.65	0.68	1.85
A322	3x	LO	OY	0	65.8 (85)	73.0 (76)	50.0 (53)	1.56	83.5	11.5	6.53	0.61	1.68
A313	3x	LO	OY	0	62.2 (83)	70.0 (79)	49.0 (43)	1.49	82.5	11.7	6.83	0.48	1.70
A35	3x	LO	OY	0	63.1 (85)	73.0 (78)	49.0 (44)	1.78	85.2	11.5	6.66	0.60	1.70

[z]LO, long ovoid; O, ovoid; R, roundish.
[y]OY, orange yellow; W, white.
[x]Maximum.
Source: Liang et al. (2011b).

Fig. 5.4. Seedless loquat: (a) of H324 (white-fleshed); (b) D425 (orange flesh) with diploid 'Jinfeng' at far right. *Source*: Liang et al. (2011b).

recovery averaging 84% as compared to 64% for the diploid seed parents. Fruit index (L/W) tended to be greater in triploids indicating that increase in length of triploid fruits was greater than width. In addition, triploids tended to have a sunken calyx. The average fruit weight of triploids averaged 48.9% higher than corresponding diploids. Flesh color ranged from white to orange reflecting color in diploid parents. Sugar content and acid were slightly lower than corresponding diploids but fruit quality of triploids reached levels required by the industry (Table 5.4). Elite strains are currently under testing in various locations in China.

The morphological characteristics and fruit quality of these super strains indicate that this breeding technique holds enormous promise for creating commercial seedless loquat. However there are a number of issues that need to be solved as follows:

1. Is pollination required for fruit set in triploids? If so, diploid pollinators would be required in solid blocks of triploids since only $(1/2)^{17}$ of pollen of triploids is expected to be haploid or diploid assuming random chromosome segregation of the 17 extra chromosomes in Division I of meiosis.
2. For seedless loquat to be successful, a range of types must be generated including a range of ripening time and various internal and external quality features (color and shape) to meet market expectations. This will require selection from literally hundreds of triploids. It is suggested that many thousands of seeds from the best diploid cultivars be grown out to identify potential triploid cultivars resulting from unreduced gametes.
3. Flow cytometry which is more efficient than chromosome counts should be adapted as a technique to evaluate triploidy,
4. The optimum levels of GA application will have to be determined for each triploid.

F. Molecular Breeding

1. Molecular Markers. Molecular markers in loquat have been used mainly for assessing genetic diversity relationships among *Eriobotrya* species and related genera using PCR-derived dominant and codominant markers.

Dominant Markers. The first type of markers used in loquat were random amplified polymorphic DNA (RAPDs) (Williams et al. 1990). These markers were used for genotyping of plant material and genetic diversity studies in loquat collections. Since they are dominant, this type of markers gives less genetic information, but they can be used when there is a lack of DNA sequence information. The first RAPD marker studies were applied to Mediterranean germplasm collections. Vilanova et al. (2001) studied cultivars with common origin, mostly local selections from Spain, and although the diversity found was low, as expected, the markers allowed the identification of cultivar homonyms and synonyms, and identification of derived natural mutations. Later, a larger collection of loquat genotypes was characterized using these markers by Badenes et al. (2004a). Although the accessions had a close origin, they were mostly local cultivars from the Mediterranean basin countries, the markers provides a high rate of cultivar identification. RAPD markers have also been used to elucidate the identity of cultivars in Chinese collections (He et al. 2007; Yang et al. 2009a, b). In all these studies, RAPD markers were useful for identification of loquat genotypes, however, the lack of reproducibility among laboratories made them less convenient for genotyping collections, since the data are hardly exchangeable. They were replaced later by amplified fragment length polymorphism (AFLPs) (Vos et al. 1995). These markers consisted of selective amplification of restriction fragments from a total digested genomic DNA, combining a restriction enzyme analysis with ligation with previous amplified adapters. In this way the PCR amplification was not based on random priming but was selective, and the reproducibility of the technique was acceptable, allowing comparison of genotyping experiments among laboratories. Yang et al. (2009c) used these markers for establishing relationships among *Eriobotrya* species. Wu et al. (2011) analyzed the genetic diversity in 34 cultivated loquat germplasm accessions using AFLP. A total of 1,091 AFLP bands were generated with eight informative primer combinations, of which 993 (91.0%) were polymorphic bands. A mean of 124.1 polymorphic bands was detected for each AFLP primer combination.

Codominant Markers. Microsatellites, simple sequence repeat or SSR markers (Tautz and Schlötterer 1994) have also been used for loquat diversity studies. Badenes et al. (2004b) used SSRs cloned from *Malus* species to study a set of loquat accessions, found a good degree of transferability between *Malus* and *Eriobotrya* genera. This transferability was confirmed by Soriano et al. (2005), who found a high percentage of transferability of the SSR markers developed from *Malus* (Gianfranceschi et al. 1998) to *Eriobotrya*, confirming the usefulness of microsatellite markers from apple species as a suitable tool for loquat cultivar identification and genetic studies.

However, microsatellites optimized from the same species may be of greater value in assessing diversity. Gisbert et al. (2009a) developed and characterized the first 21 polymorphic microsatellite loci from a CT/AG-enriched loquat genomic library. The observed heterozygosity ranged between 0.20 and 1.00, and expected heterozygosity ranged between 0.17 and 0.81. Three markers were multilocus and eight loci departed significantly from Hardy–Weinberg equilibrium, and so were excluded from further evaluations The remaining markers were used by Gisbert et al. (2009b) for studying the genetic relationships among 83 loquat accessions from different countries belonging to the European loquat germplasm collection held at the Instituto Valenciano de Investigaciones Agrarias (IVIA). A total of nine SSRs from *Malus* and

Eriobotrya revealed 53 informative alleles and S-RNases consensus primers detected 11 selfincompatibility putative alleles. The combined data allowed an unambiguously separation of 80 of the 83 accessions studied.

Another type of marker used for genotyping loquat is the internal transcribed spacer (ITS) from ribosomal DNA (Li et al. 2007). Following the study of Zhai et al. (2008), ITSs from loquat rDNA were isolated by Yang et al. (2011). The authors characterized the sequences and found 130 variable points suitable for classification of *Eriobotrya* species.

Intersimple sequence repeats (ISSR) (Zietkiewicz et al. 1994) combined with RAPD markers has been used to establish the genetic relationships among Daduhe loquat (*E. prinoides* var. *dadunensis*), oakleaf loquat (*E. prinoides*) and common loquat (*E. japonica*) by Luo et al. (2011). ISSR markers proved very useful for studying genetic relationship among loquat germplasm from China (Xie et al. 2007)

2. Marker-Assisted Selection (MAS). Development of molecular markers linked to the traits of interest has the advantage of allowing selection of desirable genotypes as seedlings, reducing the number of plants evaluated. The lack of genetic studies has made MAS of limited value in loquat breeding to date. However, the selfcompatibility trait in *Eriobotrya*, is a gametophytic mechanism mediated by RNAs, is common to other rosaceous species, and homology of conserved regions have been used for development of markers for the selfincompatibility trait in loquat. S allele fragments were determined by amplification of partial degenerated primers designed from conserved regions of S allele sequences of *Malus × domestica* and *Pyrus* spp. (Raspé and Kohn 2002). Differences among the alleles amplified in accessions allows determination of the S alleles, to identify the compatible allele Sc and establish the groups of intercompatibility in loquat (Gisbert et al. 2009b). Differences in length of S alleles amplified by primers derived from conserved regions of RNAs gene on five selfincompatible loquat cultivars were determined by Carrera et al. (2009). Both studies provided a genomic PCR-based approach useful for identifying S-RNase alleles in loquat, which allow prediction of selfcompatibility of genotypes and intercompatible groups of cultivars for planning of crosses in breeding.

3. Genetic Linkage Maps. Construction of molecular marker linkage maps, is an important tool to expedite plant breeding. Such maps are of greatest value to breeding when markers are identified which are tightly linked to traits expanding the use of marker-assisted selection (MAS), which is described above. It is to then possible to use MAS for traits in which the associated gene is unknown, and where multiple loci (quantitative trait loci, QTLs) contribute to important traits. Linkage mapping, which allows quantitative character dissection, combined with candidate genes approaches, are new genomic tools for molecular breeding (Yamamoto et al. 2004).

Gisbert et al. (2009c) reported the construction of the first genetic maps of two loquat cultivars based on AFLP and microsatellite markers from *Malus*, *Eriobotrya*, *Pyrus*, and *Prunus* (Fig. 5.5). An F_1 population

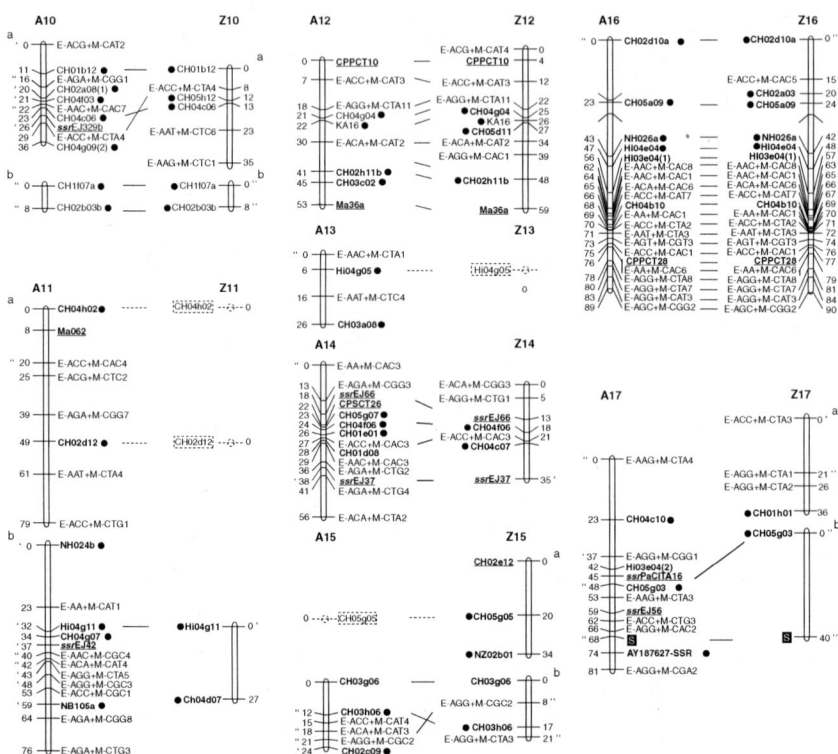

Fig. 5.5. Molecular genetic maps of 'Algerie' and 'Zhaozhong-6'. The map includes AFLP and SSR markers. The codominant markers came from *E. japonica* (16), *P. persica* (9), *Malus* × *domestica* (3), *P. armeniaca* (1), and *P. salicina* (1). Seventeen linkage groups were obtained, which agrees with the single haploid chromosome number of Pomoideae. The map of 'Algerie' contains 177 loci (83 SSRs y 94 AFLPs) expanding 900 cM. The map of 'Zaozhong-6' contains 146 loci (64 SSRs y 82 AFLPs), expanding 870 cM. In bold the codominant markers used. SSR markers from loquat underlined. The arrows indicated the markers establishing synteny between linkage groups. Solid circles indicate anchor markers with other *Malus* and *Pyrus* maps. Source: Adapted from Gisbert et al. (2009c).

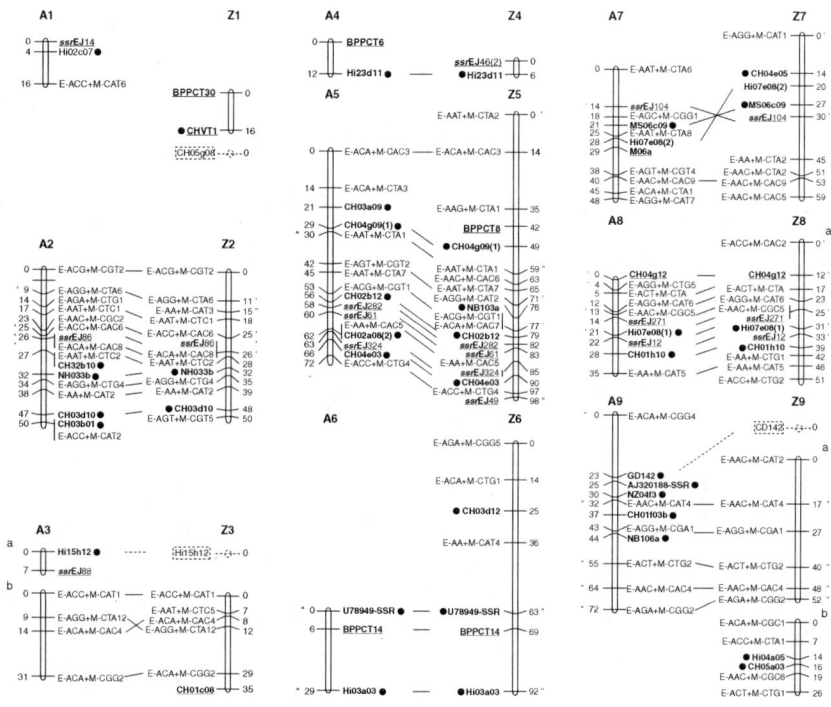

Fig. 5.5. (*Continued*)

consisting of 81 individuals, derived from the cross between 'Algerie' and 'Zaozhong-6', was used to construct both maps. A total of 111 scorable simple sequence repeat (SSR) loci were identified by testing of 440 SSR primer pairs in the analyzed progeny. The SSR transferability to *Eriobotrya* was found to be 74% from apple, 58% from pear and 49% from *Prunus* spp. In addition, 183 polymorphic AFLP were used to establish the maps. The 'Algerie' map was organized in 17 linkage groups covering a distance of 900 cM and comprising 177 loci (83 SSRs and 94 AFLPs) with an average marker distance of 5.1 cM. The 'Zaozhong-6' map covered 870 cM comprising 146 loci (64 SSRs and 82 AFLPs) with an average marker distance of 5.9 cM. The selfincompatibility trait (S-locus) was evaluated as an additionalcodominant marker, by using the primers described above, and permitting assessment of all alleles present. The segregation of the S-locus marker was consistent with a half-compatible reaction in which one S-allele was shared by both parents (1:1). Selfincompatibility trait was mapped at a LOD score of 6.0 at the distal part of LG17, agrees with maps from species from Maloideae subfamily. The high degree of synteny (retention of gene order on linkage groups) observed between *E. japonica* and other *Maloideae* species supports the potential use of the biotechnology resources from *Malus* to loquat, such as markers, sequence information (Velasco et al. 2010), and candidate genes.

4. Transformation. Loquat transformation has not been achieved. However, regeneration protocols are in progress from leaves, stems, and roots (Blasco et al. 2011)

VI. FUTURE PROGRESS

Loquat is an ancient rosaceous fruit of China with records indicating that has been grown for over 2000 years (Lin 2008). The fruit is still widely acclaimed and appreciated in China and plantings are still increasing. The fruit has possibilities of becoming an important crop in subtropical and Mediterranean climates. Loquat is an intriguing species that blooms in autumn and early winter and is one of the first gifts of spring. The delicious and attractive fruit is nonclimacteric and has the potential of long storage life and can be consumed both fresh and processed. Its main disadvantage is the many large seeds and the tender flesh. It is clear that many of the problems of this crop could be improved by breeding efforts. A survey of cultivars that have been selected from open pollinated seed indicate many with high productivity, disease resistance, large size, various flesh color, high Brix, rich flavors, attractive appearance, various ripening dates, and long keeping quality.

Most efforts at crop improvement have been achieved by evaluation and selection of existing cultivars that have been selected around the world but it is clear that vigorous programs of hybridization and recurrent selection should be able to produce superior clones that are adapted to specific areas. Efforts are underway in China (Zheng 2007), Japan (Lin 1998), Spain (Gisbert et al. 2007), and Turkey (Tepe et al. 2011). At the present time seedlessness has been achieved in China by selection of naturally occurring triploid clones which require applications of gibberellic acid for fruit set. A number of seedless triploids have been produced and are currently under test to determine which selections have a chance at commercial success. However, more efforts are needed to exploit an even wider selection of germplasm.

The production of loquat would be greatly improved by the use of dwarfing rootstocks (Janick 2011). Dwarfing in loquat has been achieved with quince rootstocks but there are compatibility problems. Evaluation efforts are underway to determine if dwarfing rootstocks could be obtained within diverse *Eriobotrya* species (Zhang et al. 2010)

Advances in genomics and synteny in the Rosaceae offer exciting possibilities for loquat improvement. (Hummer and Janick 2009; Badenes et al. 2009). The development of saturated gene maps with codominant and transferable markers suggest that MAS could be used advantageously in loquat. Many of the genes responsible for desirable quality traits within *Prunes*, *Malus*, *Pyrus*, *Cydonia*, and *Fragaria* may be relevant to loquat. These include genetic dwarfing, enhanced fruit quality, and disease resistance. Furthermore, use of transformation to incorporate critical genes from other fruit in the Rosaceae might minimize consumer resistance, which is greater with transgenesis using more distant species. Joint conventional and molecular breeding efforts involving scientists involved with loquat throughout the world would be the best strategy to achieve the desired goals of genetic improvement.

LITERATURE CITED

Badenes, M.L., T.P. Canyamás, C. Romero, E. Giordani, J. Martínez-Calvo, and G. Llácer. 2004a. Characterization of underutilized fruits by molecular markers: A case study of loquat. Genet. Resour. Crop Evol. 51:335–341.

Badenes, M.L., T. Canyamás, C. Romeron, J.M. Soriano, and G. Llácer. 2004b. Genetic diversity of an European collection of loquat based on RAPD and SSR molecular markers. Options Méditerranéennes 58:53–56.

Badenes, M.L., S. Lin., X. Yang, C. Liu, and X. Huang. 2009. Loquat (*Eriobotrya* Lindl.). p. 525–538. In: K.V. Folta and S.E. Gardiner (eds.), Genetics and genomics of Rosaceae. Springer, New York.

Blasco, M.M. Naval, and M.L. Badenes. 2011. Production of haploid and tetraploid loquat (*Eriobotrya japonica* Lindl.). Proceedings of 19th EUCARPIA General Congress. May 21–25, 2012. Budapest.

Blumenfield, A. 1995. Underutilized fruit trees in Israel. p. 31–38. In: G. Llácer U. Aksoy and M. Mars (eds.), Underutilized fruit crops in the Mediterranean region, Vol. 13. Cahiers Options (CIHEAM)

Carrera, L., J. Sanzol, M. Herrero, and J.I. Hormaza. 2009. Genomic characterization of self-incompatibility ribonucleases (*S-RNases*) in loquat (*Eriobotrya japonica* Lindl.) (Rosaceae, Pyrinae). Mol. Breed. 23:539–551.

Chan, B.G. and J.C. Crain. 1967. The effect of seed formation on subsequent flowering in apple. Proc. Am. Soc. Hort. Sci. 91:63–68.

Chen, J.S., R.Z. Chen, and J.R. Fu. 1998a. Germination character of loquat (*Eriobotrya japonica* L.) seed. Seed 99 (6):3–6.

Chen, J.S., R.Z. Chen, and J.R. Fu. 1998b. Study on the desiccation deterioration and moist storage of loquat seeds. J. Sun Yatsen Univ. Suppl. 1998 (4):11–14.

Chen, J.Y., B. Chen, L.M. Zhang, M.H. Hu, and X.D. Xu. 2003. Effects of different sampling method for scanning electron microscopic observation on pollen morphology of loquat varieties. Fujian J. Agr. Sci. 18 (2):107–111.

Chen, W.X. 1958. Observation of biological characteristics in loquat. J. Fujian Agr. Coll. 7:29–49.

Chen, Z.G., and S.Q. Lin. 1991. Some research on micropropagation of horticultural crops. p. 28–32. In: R. Chen (ed.), Advances in biotechnology in China (in Chinese). China Scientech Press, Beijing.

Collins, J.L. 1960. The pineapple: Botany, cultivation and utilization. Leonard Hill, Interscience Publ. London.

Condit, I.J. 1947. The fig. Chronica Botanica, Waltham.

DeNettancourt, D., 1977. Incompatibility in angiosperms. Monographs on theoretical and applied genetics, Vol. 3. Springer-Verlag, Berlin.

Deng, Y.Y., X.H. Yang, and D.G. Li. 2009. Anatomical study on the seedless formation of loquat (*Eriobotrya japonica*) treated by GA_3. J. Hort. Sinica. 26 (3):409–413.

Ding, C.K., Q.F. Chen, Q.Z. Xia, and T.L. Sun. 1991. The effects of mineral elements and growth regulators on the pollen germination and fruit set of loquat trees. China Fruits 1991 (4):18–20, 40.

Ding, C.K., Q.F. Chen, T.L. Sun, Q.Z. Xia, and D.W. Zhu. 1995. Germplasm resources and breeding of *Eriobotrya japonica* Lindl. in China. Acta Hort. 403:121–126.

Einset, J. 1959. Spontaneous polyploidy in cultivated apples. Proc. Am. Soc. Hort. Soc. 59:291–302.

Ellis, R.H., T.D. Hong, and E.H. Roberts. 1991. Seed moisture content, storage, viability and vigor. Seed Sci. Res. 1:275–279.

Fan, G.E., J.R. Qian, G.P. Qiu, Y.P. Wang, and X.F. Wang. 2010. Temperature and humidity of the year and the effect on the development of loquat. Zhejiang Citrus. 27 (3):30–32.

García-Legaz, M.F., E. López Gómez, J. Mataix Beneyto, A. Torrecillas, and M.J. Sánchez-Blanco. 2005. Effects of salinity and rootstock on growth, water relations, nutrition and gas exchange of loquat. J. Hort. Sci. Biotechnol. 80:199–203.

George, A. and R.E. Paull. 2008. Annona squamosa × Annona cherimola atemoya. p. 54–60. In: J. Janick and R.E. Paull (eds.), Encyclopedia of fruit and nuts. CABI, Wallingford, Oxfordshire, UK.

Gianfranceschi, L, N. Seglias, and R. Tarchini. 1998. Simple sequence repeats for the genetic analysis of apple. Theor. Genet. 96:1069–1076.

Gisbert, A.D., A. Guillem, J. Martínez-Calvo, G. Llácer, and M.L. Badenes. 2007. Contribution of biotechnology in genetic studies and breeding of loquat at IVIA, Spain. Acta Hort. 750:93–96.

Gisbert, A.D., L.I. Capuz, J.M. Soriano, G. Llácer, C. Romero, and M.L. Badenes. 2009a. Development of microsatellite markers from loquat [*Eriobotrya japonica* (Thunb.) Lindl.]. Mol. Ecol. Res. 9:803–805.

Gisbert, A.G., J. Martinez-Calvo, G. Llácer, C. Romero, and M.L. Badenes. 2009b. Genetic diversity evaluation of a loquat [*Eriobotrya japonica* (Thun.) Lindl.] germplasm collection by SSR and S-allele fragments. Euphytica 168:121–134.

Gisbert, A.G., J. Martinez-Calvo, G. Llácer, M.L. Badenes, and C. Romero. 2009c. Development of two loquat [*Eriobotrya japonica* (Thunb.) Lindl.] linkage maps based on AFLP and SSR markers from different Rosaceae species. Mol. Breed. 23:523–53.

Gorguet, B., A.W.van Heusden, and P. Lindhout. 2008. Parthenocarpic fruit development in tomato. Plant Biol. 7:131–139.

Goubran, F.H. and B.M. El-Zeftawai. 1986. Induction of seedless loquat. Acta Hort. 179:381–384.

Guo, Q.G., K. Ji, Q. Wu, Q. He, Y.K. Wu, X.L. Li, and G.L. Liang. 2010. Pollen morphology and viability of Longquan No. 1 natural triploid of loquat. J. Fruit Sci. 27:391–396.

Guo, Q., X. Li, W. Wang, Q. He, and G. Liang. 2007. Occurrence of natural triploids in loquat. Acta Hort. 750:125–127.

He, Q., X.C. Zhao, Q.G. Guo, X.L. Li, and G.L. Liang. 2007. Molecular identification of 5 loquat cultivars. Acta Hort. 750:155–158.

Hu, Z.Q., and Y.G. Lin. 2010. Effect of GA_3+ CPPU induction time on loquat pit development. Fujian J. Agr. Sci. Publ. 25:707–710.

Hartmann, H.T., and D.E. Kester. 1975. Plant propagation, 3rd ed. Prentice Hall, Englewood Cliffs, NJ.

Huang, J.S. 1984. The culture of tetraploid loquat. Ming No. 13. China Fruits 984 (2): 27–29.

Huang, J.S. 1989. The main achievements of scientific research on loquat forty years after liberation. China Fruits 1989 (2):5–8.

Huang. J.S. 2000. New cultivation techniques of Loquat, PN Fujian Science & Technology Press, Fuzhou.

Hummer, K.E. and J. Janick. 2009. Rosaceae: Taxonomy, economic importance, genomics. p. 1–17. In: K.V. Folta and S.E. Gardiner (eds.), Genetics and genomics of Rosaceae. Springer, New York.

Hussain, A., N.A. Abbasi, and A. Akhtar. 2007. Fruit characteristics of different loquat genotypes cultivated in Pakistan. Acta Hort. 750:287–292.

Igic, B., and J.R. Kohn. 2001. Evolutionary relationships among self incompatibility RNases. Proc. Natl. Acad. Sci. U S A 98:13167–13171. doi:10.1073/pnas. 231386798I.

Janick, J. 2011. Prediction for loquat improvement in the next decade. Acta Hort. 887:25–29.

Jia, D.X. 2009. Investigation in fruit development and characteristics of 'Dahongpao' loquat in green house in North China. Northern Fruits 2009 (4):18, 28.

Jiang, J.M., X.P. Chen, J.H. Xu, Y.J. Liu, H.Y. Liu, and S.Q. Zheng. 2007. Irradiation mutagenesis in loquat: fruit trait mutations. Acta Hort. 750:249–252.

Jiang, F., A.P. Huang, Z.F. Chen, C.J. Deng, X.M. Chen, X.P. Chen, X.Y. Zhang, L.J. Zhang, and S.Q. Zheng. 2009. Study on the seed traits in loquat (*Eriobotrya Japonica* Lind.) germplasm resource. J. Fujian Agr. Coll. 2009 (4):19–24.

Karadeniz, T. and M. Şenyurt. 2007. Pomological characterization of loquat selections of the Black Sea region of Turkey. Acta Hort. 750:113–116.

Kihara, H. 1951. Seedless watermelon. Proc. Am. Soc. Hort. Sci. 58:217–230.

Kihara, H. 1981. Seedless fruit. Plant Breed. Abstr. 51: 224.

Ledbetter, C.A., and D.W. Ramming. 1989. Seedlessness in grapes. Hort. Rev. 11:159–164.

Li, K.Y., and M.Q. Ding. 1984. Preliminary investigation on the embryonic development of loquat embryo: Microsporogenesis and the development of male gametophyte. J. Hubei Univ. (Nat. Sci. Ed.) 1984 (1):91–93.

Li, K.Y., M.Q. Ding, and M. Li. 1986. Preliminary investigation on the embryonic development of loquat embryo: Microsporogenesis and the development of male gametophyte. J. Hubei Univ. (Nat. Sci. Ed.) 1986 (1):90–93.

Li, N.Y. 1982. Observation on differentiation of flower bud in loquat. China Fruits 1982 (2): 12–16.

Li, P., S.Q. Lin, G.B. Hu, and Y.H. He. 2007. A preliminarily phylogeny study of the *Eriobotrya* based on the ITS sequences. Acta Hort. 750:241–24.

Li, Z.Y., and W.S. Tang. 2006. Integrated control of frozen damage of the big five-pointed Star Loquat. Fluoresc. Southwest Hort. 34 (4):65–66.

Liang, G.J., G.P. Huang, L. Deng, J.B. Zhong, and X.W. Kong. 2011a. Effects of boron, sucrose, calcium and DA-6 on growth of pollen tube of loquat. J. Zhaoqing Univ. 32 (2): 50–52, 56.

Liang, G.L., W.X. Wang, X.L. Li, Q.G. Guos, S.Q. Xiang, and Q. He. 2011b. Selection of large-fruited triploid plants of loquat. Acta Hort. 887:95–100.

Liang, G.L., W.X. Wang, S.Q. Xiang, C.G. Guo, and X.L. Li. 2011c. Genomic in situ hybridization (GISH) of natural triploid loquat seedings. Acta Hort. 887:97–99.

Liang, G.L., W.X. Wang, S.Q. Xiang, Q.G. Guo, X.L. Li, and Q. He. 2011d. Karyotype of diploid and triploid loquat. Acta Hort. 887:255–259.

Liang, G.L., W.X. Wang, S.Q. Xiang, Q.G. Guo, X.L. Li, and Q. He. 2011e. Morphological comparing between diploid and triploid loquat. Acta Hort. 887:261–263.

Lin, Q.L. 1985. A study on the formation of the plantlets from loquat endosperm *in vitro*. J. Fujian Agr. Univ. 14:117–125.

Lin, S.Q. 1991. Callus establishment from embryo callus protoplast in loquat. J. Fujian Agr. Univ. 20:179–184.

Lin, S.Q. 1992. Some observations on embryogenesis in loquat. J. Fujian Agr. Univ. 21:67–71.

Lin, S.Q. 1995. Plant regeneration from protoplast in loquat and some basic research. Ph.D. dissertation. Fujian Agr. Univ., China.

Lin, S.Q. 1998. Loquat's production and scientific research in Japan. South China Fruits 27 (5): 30–31.

Lin, S.Q., X.H. Yang, C.M. Liu, Y.L. Hu, Y.H. He, G.B. Hu, H.L. Zhang, X.L. He, and Liu, Z.L. 2004. Natural geographical distribution of genus *Eriobotrya* plants in China. Acta Hort. Sinica 31 (5): 569–573.

Lin, S.Q. 2008. *Eriobotrya japonica:* loquat. p. 642–651. In: J. Janick and R.E. Paull (eds.), Encyclopedia of fruit and nuts. CABI, Wallingford, Oxfordshire, UK.

Lin, S.Q., J.T. Ling, N. Nito, and M. Iwamasa. 1989. Isolation and culture of protoplast in loquat. J. Hort. Soc. Japan 58 (Suppl. 2): 48p–49.

Lin, S.Q., R.H. Sharpe, and J. Janick. 1999. Loquat: Botany and horticulture. Hort. Rev. 23:233–276.

Lin, S.Q., X. Huang, J. Cuevas, and J. Janick. 2007. Loquat: An ancient fruit crop with a promising future. Chronica Hort. 47 (2): 12–15.

Llácer, G, M.L. Badenes, and J.M. Calvo. 2003. Plant material of loquat in Mediterranean countries. Proc. First Int. Loquat Symp. Options Mediterraneennes 58:45–52.

Luo, N., Y.Q. Wang, Y. Fu, and Q. Yang. 2011. Genetic relationships among *Eriobotrya prinoides* var. *dadunensis*, *Eriobotrya prinoides* and *Eriobotrya japonica* using RAPD and ISSR markers. Acta Hort. 887:59–64.

Mann, G.S. and P. Sagar. 1987. Activity and abundance of flower visiting insects of loquat. India J. Hort. 44:123–125.

Manuel, F.G., L., Elvira, M.B., Jorge, N., Alejandra, and M., Jesus Sanchez-Blanco, 2008. Physiological behaviour of loquat and anger rootstocks in relation to salinity andcalcium addition. J. Plant Phys. 165:1049–1060.

Morton, J.F. 1987. Loquat. p. 103–108. In: Fruits of warm climates. Creative Resource Systems, Inc., Winterville, NC.

Muramatsu, H, T. Ikeda, and I. Ichinose. 1963. Research on flower bud differentiation of loquat (in Japanese). Kyushu Agric. Res. (Horticulture Section). 1963 (19):228–229.

Nesumi, H. 2006. Loquat (Biwa). Horticulture in Japan 2006. Shoukadoh Publication p. 85–95.

Ollitrault, P., D. Dominique, F. Luro, and Y. Froelicher. 2008. Ploidy manipulation for breeding seedless triploid citrus. Hort. Rev. 30:323–352.

Padoan, D., B. Chiancone, M.A. Germanà, P.S.S.V. Khan, I. Barany, M.C. Risueño, and P.S. Testillano. 2011. First stages of microspore reprogramming to embryogenesis through isolated microspore culture in loquat. Acta Hort. 887:285–290.

Paris, H.S., and D.N. Maynard. 2008. Cucumis spp.: cucumber, melon, West India gherkin, kiwano. p. 282–292. In: J. Janick and R.E. Paull (eds.), Encyclopedia of fruit and nuts. CABI, Wallingford, Oxfordshire, UK.

Polat, A.A. 1997. Determination of germination rate coefficients of loquat seeds and their embryos stratified in various media for different duration (in Turkish with an English summary). Turkish J. Agr. For., 21 (3):219–224.

Polat, A.A. 2007a. Loquat production in Turkey: Problems and solutions. Eur. J. Plant Sci. Biotech. 1 (2): 187–199.

Polat, A.A. 2007b. Selection studies on loquat growing in Bakras (Hatay), Turkey. Acta Hort. 750:169–174.

Polat, A.A., and N. Kaska. 1992. Investigations on the budding compatibility and forming of bud union between some loquat varieties and Quince-A rootstock (in Turkish withan English summary). Turkish J. Agr. Forestry 16 (4):773–788.

Pratt, C. 1983. Somatic selection and chimeras. p. 172–185. In: J.N. Moore and J. Janick (eds.), Methods in fruit breeding. Purdue University Press, West Lafayette IN.

Prederi, S. 2001, Mutation induction and tissue culture in improving fruits. Plant Cell Tiss. Org. Cult. 64:185–2010.

Raspé, O., and J.K. Kohn. 2002. S-allele diversity in *Sorbus aucuparia* and *Crataegus monogyna* (Rosaceae: Maloideae). Heredity 88:458–465.

Rodriguez, M.C., J. Cuevas, and J.J. Hueso. 2007. Flower development in 'Algerie: loquat under scanning electron microsopy. Acta Hort. 750:337–342.

Shih, J. 2007. Loquat production in Taiwan. Acta Hort. 750:55–60.

Simmonds, N.W. 1976. Bananas p. 211–213. In: N.W. Simmonds (ed.), Evolution of crop plants. Longman, London.

Soriano J.M., C. Romero, S. Vilanova, G. Llácer, and M.L. Badenes. 2005. Genetic diversity of loquat [(*Eriobotrya japonica* (Thunb.) Lind.] assessed by SSR markers. Genome 48:108–114.

Tautz. D. and C. Schlötterer. 1994. Simple sequences. Curr. Opin. Genet. Dev. 4 (6): 832–837.

Tepe, S., E. Turgutoglu, M.A. Arslan, and A.A. Polat. 2011. Improvement of loquat by conventional breeding. Acta Hort. 887:89–94.

Terai, O. 2002. Breeding loquat in Japan. First International Symposium on Loquat. Valencia, Abstr.

Thunberg, C.P. 1784. Flora Japonica. Linnean Society, London.

Velasco et al. 2010. The genome of the domesticated apple (*Malus* × *domestica* Borkh.). Nature Genetics 42:833–839.

Vilanova S., M.L. Badenes, J. Martínez-Calvo, and G. Llácer. 2001. Analysis of loquat germplasm (*Eriobotrya japonica* Lindl.) by RAPD molecular markers. Euphytica 121:25–29.

Vos, P., R. Hogers, M. Bleeker, M. Reijans, T. Vande Lee, M. Hornes, A. Frijters, J. Pot, J. Peleman, M. Kuiper, and M. Zabeau. 1995. AFLP: a new technique for DNA fingerprinting. Nucleic Acids Res. 23:4407–4414.

Wang, F.F. 2003. Meteorological condition analysis for loquat cultivation in Shixing town. Guangdong Geol. 2003 (1):45–46.

Wang, J.F., Y.X. Liu, X.J. Liu, and S.Q. Lin. 2004. Dry freezing cryopreservation of loquat pollen. Chinese Agr. Sci. Bul. 20 (1): 1–2, 20.

Williams J.G., A.R. Kubelik, R. Livak, J.A. Rafalski, and S.V. Tingey. 1990. DNA polymorphisms amplified by arbitrary primers are useful as genetic markers source. Nucleic Acids Res. 18:6531–6535.

Wu, D.H., Z.Z. Liu, X.J. Fei, J.H. Chen, G.M. Yu, Y.S. Li, and B.L. Sheng. 1991. A study on low-temperature effect on blossom and fruit-setting of loquat. Fujian Fruits 1991 (3):10–13.

Wu, J.C., X.H. Yang, and S.Q. Lin. 2011. Genetic diversity of loquat cultivars assessed by AFLP markers. Acta Hort. 887:43–48.

Xia, Q.Z. 1993. Pollination and fruit-setting rates of different loquat cultivars' combinations. Fujian Fruits. 1993 (2):3–4.

Xie, J.H., X.H. Yang, X.Q. Lin, and W. Wang, 2007. Analysis of genetic relationships among Eriobotrya germplasm in China using ISSR markers. Acta Hort. 750:203–208.

Yamamoto, T., T. Kimura, T. Saito, K. Kotobuki, N. Matsuta, R. Liebhard, C. Gessler, W.E. VandeWeg, and T. Hayashi. 2004. Genetic linkage maps of Japanese and European pears aligned to the apple consensus map. Acta Hort. 663:51–56.

Yan, J., Y.Q. Wang, L. Tao, and N. Luo. 2011. Meiosis of pollen mother cells of loquat. Acta Hort. 887:271–274.

Yang, X.H., K. Glakpe, S.Q. Lin, Y.L. Hu, Y.H. He, J.R. Yuanshi, Y.X. Liu, G.B. Hu, and C.M. Liu, 2005. Taxa of plants of genus Eriobotrya around the world and native of Southeastern Asia. J. Fruit Sci. 22 (1):55–60.

Yang, X.H., C.M. Liu, and S.Q. Lin. 2007. The genetic relationship among common loquat, Daduhe loquat and Oakleaf loquat, according to RAPD and AFLP analysis. Subtrop. Plant Sci. 36 (2):9–12.

Yang, X.H., P. Li, C.M. Liu, and S.Q. Lin. 2009a. Genetic diversity in Eriobotrya genus and its closely related plant species using RAPD markers. J. Fruit Sci. 26:55–59.

Yang, X.H., C.M. Liu, and S.Q Lin. 2009b. Genetic relationships in Eriobotrya species as revealed by amplified fragment length polymorphism (AFLP) markers. Scientia Hort. 122:264–268.

Yang, X.H., Y.X. Wu, and S.Q. Lin. 2009c. SEM observation on the pollen morphology of six Eriobotrya plants. J. Fruit Sci. 26:572–576.

Yang X.H., S.Q. Lin, G.B. Hu, P. Li, S.J. Xu. 2011. Preliminary study on ITS sequencing and characterization of Eriobotrya. Acta Hort. 887:85–88.

Zhai, H.C., Y.N. Song, and W.W. Zheng. 2008. ITS sequence analysis of nuclear ribosomal DNA of Prunus mume in Fujian Province. Subtrop. Plant Sci. 37 (1):12–16.

Zhang H.L., Z.K. Zhang, S.Q. Lin, and J.G. Li. 2010. Screening wild Eriobotrya species as rootstocks for loquat. J. Hort. Sci. Biotech. 85 (5):399–404.

Zhang, C.X., H.K. Wang, C.R. Chu, and J.H. Cai. 2005. The investigation on cold resistance of some main varieties of 'Baisha' loquat. Fujian Fruits 2005 (1):23–24.

Zheng, S.K., and T.G. Liang. 1989. The preliminary study of embryo and endosperm development in loquat. J. Fruit Sci. 6:229–231.

Zheng, S.Q. 2007. Achievement and prospect of loquat breeding in China. Acta Hort. 750:85–91.

Zietkiewicz, E., A. Rafalski, and D. Labuda. 1994. Genome fingerprinting by simple sequence repeat (SSR)-anchored polymerase chain reaction amplification. Genomics 20:176–183.

6

Prognostic Breeding: A New Paradigm for Crop Improvement

Vasilia A. Fasoula
Institute of Plant Breeding, Genetics and Genomics and
Center for Applied Genetic Technologies
The University of Georgia
111 Riverbend Road
Athens, GA 30602, USA

ABSTRACT

Innovative approaches are required to accelerate the average annual genetic gain through selection in breeding programs. Prognostic breeding is a methodology that embodies field phenotyping for high and stable crop yield of plants and sibling lines grown at ultrawide plant spacing. Two novel prognostic equations assess concurrently the yield potential and stability of performance. Prognostic breeding introduces the use of sibling testing where the stability of performance of a selected plant is determined by the concurrent evaluation of its siblings, in contrast to traditional breeding where the stability of performance of a selected plant is determined by the evaluation of its progenies in following generations. Prognostic breeding provides an integrated methodology where inbreeders, outbreeders, and clonally propagated crops are treated similarly, although sibling lines are selfed in inbreeders, outcrossed in outbreeders, and cloned in clonally propagated crops. The two prognostic equations maximize genetic gain by enabling concurrent and reliable selection for high productivity and stability both among sibling lines and among plants within sibling lines. Prognostic breeding leads to the development of density-neutral cultivars, that is, cultivars yielding optimally over a wider range of plant densities. Widespread application of prognostic breeding in all three categories of crops (inbreeders, outbreeders, and clonally propagated) ensures accurate whole-plant field phenotyping and offers great

---

Plant Breeding Reviews, Volume 37, First Edition. Edited by Jules Janick.
© 2013 Wiley-Blackwell. Published 2013 by John Wiley & Sons, Inc.

potential for maximizing selection efficiency, minimizing selection cost, shortening the time-interval to release cultivars, and accelerating the genetic improvement of crops.

KEYWORDS: coefficient of homeostasis; crop yield potential; density-neutral cultivars; honeycomb designs; plant breeding; prognostic equations; sibling testing; stability of performance; whole-plant phenotyping

ABBREVIATIONS
  I. INTRODUCTION
  II. GENETIC COMPONENTS OF CROP YIELD POTENTIAL
  III. A NEW GENERAL RESPONSE EQUATION
  IV. PROGNOSTIC EQUATIONS FOR SINGLE PLANTS AND SIBLING LINES
    A. Soybean Case Study
    B. Cotton Case Study
    C. Maize Case Study
  V. THE ADVANTAGES OF PROGNOSTIC BREEDING
  VI. THE MARRIAGE OF PHENOTYPING WITH GENOTYPING
  VII. OUTLOOK
LITERATURE CITED

ABBREVIATIONS

DD	Density-dependent
DN	Density-neutral
D	Prefix for honeycomb designs
EE	Emergency equation
p	Plant
s	Sibling line
p-YI	Plant yield index
p-PE	Plant prognostic equation
s-CH	Sibling line coefficient of homeostasis
s-YI	Sibling line yield index
s-PE	Sibling line prognostic equation

I. INTRODUCTION

Escalating global food demand, projections that global population will reach or exceed 9 billion by 2050, and the threat of food insecurity or famine call for a doubling in existing global food production in the next decades (FAO 2009; Grierson et al. 2011). As in the past, this critical increase in crop production globally will have to be achieved primarily by increasing crop yields per unit area (67%), the cultivation of new land (20%), and higher cropping intensity (12%) (Gregory and George 2011). Attainment of higher yields on existing agricultural land is preferable to clearing new lands if global crop demand is to be met with minimal environmental impact and in a sustainable manner (Tilman et al. 2011).

There is increasing evidence for the existence of yield gaps between farm and potential yields globally (Lobell et al. 2009; Licker et al. 2010; Edmeades et al. 2010), as well as yield plateaus and losses for major crops (Peltonen-Sainio et al. 2009; Graybosch and Peterson 2010; Brisson et al. 2010; Supit et al. 2010). Improvement of crop management practices, especially in nations that use less intense agricultural practices, can bridge part of the gap (Fan et al. 2011) and efforts toward this goal should be continuous. However, the genetic improvement of plants holds a key to the sustainable increase of both potential and farm crop yields and to the eventual overcoming of yield plateaus. To meet the challenges of genetic improvement, innovative approaches are required to accelerate the progress through selection and increase the average annual genetic gain.

Throughout the 20th century, extensive breeding efforts to improve the yield potential of crops produced an annual genetic gain ranging from 0.5 to 1.6%, depending on the crop and the conditions of experimentation, including barley (*Hordeum vulgare*), cotton (*Gossypium hirsutum*), maize (*Zea mays*), oat (*Avena sativa*), sorghum (*Sorghum bicolor*), soybean (*Glycine max*), rice (*Oryza sativa*), and wheat (*Triticum aestivum*) (Austin et al. 1980; Wych and Rasmusson 1983; Wych and Stuthman 1983; Schmidt 1984; Miller and Kebede 1984; Specht and Williams 1984; Meredith and Bridge 1984; Duvick 1984; Waddington et al. 1986; Cox et al. 1988; Feil 1992; Tollenaar et al. 1994b; Rajaram 1999; Fok 2000; Peng et al. 2009; Graybosch and Peterson 2010; Fischer and Edmeades 2010; Edmeades et al. 2010).

Notably, all of the successfully improved crops were inbreeders whose breeding efforts have been characterized by capitalizing on additive genetic variation and lacking deleterious genes, except maize that although an outbreeder, is characterized by the predominance of additive genetic variation (Duvick et al. 2004). Conversely, in crops characterized by nonadditive genetic variation, like the clonally propagated potato (*Solanum tuberosum*) (Peloquin 1995), selection for productivity has failed to establish genetic gain. Results on the yield gains in potato from 1860 to the present, suggested a yield stasis despite intensive breeding efforts (Douches et al. 1996). Alfalfa (*Medicago sativa*) is another crop that has failed to establish genetic gain through selection because of the high load of deleterious genes. The genetic contribution to yield in alfalfa over an 80-year period from 1898 to 1985 was only 3% (Hill et al. 1988; Holland and Bingham 1994; Riday and Brummer 2002). The principal cause of the reported yield stasis in both potato and alfalfa has to be searched for in the high load of deleterious genes locked in repulsion-phase linkages and responsible for the high degree of degeneration following gene fixation (Fasoula and Fasoula 2002). Unlike these crops, Rasmusson and Phillips (1997), reported that in barley and wheat, two inbreeders largely devoid of deleterious genes, genetic advance is realized even when the genetic basis is narrow.

This characteristic 1% average genetic gain per annum was interrupted by the advent of the Green Revolution, the heyday of which was from 1965 to 1985. In that period, crop yields of wheat and rice were nearly doubled. The Green Revolution in wheat owes its success to the incorporation into cultivars of four categories of genes responsible for (1) high yield potential, (2) high stability of performance, (3) broad adaptability, and (4) good quality (Borlaug 1983, 2007; Ortiz et al. 2007, 2008a,b; Trethowan et al. 2007). Historically, two separate events contributed to the realization of the green revolution in wheat (Ortiz et al. 2008a,b). In the first event between 1951 and 1962, Borlaug incorporated genes for broad adaptation, daylength insensitivity, and stability of performance through "shuttle breeding" between two environmentally contrasting sites in Mexico. The two sites not only had contrasting climates, they also differed in altitude (by about 2,200 m) and latitude (by 10°). The effect of shuttle breeding was to endow those early improved tall wheat lines with wide adaptability and resistance to stem rust and other diseases (Ortiz et al. 2007). In the second event between 1962 and 1975, Borlaug used genetic materials obtained from Dr. Orville Vogel (Washington State University) that contained dwarfing genes of the Japanese cultivar 'Norin 10'. Borlaug was successful in incorporating the high yield potential genes of 'Norin 10' to the previously incorporated stability, adaptability, and quality genes to develop the first high-yielding semidwarf wheat lines (Ortiz et al. 2008a,b). Incorporation of the four categories of genes was achieved through crossing of cultivars containing these genes, followed by extensive multisite selection among their derived advanced lines.

However, the Green Revolution of the 1960s, which created cultivars with yield potential about twice the yield of the traditional ones, slowed down steadily after 1985 (Swaminathan 2007). This is due to the fact that the higher yield frontiers of the green revolution offered by the two plant productivity packages (one of high yielding and broadly-adapted cultivars and the other of high-input cultural practices), could not keep up with the immense growth rate of the population. If global yields are to continue to increase in a sustainable manner, two new packages are essential; practices that sustain and improve soil fertility conditions and nutrient availability (outside the scope of this review), and innovative breeding technologies for accelerating the annual genetic gain to develop cultivars capable of exploiting favorable as well as marginal environments. This review discusses the main causes of selection inefficiency in breeding programs, identifies the genetic components of crop yield potential, derives a new general response equation and two prognostic equations for evaluation and selection, and presents three case studies using prognostic breeding.

Five main causes of selection inefficiency have been identified, namely, (1) density and interplant competition, (2) soil heterogeneity, (3) genotype-by-environment (GE) interaction, (4) heterozygosity, and (5) lack of exploitation of the adaptive variation released constantly by the genome in response to environmental stimuli (Fasoula 1990; Fasoula and Fasoula 1997a,b, 2000, 2002). How do these factors act to hinder efficiency of selection in a breeding program? Competition interferes with the equal sharing of growth resources among plants and as a result it decreases crop yields, reduces selection efficiency, and decreases the yields of cultivars over time due to the gradual proliferation of low yielders–strong competitors at the expense of high yielders–weak competitors (Fasoula and Fasoula 1997a, 2000, 2002). For example, Peng et al. (1999) reports that an International Rice Research Institute (IRRI) rice variety 'IR8' yields less today than it did when first released 30 years ago. The negative effects of competition are minimized by selecting at ultrawide plant spacing and adopting as the unit of evaluation and selection the individual plant during all stages of the breeding program. Soil heterogeneity, the second cause of inefficiency, interferes significantly with single-plant selection for yield. Utilization of the honeycomb selection designs that possess unique spatial properties (Fasoulas and Fasoula 1995) allows efficient selection for yield potential and stability of performance. These experimental designs accomplish two main targets: (1) they remove the confounding effect of soil heterogeneity on single-plant yields by placing every plant in the center of a moving circular complete replicate that serves as a common denominator and (2) they permit effective selection for stability of performance by placing the plants of evaluated entries in the corners of a triangular grid pattern spreading across the whole field area.

To control and exploit the remaining causes of inefficiency, that is, GE interaction, heterozygosity, and the lack of exploitation of adaptive variation, use is made of the entry or sibling line coefficient of homeostasis (s-CH) (Fasoula 2006, 2008). As will be explained, selection for stability of performance in prognostic breeding is not accomplished via the progenies of each plant that are available in the next generation, but via the siblings of each plant that are available in the generation of selection. Siblings are plants having the same seed parent (mother) and are available in the same generation of evaluation. The coefficient of homeostasis enables efficient selection for crop yield potential under ultrawide plant spacing, as it measures with accuracy the factors affecting stability of performance and allows the conversion of the plant or sibling line yield potential into crop yield potential.

Prognostic breeding refers to the crop improvement methodology, which predicts plants and sibling lines equipped with genes and mechanisms controlling high and stable crop yield potential. The unit of evaluation and selection is the plant or a line of sibling plants. Except the first cycle of selection, plants within entries in all other selection cycles are siblings, namely, plants having the same mother. The novelty is that plant evaluation for crop yield potential is approached by evaluating and predicting the crop yield potential of each plant through its siblings. In traditional breeding, entry is the progeny line, whereas in prognostic breeding entry is the sibling line. Progeny lines comprise and evaluate the progenies of selected plants available in successive generations, whereas sibling lines comprise and evaluate sibling plants available in the current generation of selection. This review explains that prognosis is accomplished though the use of two novel equations, the plant prognostic equation (p-PE) that evaluates the crop yield potential of each plant and the sibling line prognostic equation (s-PE) that evaluates the crop yield potential of each sibling line.

The stability of performance of individual plants can be assessed through either their progenies or their siblings. Compared to traditional breeding that uses progeny testing, prognostic breeding uses sibling testing to assess the stability of performance. The key difference between sibling testing and progeny testing is that sibling testing increases efficiency. This is because in the case of sibling testing, plants have their siblings available in the generation of selection and the two components of the crop yield are evaluated concurrently. In contrast, in the case of progeny testing, selected plant progenies are not available in the generation of selection, and the two components of crop yield are evaluated independently in successive generations. In progeny testing, the confounding effects of soil heterogeneity on plot yields are minimized through the utilization of specific types of field designs (e.g., RCB or lattice designs). In sibling testing, the confounding and favorable effects of soil heterogeneity on single-plant yields are minimized and exploited, respectively, by the honeycomb field designs.

The two prognostic equations are the basis of prognostic breeding. Their use maximizes genetic gain by enabling concurrent and reliable selection for high productivity and stability both among sibling lines and among plants within sibling lines. The first parameter of the equations measures the stability of the sibling line to which each plant belongs and the second parameter the yield potential of plants or sibling lines. The product of the two parameters converts plant yield potential under ultrawide spacing into crop yield potential, and predicts plants equipped with yield and stability genes and mechanisms. Selection for the first parameter (stability of performance) extends the upper limit of the optimal plant density range (cultivars yield well at high plant densities), while selection for the second parameter (high yield potential per plant) extends the lower limit of the optimal plant density range (cultivars yield well at lower plant densities) (Fasoula and Fasoula 2000). Thus, the developed cultivars tend to be density-neutral (DN), that is, yield optimally over a wider range of plant densities. The main advantage of developing DN cultivars is that these cultivars can exploit more effectively marginal environments by using lower planting rates in conformity with the availability of resources. In addition, these cultivars can overcome water fluctuations from year-to-year by using lower seeding rates, which optimize production via more efficient utilization of water.

II. GENETIC COMPONENTS OF CROP YIELD POTENTIAL

Fasoula and Fasoula (2000, 2002, 2003) partitioned the crop yield potential of an entry (which may be progeny lines, sibling lines, families, or cultivars among others) into three components encompassing productivity, stability, and adaptability that are measured by simple phenotypic parameters in plants grown at ultrawide plant spacing. Joint selection for the three genetic components leads to DN cultivars, that is, cultivars that attain optimal yields over a wide range of plant densities. The reason is that selection for stability of performance extends the upper limit of the optimal plant density range (required for cultivars to yield well at high plant densities), while selection for high yield potential per plant extends the lower limit of the optimal plant density range (required for cultivars to yield well at lower plant densities) (Fasoula and Fasoula 2000). In prognostic breeding, the entries consist of sibling lines, that is, lines of plants that have the same mother.

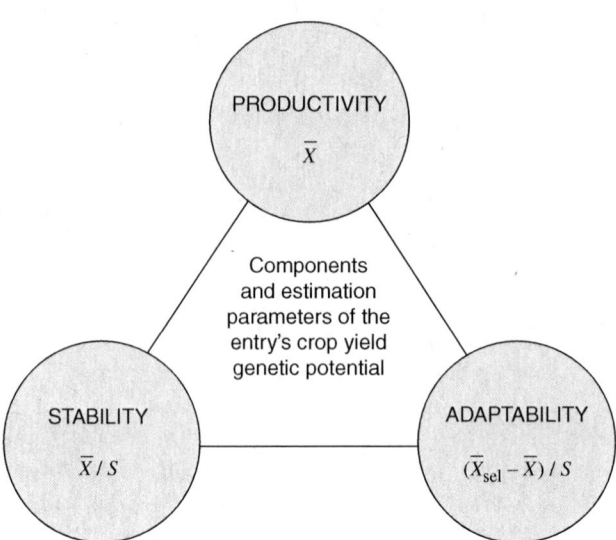

Fig. 6.1. The three components of the crop yield potential of an entry (which may be progeny line, sibling line, family, or cultivar) are (1) the productivity measured by the entry's mean plant yield $\bar{x}$, (2) the stability of performance or tolerance to stresses measured by the entry's standardized mean $\bar{x}/s$, and (3) the adaptability or responsiveness to inputs measured by the entry's standardized selection differential $(\bar{x}_{sel} - \bar{x})/s$. In prognostic breeding, the entries consist of sibling lines as explained in the text.

Prognostic breeding is an integrated methodology that treats inbreeders, outbreeders, and clonally propagated crops in a similar way. The only difference is that plants of sibling lines are selfed in inbreeders, outcrossed in outbreeders, and cloned in clonally propagated crops.

As shown in Fig. 6.1, the three genetic components of the crop yield potential are (1) the genes controlling the sibling line productivity, measured by the sibling line's mean yield per plant ($\bar{x}$). These genes ensure optimal crop yields at low plant densities; (2) the second component comprises genes controlling the stability of performance measured by the sibling line's standardized mean yield per plant ($\bar{x}/s$). These genes ensure optimal crop yields at high plant densities; and (3) the third component comprises genes controlling adaptability, expressed as responsiveness to inputs and measured by the sibling line's standardized selection differential $(\bar{x}_{sel} - \bar{x})/s$, where $\bar{x}_{sel}$ is the mean plant yield of the selected plants. Selection for the three categories of genes leads to the development of cultivars characterized by high, stable, and DN crop yields (Fasoula and Fasoula 2000). The three parameters measuring the three components of the crop yield are maximized under ultrawide plant spacing and minimized under dense stands. Therefore, they are reliably evaluated under ultrawide plant spacing.

III. A NEW GENERAL RESPONSE EQUATION

The product of the three parameters measuring the three components of the crop yield potential gives the following general response equation:

$$R = (\bar{x}/s)^2 \cdot (\bar{x}_{sel} - \bar{x})$$

that is, the product of the selection differential and the stability parameter $(\bar{x}/s)^2$ that was coined the term entry or sibling line coefficient of homeostasis (s-CH) (Fasoula 2006, 2008), since in prognostic breeding the entries consist of sibling lines. Therefore, $\bar{x}$ and s are the sibling line mean plant yield and standard deviation, respectively, and $\bar{x}_{sel}$ is the mean plant yield of the selected plants within the line. Accordingly, the parameters measuring the crop yield genetic potential are reduced to two. Efficiency is increased when the breeder can reliably measure the two parameters, that is, the selection differential and the s-CH.

Comparing the aforementioned general response equation with Falconer's (1989) general response equation:

$$R = (s_g/s)^2 \cdot (\bar{x}_{sel} - \bar{x})$$

the distinction is that the entry coefficient of heritability in Falconer's equation is replaced by the s-CH. The difference between the two coefficients is that, in field trials, the entry coefficient of heritability is estimated by progeny testing in successive generations, whereas the s-CH is estimated in the generation of selection. Moreover, the s-CH is of improved value because in comparisons among entries or sibling lines, it measures any factor affecting the sibling line stability of performance (heritability, tolerance to stresses, GE interaction, or heterozygosity among others). In Falconer's general response equation, the number of selected plants cannot be reduced below 10% since the confounding effects of heterozygosity, competition, and soil heterogeneity reduce drastically response to selection. As we will discuss later, due to the honeycomb selection designs that control effectively the confounding effects of competition, heterozygosity and soil heterogeneity, the number of selected plants in prognostic breeding can be reduced significantly, even up to the lowest-theoretically limit of one; a condition where response to selection attains maximum values.

The partitioning of the crop yield potential of an entry or sibling line into three genetic components and the subsequent development of the two-parameter general response equation are fundamental in prognostic breeding for four main reasons. The first reason is the feasibility to develop DN cultivars; that is, cultivars that yield optimally over a wide range of plant densities and minimize the genotype by density interaction (Fasoula and Fasoula 2000). The fact that joint selection for the two genetic components of the response equation leads to DN cultivars has a key impact on developing cultivars suitable for marginal environments. In such environments, the genotype by density interaction plays decisive role as harvestable yields under conditions of drought stress may be ensured only when lower to medium plant densities are used. Suitable cultivars for such areas are those yielding optimally over a wider range of plant densities; that is, DN cultivars. Marginal environments require the development of DN cultivars (Fasoula and Fasoula 2000; Tokatlidis et al. 2001). For example, the optimum plant density for maize hybrids grown under drought-prone conditions is much lower than when they grow under favorable conditions (Tokatlidis et al. 2011). This provides evidence that the drought-prone environments require lower plant densities for efficient utilization of growth resources. Modern maize hybrids, being density-dependent (DD), that is, yielding optimally under a specific high plant density, may not be highly suitable for marginal environments. To attain maximum yields at favorable as well as marginal environments breeding programs should aim at developing DN cultivars. Such a goal is realized when the breeder selects concurrently for the two genetic components of the general response equation at ultrawide plant spacing, where the two components are maximized. In comparison trials among DN cultivars, the confounding effect of the genotype by density interaction is minimal to absent.

Apart from developing DN cultivars, there are three more reasons that the partitioning of crop yield into genetic components is fundamental in prognostic breeding. The second reason is that the two parameters of the general response equation measure in unison crop yield potential and are evaluated reliably at ultrawide plant spacing. The third reason is that accurate evaluation necessitates the development and utilization of field designs allowing application of very high selection pressures. This is accomplished by (a) erasing the confounding effect of soil heterogeneity on single-plant yields and (b) capitalizing on soil heterogeneity to select for stability of performance. The fourth contribution of component partitioning is that once the confounding effects of competition and soil heterogeneity are counteracted or exploited, the selection efficiency is maximized. Higher selection efficiency is accomplished by the development of two novel prognostic equations, one ensuring whole-plant precision field phenotyping for high and stable crop yield and the other precision phenotyping of sibling lines (Fasoula 2006, 2008, 2009).

IV. PROGNOSTIC EQUATIONS FOR SINGLE PLANTS AND SIBLING LINES

The principal aim of plant breeding is the development of cultivars embodying genes for high, stable, DN, and good quality crop yield. Crop yield is maximized when all plants in the field yield the same (Fasoula and Fasoula 2002). To accomplish this ideal, a number of preconditions are essential (Fasoula and Fasoula 1997a, 2002; Fasoula and Tollenaar 2005): (1) growth resources must be ample and shared equally among plants to reduce acquired competition that interferes with the equal sharing of resources, (2) plants must be genetically identical to eliminate genetic competition, (3) the derived monogenotypic cultivars must be highly homeostatic to reduce acquired competition, (4) genetically identical plants, as well as growth resources, must be evenly distributed across the field, and (5) seeds must be of equal size and of good physiological maturity to ensure even germination and growth (Janick 1999; Tollenaar and Wu 1999). The necessity of genetically identical plants to maximize crop yield imposes as unit of selection the individual plant assessed for high and stable crop yield at ultrawide plant spacing and, ultimately, the development of monogenotypic cultivars.

Identifying in each generation plants carrying the largest possible numbers of yield and stability genes requires discovering ways that enable the breeder to predict among the thousands of plants, the few carrying the highest number of yield and stability genes. This is done by the honeycomb designs (Fasoulas and Fasoula 1995). Honeycomb field designs accomplish these tasks in two ways: (1) by assessing the yield potential of single plants after reducing the confounding effects of competition and soil heterogeneity and (2) by taking advantage of soil heterogeneity to select for stability of performance. As an example, the D-19 and D-31 replicated honeycomb designs, evaluating 19 and 31 sibling lines, respectively, are shown in Fig. 6.2 and Fig. 6.3. As illustrated, every plant in the design occupies the center of a moving circular complete replicate, which is used as a common denominator (Fig. 6.2 and

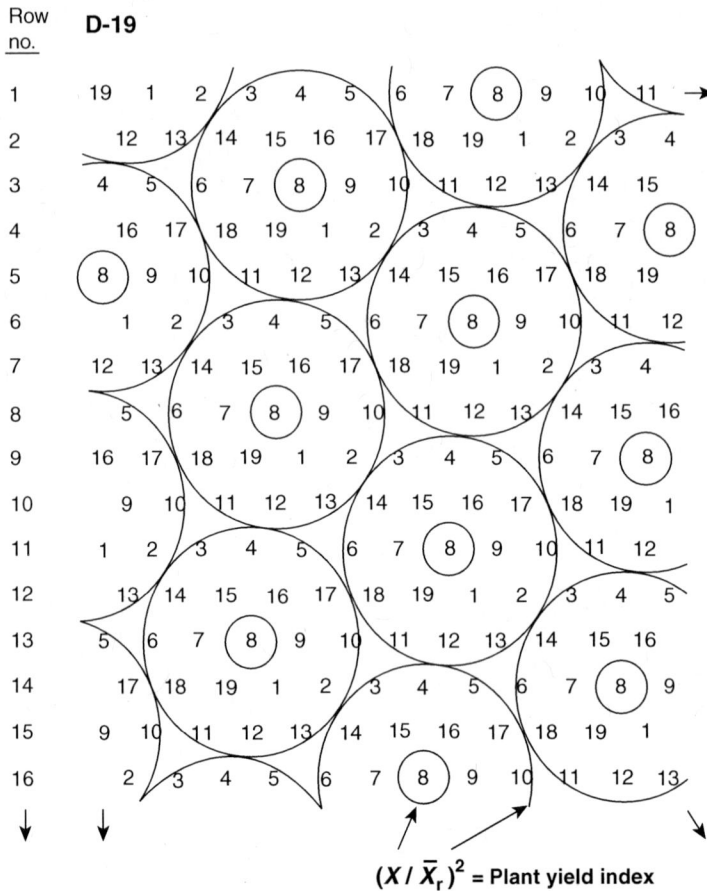

Fig. 6.2. Plants of the 19 sibling lines of the replicated D-19 honeycomb design are arranged in horizontal rows in an ascending order and the whole set is repeated regularly. The starting number of each row is different and this is essential for the formation of moving circular replicates. Namely, of replicates ensuring that every plant lies in the center of a complete replicate or block. This property permits use of the ratio $x/\bar{x}_r$ of the plant yield x to the mean yield $\bar{x}_r$ of the surrounding plants forming a moving circular complete replicate. The common denominator reduces the masking effect of soil heterogeneity on single-plant yields.

Fig. 6.3). Therefore, the number of moving replicates is equal to the number of plants in the trial. Moving replicates are exemplified by the plants of the randomly chosen sibling line 8 in Fig. 6.2 and by the plants of line 18 in Fig. 6.3. These properties apply to the plants of all the evaluated entries or sibling lines and to all honeycomb designs.

6. PROGNOSTIC BREEDING: A NEW PARADIGM FOR CROP IMPROVEMENT

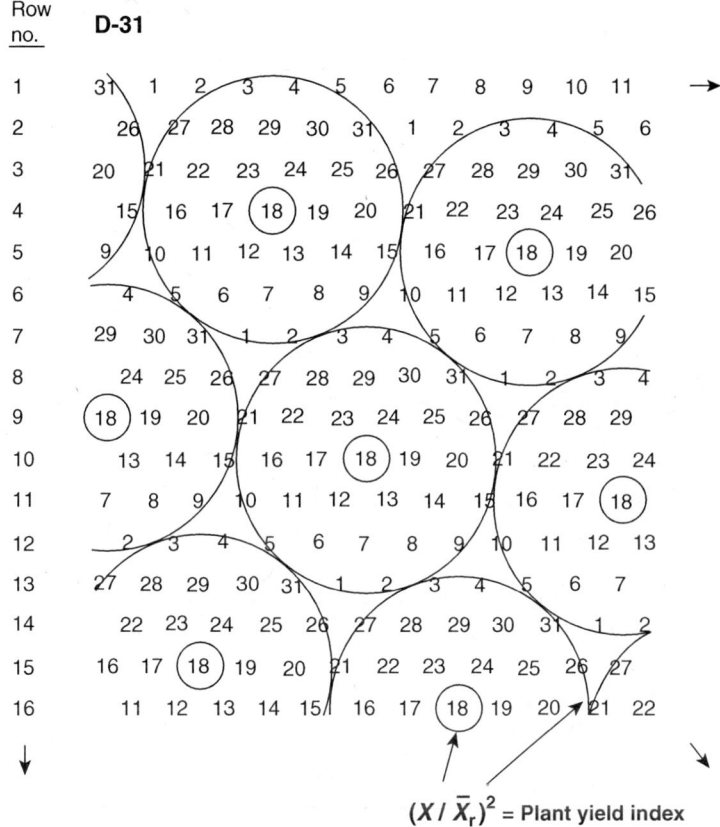

Fig. 6.3. The replicated D-31 honeycomb design evaluates 31 sibling lines. The moving replicate is illustrated for plants of line no. 18. The plant yield index p-YI = $(x/\bar{x}_r)^2$ removes the confounding effects of soil heterogeneity on single-plant yields, allowing objective ranking of plants from top to bottom. The property of every plant to occupy the center of a moving complete replicate is valid for all the honeycomb designs depicted in Fig. 6.4.

The possible numbers N capable to form honeycomb designs are given by the formula $N = X^2 + XY + Y^2$, where X and Y are whole numbers depicted in Fig. 6.4 (Fasoulas and Fasoula 1995). The honeycomb field designs minimize the confounding effects of soil heterogeneity on single-plant yields by expressing the yield of every plant x as a ratio $x/\bar{x}_r$, where x is the plant yield (g) and $\bar{x}_r$ is the mean yield (g) of the surrounding plants forming moving circular complete replicates (Fig. 6.2 and Fig. 6.3). This ratio squared $(x/\bar{x}_r)^2$ to increase its resolving

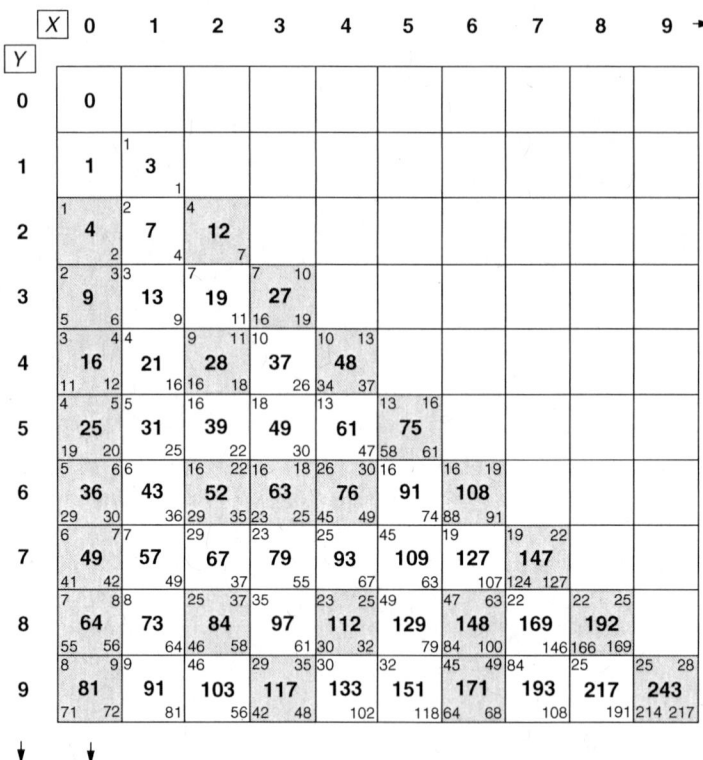

Fig. 6.4. The possible numbers N capable to form honeycomb selection designs are given by the formula $N = X^2 + XY + Y^2$, where X and Y are whole numbers. The white squares reflect the ungrouped honeycomb designs and the gray squares the grouped designs. The numbers in the corners help to construct the field rows of the honeycomb designs. *Source:* Fasoulas and Fasoula (1995).

power constitutes a unitless plant yield index (p-YI) that is devoid of the masking effects of soil heterogeneity and allows ranking plants objectively according to their yield potential. Taking the mean p-YI values of all the plants of a sibling line, the sibling line yield index (s-YI) s-YI = $(x/\bar{x}_r)^{2*}$ is obtained. The plant and sibling line yield indices take values either above or below 1. The larger the value of the index, the more superior is the yielding ability of the plant or sibling line in question.

However, since the p-YI and s-YI assess ordinary yield potential involving yield genes only and not crop yield potential involving both yield and stability genes, a parameter is required that converts ordinary yield potential into crop yield potential. This parameter is the s-CH = $(\bar{x}/s)^2$

6. PROGNOSTIC BREEDING: A NEW PARADIGM FOR CROP IMPROVEMENT

(Fasoula 2006, 2008), discussed in Chapter 3 and explained in Fig. 6.5 and Fig. 6.6, where $\bar{x}$ and s are the sibling line mean plant yield and standard deviation, respectively. Thus, if we replace the selection differential in the general response equation $R = (\bar{x}/s)^2 \cdot (\bar{x}_{sel} - \bar{x})$ with the p-YI and the s-YI, the following two prognostic equations referring to single plants and sibling lines, respectively, are defined

Plant prognostic equation

$$\text{p-PE} = (\bar{x}/s)^2 \cdot (x/\bar{x}_r)^2$$

Sibling line prognostic equation

$$\text{s-PE} = (\bar{x}/s)^2 \cdot (x/\bar{x}_r)^{2*}$$

where * is the mean p-YI values of all plants of a sibling line.

As shown in Fig. 6.5 and Fig. 6.6, the s-CH = $(\bar{x}/s)^2$, which constitutes the first and common parameter of the two prognostic equations and measures the sibling line stability of performance, is evaluated accurately due to the triangular grid pattern ensured by the honeycomb field designs. The triangular grid pattern is illustrated for plants of the sibling line 8 in Fig. 6.5 and for plants of the sibling line 18 in Fig. 6.6. The accurate measurement of the two parameters of the prognostic equations insures the following properties for the equations:

1. The high predictive power, that is, the ability to identify plants and sibling lines carrying yield and stability mechanisms.
2. The ability to convert ordinary plant and sibling line yield potential into crop yield potential.
3. The ability to assess crop yield potential under ultrawide plant spacing.
4. The ability to apply ultrahigh selection pressures.
5. The ability to produce DN cultivars.

The two parameters of the p-PE and the s-PE, which are maximized only when plants grow under ultrawide plant spacing, give optimum results when assessed in the absence of competition for growth resources. The use of ultrawide plant spacing is further imposed by the negative correlation between the yielding and the competitive ability (Kyriakou and Fasoulas 1985; Fasoula 1990; Fasoula and Fasoula 1997a). However, selection at ultrawide plant spacing places two dilemmas. The first dilemma is how we can eliminate the strong confounding effect of soil heterogeneity on single plant evaluation for yield. Many studies suggest that the problem of soil heterogeneity

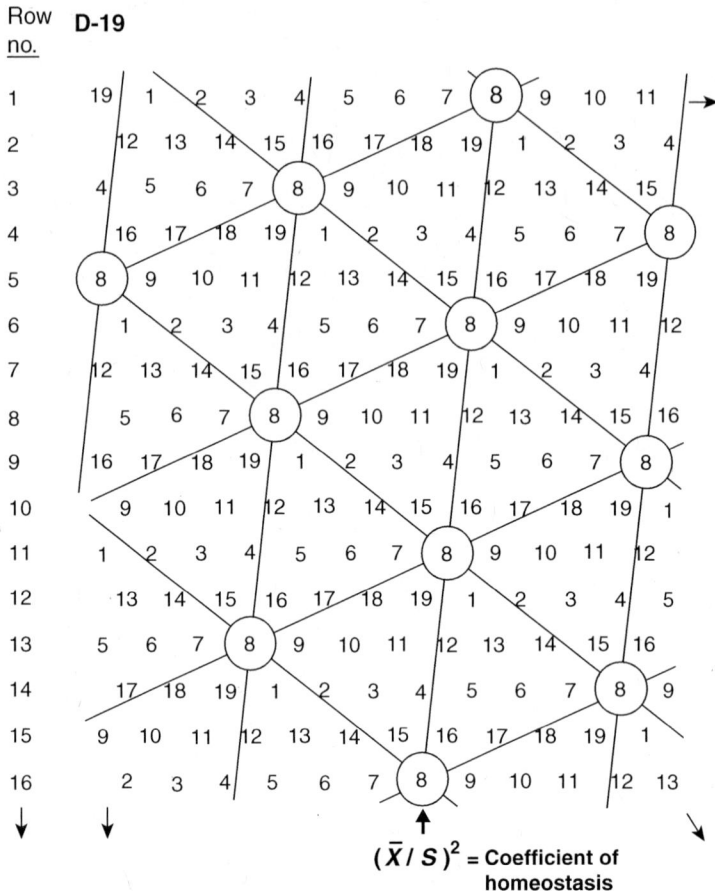

Fig. 6.5. The formation of moving circular complete replicates is a key property to the honeycomb designs. Plants of each sibling line (shown here for line no. 8) are evenly distributed across the field in the corners of a triangular grid pattern, which samples soil heterogeneity more efficiently than the random allocation. This allows the accurate measurement of the sibling line stability of performance using the sibling line coefficient of homeostasis s-CH = $(\bar{x}/s)^2$. The triangular grid pattern is formed for each of the 19 lines of the design.

or spatial variation within experimental fields affects agronomic traits and reduces the selection efficiency (Vollmann et al. 2000). The second dilemma is the extent to which evaluation of plants under wide spacing reflects evaluation under dense stands.

Regarding the masking effect of soil heterogeneity on single-plant yields, this is effectively controlled with the use of honeycomb field

6. PROGNOSTIC BREEDING: A NEW PARADIGM FOR CROP IMPROVEMENT

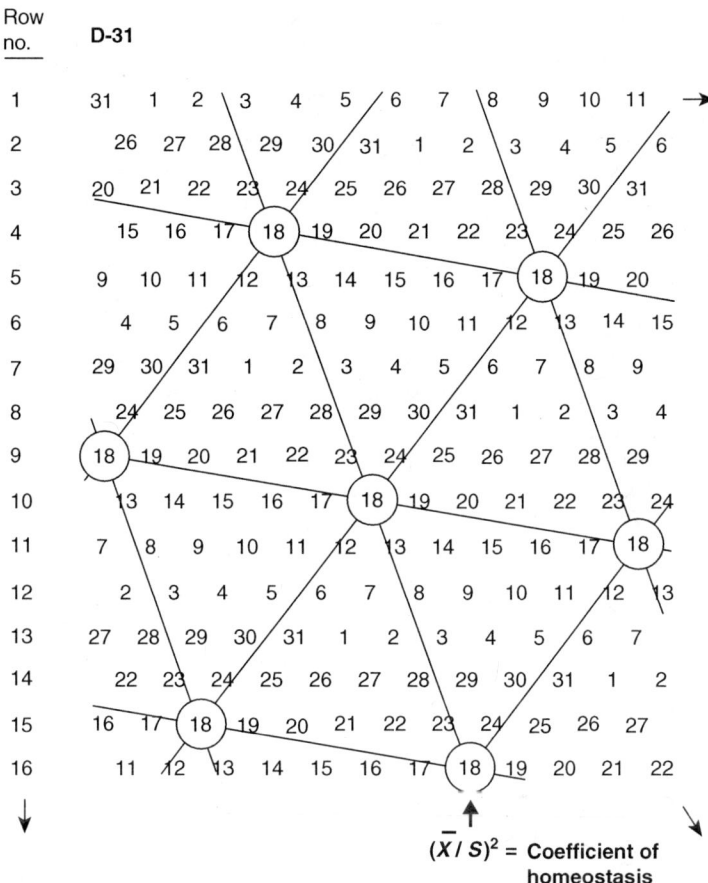

Fig. 6.6. The triangular grid pattern is shown here for plants that belong to sibling line no. 18 of the replicated D-31 honeycomb design. This triangular grid is formed for each of the 31 sibling lines that the design evaluates.

designs that accomplish two objectives. The first objective is the removal of the confounding effect of soil heterogeneity on single-plant yields by placing each plant in the center of a moving complete replicate (Fig. 6.2 and Fig. 6.3). The second objective is the efficient selection for stability of performance through effective exploitation of soil heterogeneity, by placing plants of each sibling line in a triangular grid pattern extending across the whole field (Fig. 6.5 and Fig. 6.6).

Regarding the extent to which selection under wide plant spacing reflects selection for crop yield potential, this is accomplished by the s-CH, which converts ordinary plant and sibling line yield potential into crop yield potential. The s-CH, which measures any factor interfering with the stability of performance (i.e., heritability or GE interaction), is used in the two equations to convert ordinary plant and sibling line yield potential into crop yield potential.

It is notable that in traditional breeding, the assessment of single plants for stability of performance is accomplished through progeny testing under dense stands. In prognostic breeding, the stability of performance is assessed through sibling testing. The principal difference is that progeny testing evaluates plant progenies and is realized in successive generations following the visual selection of plants, while sibling testing evaluates sibling plants and is completed in the current generation. Thus, progeny testing requires more years of evaluation than sibling testing. In addition, progeny testing requires densely-grown field trials and such trials cannot be grown in early generations due to the scarcity of the seed. Conversely, sibling testing is accomplished by the s-CH and the p-YI that assess concurrently stability of performance and yield potential in the generation of selection in individual plants grown at ultrawide plant spacing.

The p-PE and the s-PE render cultivars DN and secure efficient selection for high and stable crop yield over a wide range of sites, years, and plant densities. DN cultivars are coupled with two more advantages: (1) they compete effectively against the weeds by having early fast growth that covers the field before the majority of weeds emerge and (2) they possess deep and extensive root system that adds tolerance to drought (Fasoula and Fasoula 1997a, 2000). DN cultivars are better equipped to exploit marginal environments and may also be suitable for alternative cultural systems like the system of rice intensification (SRI) described by Sato and Uphoff (2007).

The transition from the three-component crop yield potential (Fig. 6.1) to the two-component p-PE and s-PE has comprised the refinement of the honeycomb breeding. This finishing point by the two prognostic equations may lead to the transformation of conventional into prognostic breeding and the maximization of selection efficiency. In other words, honeycomb breeding becomes synonymous to the meticulous application of the two prognostic equations. Ranking of all plants and all sibling lines according to the values of the two prognostic equations identifies the top plants and sibling lines that carry the best yield and stability mechanisms and increases advance through selection. Growing the same honeycomb trial in sites representing production environments and analyzing the trial by the two prognostic equations within and across sites ensures efficient selection for local and broad adaptation. Selection for local adaptation is performed in environments with extreme or unique conditions, whereas selection for broad adaptation is performed across a target number of environments.

Notably, a prerequisite for the successful application of the two prognostic equations is the successful establishment of plants in the field. In case of failure to achieve a satisfactory stand in the field, the following unitless emergency equation (EE) may be used:

$$EE = (\overline{x}_b/s)^2 \cdot (\overline{x}_b/\overline{x}_g)^2$$

where $(\overline{x}_b/s)^2$ is the coefficient of homeostasis of the 10–30 best plants selected by truncation, the same number of best plants for all the evaluated sibling lines, and $(\overline{x}_b/\overline{x}_g)^2$ is the squared ratio of the mean yield of the 10–30 best plants in each sibling line ($\overline{x}_b$) to the grand mean yield of all sibling lines ($\overline{x}_g$). Note that selection according to the EE becomes more efficient if confined to the very top sibling lines in the rank. Among sibling lines having similar EE values, the one with the largest coefficient of homeostasis is preferred as it predicts stability of performance more accurately. The EE should not be used in the place of the two prognostic equations, but only in extreme and difficult conditions.

The resolving and prognostic power of the p-PE and the s-PE that maximize response to selection, will be exemplified using data from three studies, one with soybean, the second with cotton, and the third with maize.

A. Soybean Case Study

Despite its importance, exploitation of intracultivar variation has not been widely investigated due to the belief that elite cultivars are highly homogenous. Evidence from selection experiments within homogenous gene pools (Sprague et al. 1960; Russell et al. 1963; Higgs and Russell 1968; Dudley and Lambert 2004) suggests that the genome is more flexible and plastic than once assumed. McClintock (1984) suggested that the genome is dynamic and that it can modify itself in response to environmental stresses. Although elite cultivars are fairly homogenous, latent genetic variation among the single plants of a cultivar exists and mechanisms that generate *de novo* variation are also present (Fasoula 1990; Fasoula and Fasoula 2000). Rasmusson and Phillips (1997) reported that elite gene pools have inherent mechanisms to provide a continuing source of new genetic variation and hypothesized that selection gain occurs because of variation present in the original gene pool as well as variation generated *de novo*.

Intracultivar selection was applied within the elite soybean cultivar 'Haskell' released in the Southeast for high productivity and tolerance to various pests and diseases (Boerma et al. 1994). A total number of 350 plants were grown under ultrawide plant spacing (1.4 plants m^{-2}) to select for yield potential, seed protein and oil content and other agronomic traits (Fasoula and Boerma 2005a,b; Fasoula and Boerma 2007). The 20 highest-yielding plants were selected by truncation and were evaluated in a D-21 honeycomb design (Fasoulas and Fasoula 1995) the next year. The D-21 honeycomb trial evaluated the 20 selected sibling lines plus the check, using 32 plants per line at a spacing of 1.4 plants m^{-2} (Fasoula and Fasoula 2000; Fasoula and Boerma 2005b). Codes 1–20 of the design were assigned to the sibling lines and code 21 was assigned to the check cultivar 'Haskell'. As discussed before, sibling lines evaluate plants that are siblings.

Table 6.1 gives the results of the 20 selected lines plus 'Haskell', ranked on the basis of the three genetic components of crop yield as described in Chapter 2. Mean yield per plant ranged from 150 to 212 g, tolerance to stresses from 4.36 to 8.40 and responsiveness to inputs from 1.19 to 1.72. Because percentages are common measuring units to all three components, lines were ranked according to the grand percentage mean of the three components. Among the five top lines with similar grand percentage mean (80%–85%), the best three lines are lines 12, 18, and 10. The criterion to discriminate among the best five lines is the specific weight of each of the three components participating in the grand percentage mean. In other words,

Tolerance to stresses > responsiveness to inputs > yield per plant

If we consider the specific weight of each component, that is, tolerance to stresses is more essential than the other two components, the best lines are line 12 (holding the top position), followed by lines 10 and 18. This is because line 12 has the largest value for stress tolerance (8.4), whereas lines 10 and 18 have values of 6.5 and 6.32 for stress tolerance, respectively (Table 6.1). Subsequently, the two lines selected as superior to 'Haskell', that is, lines 12 and 18, were evaluated in randomized complete block (RCB) trials across years and sites. The results are averaged across 4 years and 17 locations and are presented in Table 6.2. They demonstrate that line 12 exhibited a 5% significant seed yield superiority over 'Haskell' in randomized trials across 4 years, while line 18 exhibited a 4% significant yield superiority. Therefore, the top lines 12 and 18 selected on the basis of the three components of crop yield were confirmed to be superior when these lines were grown in RCB trials across years.

6. PROGNOSTIC BREEDING: A NEW PARADIGM FOR CROP IMPROVEMENT

Table 6.1. Ranking of the 20 lines selected from the soybean cultivar 'Haskell' on the basis of the % grand mean of the three components of the crop yield potential.

Line code no.	Grand mean (%)	Tolerance to stresses ($\bar{x}/s$)	(%)	Responsiveness to inputs ($\bar{x}_{sel} - \bar{x})/s$	(%)	Mean yield per plant (g) ($\bar{x}$)	(%)
9	85	5.82	69	1.50	87	212	100
10	85	6.50	77	1.54	90	185	87
18	84	6.32	75	1.53	89	189	89
12	83	8.40	100	1.22	71	168	79
19	80	5.14	61	1.44	84	204	96
8	79	5.71	68	1.52	88	173	82
17	79	4.72	56	1.72	100	172	81
15	78	4.71	56	1.61	94	178	84
13	78	5.57	66	1.55	90	167	79
14	77	5.42	65	1.31	76	189	89
7	77	5.71	68	1.33	77	179	85
6	76	5.94	71	1.20	70	182	86
16	75	5.27	63	1.22	71	193	91
1	75	5.11	61	1.47	85	170	80
3	75	5.62	67	1.36	79	168	79
20	74	5.13	61	1.37	80	171	81
'Haskell'	73	4.36	52	1.45	84	173	82
4	72	4.82	57	1.29	75	177	84
2	71	4.94	59	1.29	75	169	80
11	69	4.88	58	1.19	69	168	79
5	69	4.61	55	1.40	81	150	71
Resolution range (%)	81–100		52–100		69–100		71–100

The $\bar{x}_{sel}$ in the responsiveness to inputs was calculated using the mean yield of the five top-yielding plants per line. Because the contribution of each of the three components is not equal, that is, tolerance to stresses > responsiveness to inputs > mean yield per plant, this has to be considered in the final selection of the top lines. For example, line no. 12 with the best tolerance to stresses (8.40) should be given top priority.

At this point, it is essential to proceed into a meta-analysis of the same results using as a prediction criterion the newly formed and unknown at that time prognostic equations. The data have been reanalyzed and the s-CH along with the s-YI and the p-YI have been calculated for all 21 lines of the experiment. The p-YI $= (x/\bar{x}_r)^2$ was calculated using the available computer program (Mauromoustakos et al. 2006) and a moving ring of 19 plants, where x is the yield of each plant and $\bar{x}_r$ is the mean plant yield of its 18 surrounding plants. Tables 6.3 and 6.4 give the results of this meta-analysis on the basis of the s-PE and the p-PE values, respectively. Table 6.3 presents the results

Table 6.2. Randomized complete block (RCB) trials across 4 years and 17 locations evaluating lines 12 and 18 selected from 'Haskell' on the basis of the % grand mean of the three components of the crop yield potential.

Line	Grand mean of the three components (%)	Seed yield (kg ha^{-1})^z	'Haskell' (%)
12	83	2882 a	105
18	84	2862 a	104
'Haskell'	73	2741 b	100

Selection on the basis of the three components predicts RCB performance across several sites and years.
^{z}Means with different letters in a column are significantly different at $P = 0.05$.

of the s-PE, the s-YI, the s-CH and the mean yield per plant for the 21 lines evaluated at 1.4 plants m^{-2} in the D-21 honeycomb design. Table 6.4 presents the top 30 out of the 672 soybean plants ranked by their p-PE value, along with their p-YI and the homeostasis values of the line they belong to.

According to Table 6.3, the best lines are 12, 10, and 18 with s-PE values of 67, 44, and 43, respectively, whereas 'Haskell' has a much lower s-PE value of 18. Table 6.4 shows the top 30 plants of the experiment based on their p-PE values that range from 66 to 101. Most of the top 30 plants belong to lines 12, 10, and 18 and these represent the 'champion' plants that the breeder selects for the next cycle of selection. Note that the check 'Haskell' is not represented in the top 30 plants. Line 12, with a s-PE value of 67, is the best line as predicted by the equations, and exhibited a 5% significant seed yield superiority across 4 years and 17 locations in RCB trials (Table 6.2). As illustrated in Table 6.3, line 12, with a relatively low s-YI (s-YI = 0.95), would not have been selected if it did not have a very high coefficient of homeostasis (s-CH = 70.6) that resulted in the high s-PE = 67. This demonstrates the importance of selecting concurrently for high yield potential and stability of performance as this is accomplished by the use of the two equations.

Comparing the data of Tables 6.1–6.4, a number of important inferences may be drawn.

1. Prognosis by the grand percentage mean of the best lines tested in RCB trials across years and sites proved to be reliable except for one limitation, which is the fact that each of the three components contributes equally to the % grand mean. As discussed before though, the contribution of the three components to the crop yield is not equal, that is, tolerance to stresses is more important than

6. PROGNOSTIC BREEDING: A NEW PARADIGM FOR CROP IMPROVEMENT

Table 6.3. Meta-analysis of the results of Table 6.1 to calculate the sibling line prognostic equation, the sibling line coefficient of homeostasis, and the sibling line yield index of the soybean lines.

Line code no.	s-PE value	s-CH $(\bar{x}/s)^2$	s-YI $(x/\bar{x}_r)^{2*}$	Mean plant yield per line (g)
12	67	70.6	0.95	168
10	44	42.3	1.04	185
18	43	39.9	1.08	189
9	40	33.9	1.19	212
6	36	35.3	1.02	182
7	33	32.6	1.01	179
8	32	32.6	0.97	173
14	31	29.4	1.06	189
19	30	26.4	1.14	204
16	30	27.8	1.08	193
3	30	31.0	0.95	168
13	29	31.0	0.94	167
20	25	26.3	0.96	171
1	25	26.1	0.95	170
4	23	23.2	1.00	177
2	23	24.4	0.95	169
15	22	22.2	1.00	178
17	22	22.3	0.97	172
'Haskell'	18	19.0	0.97	173
5	18	21.3	0.84	150
Resolution range (%)	27–100	27–100	71–100	71–100

*Mean of all p-YI values of a sibling line.
Ranking of the 20 lines along with the check 'Haskell' is based on the unitless s-PE values that range from 18 (line 5) to 67 (line 12). Line 12 has the best homeostasis (s-CH = 70.6) of all the lines and the best s-PE = 67.

responsiveness to inputs and the mean yield per plant. Therefore, this limitation should be taken into consideration when selecting among the top lines.

2. The use of the two prognostic equations offsets this limitation. To select the top lines and the best plants within the lines on the basis of their s-PE and p-PE values, the breeder evaluates the results of Table 6.3 and 6.4 and makes the best selections. Since Table 6.4 presents the plants ranked on the basis of the p-PE values, the breeder can select real 'champion' plants belonging to 'champion' lines and maximize genetic gain. As explained before, the s-PE and p-PE values estimate the sibling line crop yield potential and the plant crop yield potential, and this is the reason that breeders can accurately select the top plants within the best lines.

Table 6.4. Meta-analysis to present the top 30 of the 672 soybean plants belonging to the 21 lines of Table 6.1 along with their p-PE, s-CH, p-YI values, and plant yield.

Top plants and line code no.		p-PE value	s-CH $(\bar{x}/s)^2$	p-YI $(x/\bar{x}_r)^2$	Plant yield (g)
1	18	101	39.9	2.54	254
2	9	98	33.9	2.90	301
3	18	98	39.9	2.45	250
4	12	95	70.6	1.34	181
5	12	93	70.6	1.31	174
6	10	88	42.3	2.08	255
7	9	87	33.9	2.58	239
8	12	85	70.6	1.21	202
9	12	82	70.6	1.16	183
10	14	80	29.4	2.71	285
11	12	78	70.6	1.11	164
12	13	77	31.0	2.48	239
13	10	76	42.3	1.80	233
14	7	76	32.6	2.33	261
15	9	76	33.9	2.24	265
16	12	72	70.6	1.02	183
17	12	71	70.6	1.01	170
18	9	71	33.9	2.09	254
19	12	69	70.6	0.97	180
20	10	67	42.3	1.58	214
21	8	67	32.6	2.04	238
22	12	66	70.6	0.94	200
23	9	66	33.9	1.96	263
24	16	66	27.8	2.39	270
25	12	66	70.6	0.94	176
26	12	66	70.6	0.94	188
27	12	66	70.6	0.94	170
28	10	66	42.3	1.56	206
29	12	66	70.6	0.93	180
30	12	66	70.6	0.93	191
Resolution range (%)		65–100	39–100	32–100	54–100

The 30 top plants are ranked on the basis of their p-PE values that range from 66 to 101 units. The large differences on the p-PE among the top plants reflect the high resolving and predictive power of the p-PE, which allows application of ultrahigh selection pressures to select only the very best plants. It is notable that 14 of the 30 top plants belong to line 12 that has the best homeostasis (s-CH = 70.6).

3. Since the p-PE chooses 'champion' plants according to the demands of the growing environment, plants are selected across a wide array of environments and the role of the breeder should be focused on the successful establishment of the honeycomb trials

on sites representing production environments. Honeycomb trials grown across different environments or years are readily comparable because the parameters of the two prognostic equations are unitless.

4. The resolving power of the s-PE expressed by the resolution range of the % values (27%–100%) (Table 6.3), is larger compared to the resolving power of the grand percentage mean (81%–100%) (Table 6.1), suggesting that s-PE has a better predictive and resolving value as shown in the data of Table 6.3.

5. If we look at the top 30 plants of Table 6.4, we find that 14 plants belong to the top line 12, and 8 of them have a p-PE value that ranges from 69 to 95, thus, surpassing the mean s-PE value of 67 for line 12 (Table 6.3). This demonstrates the possibility to further improve line 12 by selecting these top plants. The same holds true for the top two plants of line 18, with p-PE values of 98 and 101, respectively (Table 6.4), compared to the mean s-PE value of 43 for line 18 (Table 6.3). The constant improvement of cultivars by nonstop selection for important quality and agronomic traits can lead to the development and release of promising and superior lines as shown by Fasoula et al. (2007a,b,c).

Notably, this meta-analysis demonstrates the advantages of using the two prognostic equations for evaluation and selection. All the plants of each soybean line were evaluated for crop yield potential. This is due to the utilization of the two prognostic equations that allow estimating concurrently and in the same generation the two components of the crop yield potential, that is, the plant yield potential and the plant stability of performance. Conversely, in traditional breeding, the two components of the crop yield are evaluated independently in two successive generations and under different growing conditions. In this study, the two components of the crop yield potential of the top line 12 were evaluated in the same generation by the s-CH = 70.6 (stability of performance) and s-YI = 0.95 (yield potential) (Table 6.3). Therefore, stability of performance and yield potential were concurrently measured in the same year. The superiority of line 12 was confirmed when grown in RCB trials across 4 years and 17 locations under high plant densities (Table 6.2), demonstrating that the use of the two equations predicted accurately the high crop yield potential of this top line.

Molecular analysis of the intracultivar phenotypic variation discovered within the elite soybean cultivar 'Haskell' was performed using 144 simple sequence repeat (SSR) markers in seven lines selected within

'Haskell' with the objective to determine the source of the variation (Yates et al. 2012). The data showed that most of the intracultivar SSR variation discovered in 'Haskell' was the result of residual heterozygosity in the initial plant selected to become the cultivar. More specifically, 93% of the SSR variant alleles were traced in the Haskell Foundation seed source where selection was originated. The remaining 7% variant bands could not be detected in the Foundation source and were likely the result of mutation or some other mechanism generating *de novo* variation. Similar results were obtained for the intracultivar variation discovered within another two elite soybean cultivars, 'Benning' and 'Cook' (Yates et al. 2012). These results are in agreement with those of Haun et al. (2011) who explored the genetic basis of intracultivar variation in soybean by investigating the nucleotide, structural, and gene content variation of different individuals from the reference cultivar 'Williams 82'. They concluded that soybean haplotypes can possess a high rate of structural and gene content variation and the impact of intracultivar genetic heterogeneity may be significant. These notable results provide molecular evidence that inbred lines and other homogenous gene pools are not permanent genetic stocks with all plants being genetically identical, but material that can be constantly improved and upgraded by nonstop selection for important quality and agronomic traits.

B. Cotton Case Study

The genetic material used in this study was the elite cotton cultivar 'Celia', a leading cultivar in Greece for several years and the research lasted 4 years (Fasoula and Lithourgidis 2008; Fasoula 2012). In 2004, random seeds of the cultivar 'Celia' were grown in a honeycomb design at the Aristotelian University Farm, near Thessaloniki. Plants were spaced 1 m apart to eliminate the confounding effect of competition on single-plant yields and permit the full expression of their yield potential (Fasoulas and Fasoula 1995). After the first year of experimentation, the 30 top-yielding plants were selected on the basis of the second parameter of the p-PE, that is, the p-YI, using the available computer software (Mauromoustakos et al. 2006). In 2005, the 30 selected plants together with 'Celia' were grown in a replicated D-31 honeycomb design (Fig. 6.3 and Fig. 6.6) with 28 plants per line for a total number of 868 plants (Fasoula 2012). Plants were spaced 1 m apart. After harvest, the s-PE, s-YI and s-CH values were calculated for the 31 sibling lines. Each of the 31 sibling lines evaluated sibling plants.

The results are presented in Table 6.5 and show the ranking of the 31 sibling lines on the basis of the s-PE values that ranged from 3 to 30.

6. PROGNOSTIC BREEDING: A NEW PARADIGM FOR CROP IMPROVEMENT

Table 6.5. Ranking of the 31 cotton sibling lines of the replicated D-31 honeycomb design on the basis of their unitless s-PE values.

Line code no.	s-PE value	s-CH $(\bar{x}/s)^2$	s-YI $(x/\bar{x}_r)^{2*}$	Mean seed yield per plant (g)
3	30	20.8	1.42	400
5	30	23.7	1.28	402
16	23	20.3	1.14	379
6	19	16.3	1.17	392
19	18	14.5	1.23	419
8	17	16.1	1.04	376
1	15	14.4	1.02	370
18	13	7.8	1.64	427
23	13	12.7	1.04	371
28	12	12.1	1.02	358
7	11	10.0	1.10	381
29	10	9.2	1.12	379
15	10	11.2	0.92	346
30	10	8.4	1.23	417
4	10	9.3	1.02	356
9	10	9.4	1.04	392
21	10	9.7	1.00	365
22	9	11.2	0.83	315
12	8	7.1	1.17	395
25	8	8.8	0.90	343
27	7	6.8	1.00	338
10	7	7.2	0.90	347
'Celia'	6	7.5	0.81	341
24	6	4.8	1.20	359
26	6	5.8	1.02	352
13	5	5.2	1.02	383
11	5	6.1	0.88	355
14	5	7.2	0.64	313
17	4	4.9	0.85	351
20	3	3.2	0.96	339
2	3	3.8	0.69	286
Resolution range (%)	10–100	14–100	49–100	67–100

*Mean of all p-YI values of a sibling line.

The s-CH values range from 3.2 to 23.7, with line 5 exhibiting the best tolerance to stresses (s-CH = 23.7). The s-PE values range from 3 to 30 and demonstrate their high resolving and predictive power for selection purposes. Lines 3 and 5 with s-PE = 30 separate themselves in the ranking as top lines.

The two best sibling lines were no. 3 and no. 5 with a s-PE = 30, which is five times larger than the s-PE value of 'Celia' (s-PE = 6). Table 6.6 gives the ranking of the top 30 out of 868 plants on the basis of their p-PE values that range from 26 to 65 units. As we can see, 'Celia' is not represented in the top 30 plants, while lines 3 and 5 are represented by

Table 6.6. Ranking of the top 30 of 868 cotton plants of the replicated D-31 honeycomb design on the basis of their p-PE values that range from 26 to 65 units.

Top plants and line code no.		p-PE value	s-CH $(\bar{x}/s)^2$	p-YI $(x/\bar{x}_r)^2$
1	3	65	20.8	3.13
2	5	55	23.7	2.31
3	3	55	20.8	2.66
4	3	55	20.8	2.62
5	6	52	16.3	3.17
6	5	48	23.7	2.04
7	5	46	23.7	1.96
8	16	46	20.3	2.28
9	16	40	20.3	1.96
10	5	39	23.7	1.63
11	3	37	20.8	1.80
12	5	36	23.7	1.53
13	5	36	23.7	1.51
14	5	36	23.7	1.51
15	16	36	20.3	1.77
16	16	35	20.3	1.74
17	3	34	20.8	1.61
18	8	33	16.1	2.04
19	16	33	20.3	1.61
20	19	33	14.5	2.25
21	19	31	14.5	2.16
22	6	31	16.3	1.90
23	6	28	16.3	1.71
24	6	28	16.3	1.69
25	3	28	20.8	1.32
26	3	27	20.8	1.28
27	19	27	14.5	1.88
28	8	27	16.1	1.69
29	1	27	14.4	1.90
30	1	26	14.4	1.80
Resolution range (%)		40–100	69–100	58–100

Notably, only 7 of the 31 lines are represented in the top 30 plants (lines 5, 3, 16, 6, 19, 8, and 1) and only 4 of 31 lines are represented in the top 15 plants (lines 5, 3, 16, and 6).

seven plants each. Although lines 3 and 5 have equivalent s-PE values, line 3 is superior to line 5 in yield potential (s-YI = 1.42) and inferior in homeostasis (s-CH = 20.8), whereas line 5 is superior in homeostasis (s-CH = 23.7) and inferior in yield potential (s-YI = 1.28) (Table 6.5). Line 5 with the largest coefficient of homeostasis (s-CH = 23.7) has a better prognostic power since higher homeostasis denotes better stability of performance. This demonstrates the importance of selecting concurrently for (1) high yield potential and (2) stability of performance in order to assess the crop yield potential, as this is realized by the use of the two prognostic equations.

Based on these results of Tables 6.5 and 6.6, the cotton lines 3 and 5 were selected for their high crop yield potential and were further evaluated in RCB trials. Seed from the five best plants of lines 3 and 5 (Table 6.6) was mixed to compose the material for the RCB trials of the years 2006 and 2007. Trials in all 4 years were grown in the same field to eliminate the confounding effect of the genotype by field interaction. The results are presented in Table 6.7. As shown, line 5 with the largest coefficient of homeostasis (s-CH = 23.7) significantly outyielded 'Celia' in 2006 by 19% and line 3 outyielded 'Celia' by 10%. In 2007, line 5 outyielded 'Celia' by 22%, whereas line 3 outyielded 'Celia' by 2%. The results of Tables 6.5–6.7 permit to arrive at the following inferences.

1. Large s-PE values can predict lines with high and stable crop yield potential, although evaluation of single plants is performed at ultrawide plant spacing (Fasoula 2012).
2. The prognostic power of the p-PE is due to its ability to (1) convert plant yield potential into crop yield potential, (2) identify plants carrying yield and stability mechanisms, (3) assess the crop yield potential of plants grown under ultrawide plant spacing, and (4) apply ultrahigh selection pressures.
3. Exploitable genetic variation is always present due to the high plasticity of the genome, meaning the presence of highly tuned endogenous mechanisms that keep the genome in a constant flux (Cullis 1990; Rasmusson and Phillips 1997; Lolle et al. 2005). A great example of genome plasticity in cotton is the gradual building up of resistance to *Verticillium wilt* from genetic material that was originally susceptible to the disease (Fasoulas 2000). This phenotypic variation, either genetic or epigenetic in nature and outcome of the GE interaction, is not random but rather adaptive to the constantly changing climatic conditions. Hence, the need of nonstop selection in promising materials to meet the challenges of the continuously released adaptive variation.
4. Prognostic equations predict with precision the crop yield potential of every plant and every sibling line and allow the identification of very few 'champion' plants that maximize genetic gain by exploiting effectively both favorable and stress-prone growing environments.
5. The efficiency of the prognostic equations is linked to two prerequisites. The first prerequisite is that the units of evaluation and selection are the individual plants grown at ultrawide plant spacing. The second prerequisite is the utilization of the

Table 6.7. Comparison of the selected lines 5 and 3 with the original cotton cultivar 'Celia' across three years.

	D-31 2005			RCB 2006		RCB 2007	
Line	s-PE	s-CH	s-YI	Yield (t/ha)^z	Gain (%)	Yield (t/ha)^z	Gain (%)
5	30	23.7	1.28	6.67 a	119	6.07 a	122
3	30	20.8	1.42	6.18 ab	110	5.07 b	102
'Celia'	6	7.5	0.81	5.61 b	100	4.99 b	100

The first year shows the crop yield potential of the two top lines and 'Celia' as measured by their s-PE values in the replicated D-31 honeycomb trial grown at ultralow plant density. The second and third year show the crop yield of these top lines and 'Celia' in randomized complete block (RCB) trials grown at high plant density.
^{z}Means with different letters in a column are significantly different at $P = 0.05$.

honeycomb selection designs that ensure that the two parameters of the equations are assessed accurately.
6. The contribution of the honeycomb designs is that they maximize the reliability of the two parameters of the prognostic equations: (1) the sibling line homeostasis, by allocating the plants of every line in the corners of triangular grids that sample soil heterogeneity more effectively than random allocation (Fig. 6.5 and Fig. 6.6), and (2) the p-YI and the s-YI, by allocating every plant in the center of a moving circular complete replicate that enables expressing its yield as a ratio to the mean yield of the common denominator (Fig. 6.2 and Fig. 6.3).

Notably, in this study, the two prognostic equations were utilized to predict and select concurrently for high and stable crop yield potential the best lines 3 and 5 at ultralow plant density and the best plants within lines 3 and 5. These lines confirmed their superiority when grown in RCB trials at high plant densities across 2 years.

Compared to progeny testing used in traditional breeding, the advantages of sibling testing used in prognostic breeding along with the equations are the following: (1) every year, a large number of sibling lines and plants per line are screened concurrently for yield potential and stability of performance across several sites, allowing application of high selection pressures; (2) due to the utilization of high selection pressures, gene fixation is rapid and response to selection is high; (3) genetically uniform cultivars with high and stable crop yield and other agronomically important traits may be released as early as after 3–5 years of selection; (4) due to the high selection pressures (1%–0.5%) used, the breeding methodology for self- and cross-pollinated crops is comparable. The difference is that outbreeders enjoy a better potential for broader adaptation than inbreeders due to the free exchange of genes among selected plants; (5) ensuring an endless improvement of cultivars by nonstop selection across the changing climatic conditions and release of superior lines (Fasoula et al. 2007a,b,c); (6) selection by the prognostic equations is devoid of one limitation of conventional breeding, that is, visual selection that increases the chances of loosing superior genes irretrievably, thus reducing significantly the chances for large genetic gain; (7) released cultivars, being density neutral, are better equipped to exploit both favorable and marginal environments by adjusting properly seeding rates.

It is essential to predict with accuracy the presence of yield and stability genes earlier in the breeding program, as opposed to relying on progeny testing across sites in later generations to identify the yield and stability genes. Postponing selection for stability across sites and years reduces efficiency as stability genes may be irretrievably lost. Prognostic breeding relies on sibling testing to identify and select concurrently yield and stability genes by evaluating sibling plants. Sibling testing is more efficient than progeny testing because yield potential and stability are evaluated concurrently in the same generation. The advantages from using the prognostic equations for efficient selection among sibling lines and among plants within selected lines may be summarized as follows: (1) use of high selection pressures (1%–0.5%), (2) maximization of genetic gain through selection, (3) reduced time required to release a cultivar, (4) lower cost for cultivar development, (5) release of DN cultivars, (6) selection for quality on a single-plant basis, (7) selection for tolerance to biotic and abiotic stresses, (8) exploitation of the adaptive variation resulting from the GE interaction, (9) simultaneous selection for high and stable crop yield on a single-plant basis, (10) selection for local adaptation when extreme environments are used, and (11) mechanization and computerization of the selection process.

C. Maize Case Study

In maize, the increase of the crop yield per unit area during the transition period from populations to double-cross hybrids and finally to single-cross hybrids, was accompanied by a parallel increase of the plant density from 3 to almost 9 plants m^{-2}. This event drove scientists to regard plant density as an essential component of the crop yield potential (Yan and Wallace 1995). Another event strengthening this view has been the preference by farmers for high plant densities, owing to their observation that high plant densities suppress weeds more effectively. It is apparent that any attempt to unravel the relation between crop yield and plant density is rational.

Fasoula and Fasoula (1997a, 2000, 2002) and Fasoula and Tollenaar (2005) investigated this relation and identified the two causes that act in unison and explain the parallel increase of crop yield and plant density in maize. The first cause was the use of the monogenotypic single-cross hybrids that tolerate high plant densities due to the lack of genetic competition. The second cause was the constant improvement of single-cross hybrids for tolerance to the biotic and abiotic stresses that adds stability of performance and, by extension, tolerance to high plant densities. Many studies have showed that maize yield gains over the years are due to increases in stress tolerance (Bänziger et al. 1999; Tollenaar and Wu 1999; Tollenaar et al. 1994a, 2000). According to Duvick et al. (2004) increases in stress tolerance and in production efficiency increase the ability of the maize hybrids to tolerate ever-higher plant densities.

Referring to the causes limiting efficiency in plant breeding, Fasoula and Fasoula (1997a, 2000, 2002), elucidated that density is mistakenly considered a component of the crop yield, since the true components of the crop yield as explained in Chapter 2 are (1) the yield potential per plant measured by the mean yield ($\bar{x}$), (2) the yield stability measured by the standardized mean yield ($\bar{x}/s$), and (3) the adaptability or responsiveness to inputs measured by the standardized selection differential $(\bar{x}_{sel} - \bar{x})/s$. Because the three parameters measuring the three components are maximized in the absence of competition and minimized in its presence, their measurement should be realized at ultrawide plant spacing to ensure reliable estimates. An equally important reason for evaluating at ultrawide plant spacing is the negative correlation between yielding and competitive ability, which reduces advance through selection when plants are grown under competition (high plant densities) (Fasoula 1990; Fasoula and Fasoula 1997a).

Overcoming hybrid yield barriers in maize is one of the principal concerns of breeders and strategies for enhancing grain yield in maize have been reviewed by Tollenaar and Lee (2011). The productivity gap between maize inbred lines and hybrids has remained high in the last quarter of the 20th century (Fasoula and Fasoula 2000, 2005). The immediate impact of the large productivity gap is the high productivity cost of the hybrid seed. It is possible that the productivity gap has remained high due to the emphasis on selection for combining ability that piled up a large load of deleterious genes in the inbred lines. Selection for heterosis and combining ability retains deleterious genes, whereas continuous exploitation of the additive genetic variation reduces gradually the productivity gap between inbreds and hybrids (Fasoula and Fasoula 1997b, 2000, 2005). After Fasoula (2006, 2008, 2009) formulated the two prognostic equations for high and stable crop yield discussed previously, she proposed their use to bridge the productivity gap between maize inbreds and hybrids in order to overcome yield barriers in hybrid productivity. The best material to start this selection is the F_2 of elite commercial hybrids grown under conditions of open pollination. The principal idea is that to overcome crop yield barriers of maize hybrids, the improvement of crop yield potential of inbreds should precede that of hybrids. The preliminary results reported by Greveniotis et al. (2010, 2011) are presented below.

In 2007, 2,350 F_2 plants (C_0 cycle) of the maize hybrid 'Costanza' were grown under open pollination in a nonreplicated honeycomb trial (Fig. 6.7) in one location with plants spaced 1.25 m apart (Greveniotis et al. 2010). All plants were harvested individually and ranked on the basis of their p-YI only, p-YI = $(x/\bar{x}_r)^2$ since this was the first year of experimentation. A moving-ring size of 31 plants was used (Fig. 6.7) to measure the p-YI = $(x/\bar{x}_r)^2$, where x is the yield of each plant and $\bar{x}_r$ is the mean plant yield of its 30 surrounding plants. A very high selection pressure was used (1.2%) and the 29 top plants were selected based on their p-YI values. These plants were evaluated along with 'Costanza' and the initial F_2 in the next cycle (C_1) in a replicated D-31 honeycomb trial (Fig. 6.6) under open pollination in two contrasting locations. All plants were harvested individually and ranked according to the two prognostic equations in order to select the best plants. A similar procedure was repeated in the years 2009 (C_2 cycle) and 2010 (C_3 cycle). The results of these trials are presented in Fig. 6.8 and show how the productivity gap, as determined by the crop yield potential, was reduced from the C_0 to the C_3 cycle. Three cycles of selection by the two prognostic equations reduced the productivity gap from 80% to 8% across the two locations (Greveniotis et al. 2011). These results are very promising and suggest that the productivity gap between maize inbreds and hybrids can be significantly reduced to effectively overcome yield barriers.

Another notable event is the correlations calculated between the p-YI and the days measuring the anthesis-to-silking interval (ASI) shown in Fig. 6.9. These high and negative correlations indicate that when selecting plants with high p-YI values, the ASI is reduced to almost zero and self-fertilization responsible for gene fixation and advance through selection is maximized. For example, plants of family 12 with low yield index (p-YI = 0.1) have an ASI of 6 days, whereas plants with high yield index (p-YI = 5.0) have an ASI of only 1 day (Fig. 6.9).

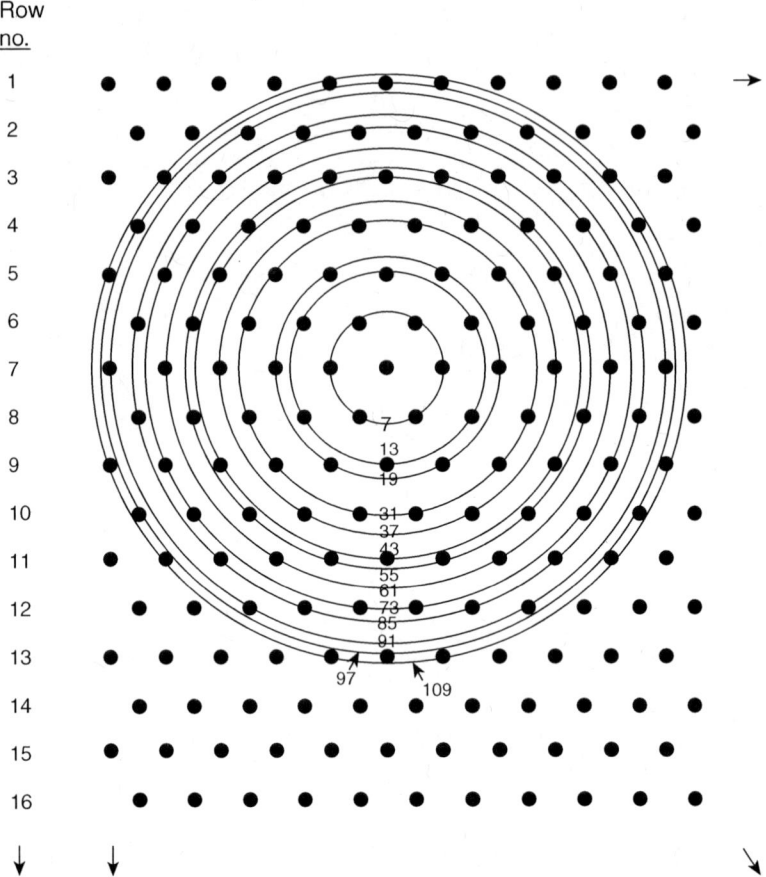

Fig. 6.7. The nonreplicated honeycomb design used in the first year of experimentation. Plants are laid out in the field randomly and evaluated only on the basis of the second parameter of the p-PE, that is, the plant yield index (p-YI), using a chosen moving-ring size; from 7 plants (smallest ring) to >109 plants (largest ring depicted here).

This correlation is also visible in the hybrid 'Costanza.' 'Contanza' plants with a high p-YI (p-YI = 5.7) have an ASI of only 2 days, while 'Contanza' plants with a low p-YI (p-YI = 0.1) have an ASI of 5 days. These results suggest that when the breeder selects for high p-YI values, he/she selects simultaneously for reduced ASI.

According to the two prognostic equations, the crop yield potential of every plant or every sibling line consists of two components: (1) the p-YI or s-YI and (2) the s-CH, which converts the p-YI or s-YI into crop yield potential. In this way, the two parameters of the crop yield

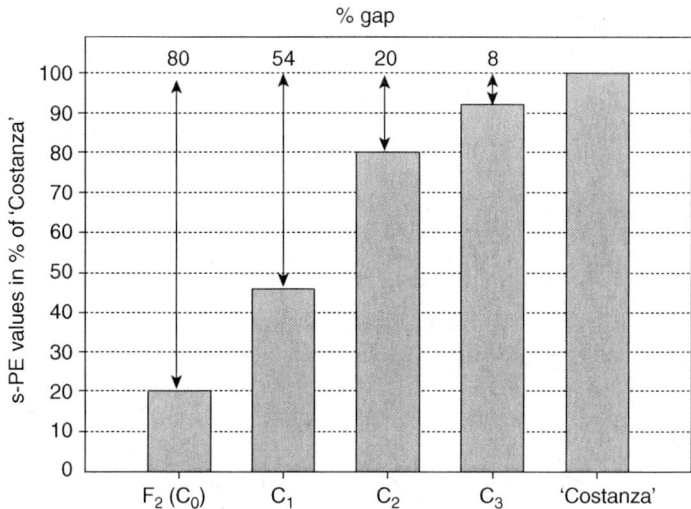

Fig. 6.8. Three selection cycles, starting from the F_2 (C_0 cycle) of the commercial maize hybrid 'Costanza', on the basis of the p-PE and s-PE, reduced the productivity gap between maize inbreds and hybrids from 80% to 8%. *Source:* Based on data from Greveniotis et al. (2011).

potential are measured concurrently and in the absence of the masking effect of competition and soil heterogeneity. Their use allows application of ultrahigh selection pressures (1%–0.5%) that accelerate gene fixation, maximize genetic gain and reduce significantly the number of selected lines participating in the next year's selection. Selection by the prognostic equations is devoid of one limitation of conventional breeding, that is, visual selection that increases the chances of loosing superior genes irretrievably, thus reducing significantly the chances for large genetic gain. The equations evaluate the stability of performance and yield potential in the same generation via the use of sibling testing, in contrast to the progeny testing that evaluates the stability of performance in successive generations and when enough seed is available to grow RCB trials. The problem is that postponing selection for stability across sites and years reduces efficiency as stability genes are irretrievably lost.

The aforementioned advantages help us to understand the superiority of the modified ear-to-row selection method suggested by Lonnquist (1964), which improves the efficiency of the ear-to-row selection by evaluating families in four sites synchronously, three testing sites and one selection site. Once the results for yield across the three testing sites

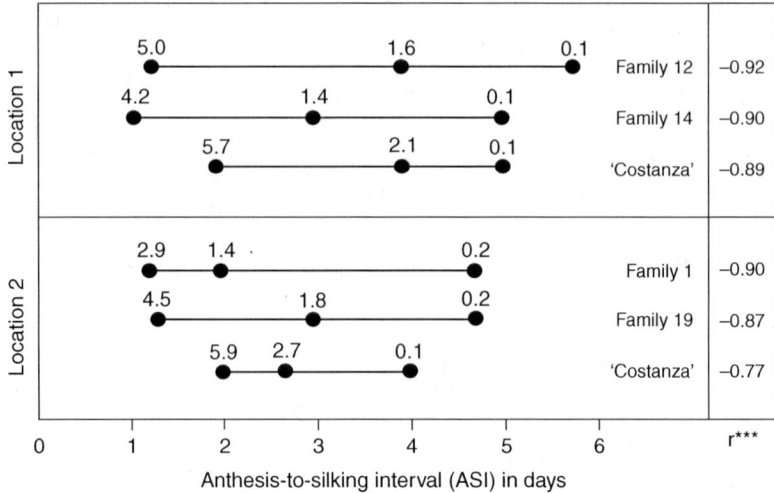

Fig. 6.9. Each black circle represents two values; one given by the horizontal axis (anthesis-to-silking interval) and the other on the top represents the mean p-YI values of three plants (top, intermediate, and bottom in the rank) of the best families and 'Costanza'. As illustrated, these pairs of values are correlated highly and negatively, meaning that selection for high p-YI reduces the ASI to almost zero. Plants with high p-YI values (>4) have an ASI of only 1–2 days, whereas plants with low p-YI values (<0.5) have an ASI of 4–6 days. *Source:* Based on data from Greveniotis et al. (2011).

are known, only the ears from outstanding lines across sites are selected. In this way, only superior families are selected in the selection site. The modified ear-to-row selection method, by selecting for yield and stability genes across three sites, has been the most efficient method for population improvement in maize. In addition, Lonnquist (1967) provided evidence that by spacing maize plants 0.75 m apart to enhance prolificacy and select for prolificacy for 5 years, the yield of the variety 'Hays Golden' was improved by 6.3% per year. These notable results substantiate the importance of predicting with accuracy the presence of yield and stability genes earlier in the breeding program, as opposed to relying on progeny testing across sites in later generations to identify the yield and stability genes.

Concurrent selection for high and stable crop yield by the two prognostic equations has an additional benefit, that is, the development of DN cultivars, namely, cultivars yielding optimally over a wider range of plant densities. This broadening of the density range of the optimum crop yield by the two prognostic equations is realized because the first component of the equations (s-CH) selects for optimal performance at higher plant densities, whereas the second component (s-YI or p-YI) selects for optimal performance at lower plant densities. The s-CH, as a parameter measuring any factor affecting the stability of performance (heritability, GE interaction, or tolerance to stresses), quantifies tolerance to biotic and abiotic stresses, which in turn ensures tolerance to high plant densities.

One of the main advantages of DN cultivars is that, by reducing the genotype by density interaction, they can exploit most profitably drought-stressed fields via proper adjustment of the seeding rates and avoidance of the damage from overpopulation. Data by Tokatlidis et al. (2011) show that the optimum plant density for maize hybrids grown under drought-prone conditions is much lower than when grown under favorable conditions, suggesting that the drought-prone environments require lower plant densities for efficient use of growth resources. In contrast, the modern maize hybrids are density-dependent (DD) since their productivity depends on a specific high plant density and may not exploit effectively drought-prone environments like those occurring in the sub-Saharan region (Fasoula and Fasoula 2000; Tokatlidis et al. 2011). Li et al. (2011) reported that modern US maize hybrids show a significant interaction with planting density, which is characteristic of DD cultivars. The development of DN maize hybrids is feasible once concurrent selection for high plant yield and stability is applied at ultralow plant densities. For example, maize hybrids improved concurrently for high plant yield and stability of performance had significantly higher yield than the check hybrid B73 × Mo17 at three different densities (0.74, 2.5, and 4.2 plants m^{-2}), while maintaining the same productivity at the high plant density of 8.4 plants m^{-2} (Tokatlidis et al. 2011).

The development of DN hybrids enables breeders to predict crop yield performance across the multitude of farmers' fields and plant densities, although selection is realized at ultrawide plant spacing on a single-plant basis. Another essential advantage of the DN cultivars is their capacity to reduce drastically the number of surviving weeds in the field. This advantage is due to the earlier emergence and rapid growth as well as the strong and extensive root system characterizing selected plants (Fasoula and Fasoula 1997a). As a result, the development of DN cultivars constitutes an effective means to select for tolerance to stresses and specifically for tolerance to drought and to increase the competitive ability of crops against the weeds. In the latter case, what is improved is not the genetic competitive ability but the ability of the DN cultivars to have rapid early growth and cover the field before the majority of weeds emerge.

Application of selection using the two prognostic equations in the F_2s of elite maize hybrids across ecological niches representing production environments, leads to high yielding and genetically homogenous DN prognostic lines of maize. A prognostic line of maize is an open-pollinated, near true-to-type breeding cultivar, developed through selection by the two prognostic equations. Such maize lines breed true-to-type due to the enhanced gene fixation induced by the high selection pressures and are ideal to be used in the production of prognostic maize hybrids. Hybrids between maize prognostic lines, will tend to be DN and capable of exploiting both favorable and marginal environments. The building up of prognostic hybrids by the two equations is an interesting topic for further academic research and could help alleviate some of the current limitations in maize breeding. The advantages of hybrid production in maize through prognostic lines may be summarized as follows.

1. Maize prognostic lines, having increased crop yield potential, reduce drastically the cost of hybrid seed.
2. Nonstop selection of maize lines by the two prognostic equations enables to build up resistance to biotic and abiotic stresses. Crossing these lines allows incorporation of this resistance to their developed hybrids, therefore, keeping ahead of mutating pathogens.
3. Hybrids between prognostic lines, being DN, exploit equally effectively favorable and marginal environments, and they do not exclude the use of high plant densities to minimize damage from weeds.
4. In case of unsatisfactory germination, DN prognostic hybrids may not require reseeding, in particular if they are herbicide tolerant.
5. Hybrids among maize prognostic lines, being endlessly improved for high and stable crop yield, exploit effectively the constantly released adaptive variation.
6. Constant improvement of maize prognostic lines under extreme environmental conditions or stresses leads to the production of prognostic hybrids with broader adaptation to these stresses.

In conclusion, the replacement of DD with DN prognostic hybrids in maize could help increase further hybrid productivity and broaden the capabilities of maize to exploit a vast category of growing environments, ranging from extremely favorable to extremely marginal. It offers also the possibility to reduce drastically the cost of hybrid seed, benefiting the farmers and the seed producers. An efficient way to meet the requirements of a sustainable agriculture is the development of DN hybrids, by replacing conventional inbred lines with prognostic lines via the utilization of the two prognostic equations discussed in detail in the previous chapters. Compared to current hybrids, DN hybrids are expected to exploit limited growth resources in marginal environments more efficiently. Furthermore, with the advent of the genetically modified hybrids, seed costs for the farmers have risen sharply and the production of DN hybrids may help reduce seed cost by using lower plant densities.

V. THE ADVANTAGES OF PROGNOSTIC BREEDING

Prognostic breeding is the methodology that utilizes two novel equations, applied on plants grown in the field at ultralow plant density, to predict plants and sibling lines that are equipped with yield and stability mechanisms. In addition, the two prognostic equations lead to monogenotypic cultivars that are constantly improved for high, stable, DN, and good quality crop yield. This is realized by ensuring in each generation reliable ranking of all plants and of the sibling lines to which plants belong. The reliable ranking allows application of ultrahigh selection pressures and identification of the top 1%–0.5% plants, in order to compose the sibling lines of the next cycle. Plants or lines selected on the basis of the two prognostic equations are coupled with three very important advantages: (1) they lead to DN cultivars that yield optimally over a wider range of plant densities because the first parameter of the equations selects for optimum yield at higher plant densities, whereas the second parameter selects for optimum yield at lower densities, (2) they have tolerance to biotic and abiotic stresses and especially to drought, due to the deep and extensive root system essential to confront the frequent scarcity of water, and (3) they compete effectively against weeds by shading the field before the majority of weeds have emerged.

Prognostic equations maximize selection efficiency by two powers, one resolving and one predictive. The resolving power of the prognostic equations stems from the fact that when plants are evaluated at ultrawide plant spacing, the differences among traits are magnified, while the prohibitive effect of competition on response to selection is eliminated. The predictive power of the prognostic equations consists in their ability to anticipate performance for high, stable, DN, and of good quality crop yield across sites and years, from evaluation at only few target production environments. Utilization of honeycomb field designs empowers the equations by accomplishing two objectives. The first objective is the elimination of the confounding effect of soil heterogeneity on single-plant yields by allocating plants in the center of moving circular complete replicates that serve as a common denominator. The second objective is the assessment of the sibling line homeostasis by allocating the plants of each sibling line in the corners of a triangular grid pattern that samples soil heterogeneity more efficiently than random allocation (Fasoulas and Fasoula 1995). This layout permits to select for stability of performance by assessing the stability of the plants of each sibling line in each triangular grid by the coefficient of homeostasis. The coefficient of homeostasis, by measuring any factor affecting stability of performance, like heritability, GE interaction, or tolerance to stresses, converts plant yield potential into crop yield potential.

In prognostic breeding, all plants of a sibling line are evaluated simultaneously for the two components of crop yield potential, that is, the plant yield potential and the plant stability of performance, in the same generation by using the two prognostic equations. In contrast, in traditional breeding, the two components of crop yield potential are evaluated independently in successive generations through progeny testing. The key novelty in prognostic breeding is that progeny testing is replaced by sibling testing, where plants have their siblings available in the generation of selection and the two components of the crop yield are evaluated concurrently.

Prognostic equations are the basis of prognostic breeding and capable of assessing the crop yield potential of individual plants and maximizing selection efficiency. Furthermore, they could help increase productivity in maize, through the development of constantly improved prognostic inbreds and hybrids. As indicated in the previous chapter, preliminary results concerning the production of prognostic inbreds, starting from F_2s of elite maize hybrids (C_0 selection cycle), show that the highly increased response to selection is accompanied by elimination of the protandry (Greveniotis et al. 2011). This reduced ASI in combination with the ultrawide plant spacing used, accelerate gene fixation and by extension advance through selection.

Overcoming hybrid yield barriers in maize is one of the principal concerns of maize breeders and strategies for enhancing grain yield in maize have been reviewed by Tollenaar and Lee (2011). The key question is why inbreds lag significantly behind maize hybrids in yield, adaptability, and stress tolerance when these traits are controlled mainly by additive genes. The answer is that selfing, by fixing not only additive, but deleterious alleles as well, preserves these alleles and causes inbreeding depression. As selfing fixes deleterious alleles, exploitation of heterosis via selection for combining ability is successful only when ensuring an effective mutual masking of these alleles. This process, however, being random, interferes with efficiency. Taking for granted the insurmountable obstacles set by selfing, the solution could be the replacement of selfing with another method. This method is accomplished by the p-PE and s-PE that, by enhancing fixation of favorable genes and eliminating the deleterious ones, allow application of ultrahigh selection pressures (1%–0.5%) under open pollination that lead to near-homozygous lines characterized by high, stable, DN and of good quality crop yield. These prognostic inbreds have higher productivity and stability compared to current inbreds and are expected to further increase maize yields through the development of superior and prognostic hybrids with lower seed cost. Prognostic inbreds, being constantly and concurrently improved for high and stable crop yield, are devoid of defective genes since these are removed during the selection process.

Another significant advantage of prognostic breeding is its ability to run selection within clonally propagated crops where exploitable genetic variation is assumed to be very small. It is well known though that somatic mutations as well as epigenetic variation can be present. In prognostic breeding, this variation can be exploited effectively by evaluating large numbers of sibling lines, from 200 to 300 with 10–15 plants per line across several environments, and exposing them to different biotic and abiotic stresses with the objective of building up tolerance to major stresses in clonally propagated crops.

Various studies have shown that the genome is more plastic than previously assumed and in constant flux (Cullis 1990; Rasmusson and Phillips 1997; Brunner et al. 2005; Lolle et al. 2005) and other studies have reported that selection within homogenous gene pools can be very effective (Fasoula 1990, 2012; Rasmusson and Phillips 1997; Fasoula and Fasoula 2000; Fasoula and Boerma 2005a,b, 2007; Tokatlidis et al. 2006). If we could analyze extensively a cross-pollinated, a self-pollinated, and a clonally propagated crop, it is unlikely that we find two plants that are completely identical genetically and epigenetically. Recent molecular data corroborate the fact that variation is present even in homogenous gene pools (Haun et al. 2011; Yates et al. 2012). Since monogenotypic cultivars are in constant flux due to latent and/or constantly released variation in response to environmental stimuli, application of nonstop prognostic breeding within these three categories of cultivars becomes essential. Prognostic breeding provides an integrated methodology where inbreeders, outbreeders, and clonally propagated crops are treated similarly. The difference is that plants of sibling lines are selfed in inbreeders, outcrossed in outbreeders, and cloned in clonally propagated crops.

VI. THE MARRIAGE OF PHENOTYPING WITH GENOTYPING

In molecular studies, accurate phenotyping is essential since genotypic characterization at the molecular level should be coupled with phenotypic predictability. Molecular techniques include identification of quantitative trait loci (QTL), marker-assisted selection, genome-wide selection and genetic transformation. The most important factor determining the success of marker-assisted strategies is the precision of the phenotyping (Peleman and van der Voort 2003; Edmeades et al. 2004). Mishra et al. (2010) reported that success in improving drought tolerance using molecular and physiological tools will largely be dependent on the accuracy of plant phenotyping and the capacity to determine the GE interactions. The accurate whole-plant field phenotyping for high and stable crop yield, accomplished by the two prognostic equations, paves the way to a mutually beneficial marriage between phenotyping and genotyping. For example, breeders can use the extreme selections of the prognostic equations, that is, the top- and bottom-ranking plants or sibling lines, to identify differences at the protein, gene, methylation, or genome levels, and use them subsequently as indices to enhance efficiency and speed up selection. Another use of the p-PE and s-PE can be the testing and further improvement of genetically modified plants through nonstop selection for high and stable crop yield under the heavy pressure of biotic and abiotic stresses (broad spectrum herbicides, damaging pests, or drought). The equations can be used to exploit effectively the constantly released adaptive variation in response to environmental stimuli. As has been discussed before, the genome, being in a state of a constant flux, constitutes an inexhaustible source of genetic variation that may ensure a lasting response to selection.

In addition, the marriage between phenotyping and genotyping allow probing into the genetic basis of important phenomena, such as heterosis, cultivar yield decline over time, yielding ability, competitive ability, or stability of performance. For example, breeders can convert F_2s of elite hybrids into prognostic lines that have improved crop yield and stability, and compare their molecular profile to that of the hybrids to discover answers concerning the genetic basis of heterosis. Similarly, selecting the top plants by the prognostic equations under ultralow versus high plant density within the same genetic material, breeders can assess the genetic difference between yielding and competitive ability by comparing the two categories of high yielders (selected at low vs. high plant density). The marriage between phenotyping and genotyping can be accomplished successfully since both phenotyping and genotyping use the individual plant as the unit of evaluation and selection.

VII. OUTLOOK

The maximization of the genetic gain through application of prognostic breeding ensured by the two prognostic equations, is founded on a number of fundamental principles that may be epitomized as follows.

1. Crop yield is maximized under one condition, that is, when all plants in the field yield the same. Maximization of crop yield is accomplished when ensuring equal sharing of growth resources among plants by eliminating the principal cause of unequal sharing, that is, competition. This is realized through (i) use of monogenotypic cultivars to eliminate genetic competition, and (ii) adoption of agronomic measures that reduce acquired competition.
2. Use of monogenotypic cultivars imposes as unit of evaluation and selection the single plant into which yield, quality, and stability mechanisms are incorporated. It imposes also the growing of plants at ultrawide plant spacing to remove genetic competition, which, being correlated negatively with yielding ability, hinders advance through selection due to the preferential selection of plants with high competitive-low yielding ability at the expense of plants with low competitive-high yielding ability.
3. The confounding effect of soil heterogeneity on single-plant yields must be reduced to incorporate yield mechanisms into plants grown at ultrawide plant spacing. Conversely, it is important to capitalize on soil heterogeneity in order to select for stability of performance to incorporate stability mechanisms. These seemingly contrasting objectives are accomplished by the honeycomb field designs that sample effectively soil heterogeneity and enable to predict with precision high yield and stability mechanisms among plants as well as among the sibling lines to which they belong.
4. The identification by the honeycomb designs of plants with high yield potential is accomplished by the p-YI $= (x/\bar{x}_r)^2$, where x is the plant yield (g) and $\bar{x}_r$ is the mean yield (g) of the surrounding plants encircled within a moving ring. The identification of

plants carrying stability mechanisms is realized by the s-CH $= (\bar{x}/s)^2$, where $\bar{x}$ and s are the mean yield per plant and standard deviation, respectively, of the sibling line to which each plant belongs. The product of the two parameters form the p-PE $= (\bar{x}/s)^2 \cdot (x/\bar{x}_r)^2$, which enables to: (1) convert plant yield potential into crop yield potential, (2) predict plants carrying yield and stability mechanisms, (3) assess crop yield potential under ultrawide plant spacing, (4) apply ultrahigh selection pressures (1%–0.5%), and (5) produce DN cultivars, that is, cultivars yielding optimally over a wider range of plant densities.

5. Replacing in the p-PE the p-YI $= (x/\bar{x}_r)^2$ with the s-YI $= (x/\bar{x}_r)^{2*}$, that is, the mean p-YI values of the plants of a sibling line, the following s-PE is defined; s-PE $= (\bar{x}/s)^2 \cdot (x/\bar{x}_r)^{2*}$. The s-PE enables to rank sibling lines according to their crop yield potential, assessed when plants grow at ultrawide plant spacing.

6. The capability of the p-PE and s-PE to predict plants and sibling lines carrying yield and stability mechanisms allows the application of ultrahigh selection pressures that maximize genetic gain and reduce drastically the number of selected plants. Furthermore, it enables to exploit plants carrying yield, tolerance to stresses, and stability mechanisms before these plants are irretrievably lost as this happens when advancing generations under dense stands.

7. Nonstop selection within elite cultivars by the two prognostic equations accomplishes three tasks: (i) it avoids cultivar yield decline over time, (ii) it improves productivity and stability by exploiting the latent and/or constantly released by the genome adaptive variation stemming from the GE interaction, and (iii) it ensures cultivar longevity through a continuous adaptation of cultivars to the challenges arising from the changing environmental conditions.

8. The applied ultrahigh selection pressures (1%–0.5%) lead to a rapid fixation of genes that unifies the breeding methodology for inbreeders and outbreeders. Regardless of the breeding system involved, application of ultrahigh selection pressures fixes additive alleles that are indispensable for advance through selection. Unifying the breeding methodology is powerful and makes prognostic breeding efficient, clear-cut, and science-based.

9. From the three possible categories of monogenotypic cultivars; that is, hybrids, clones, and those exploiting homozygote

advantage (inbred lines), the latter constitute the main target of a breeding program. These lines, being constantly improved for crop yield potential, serve to elevate further the yield potential of hybrids and can also serve as a lasting reservoir of genetic variation.

10. In prognostic breeding, selection for high and stable crop yield is not visual, thus it has the potential to become fully mechanized and computerized, adding new possibilities to maximize selection efficiency, minimize selection cost, and accelerate the genetic improvement of crops.

In conclusion, prognostic breeding is the methodology that embodies field phenotyping for high and stable crop yield of plants grown at ultrawide plant spacing. This prognosis is accomplished through utilization of the two prognostic equations that assess the plant yield potential and stability of performance. There are two ways to evaluate stability of performance; (1) by using the progenies of the plant and (2) by using the siblings of the plant. Compared to traditional breeding that uses progeny testing, the key novelty in prognostic breeding is that it uses sibling testing to assess the stability of performance. As has been explained, the difference between them is that in sibling testing, plants have their siblings available in the generation of selection and the two components of the crop yield are evaluated concurrently. In contrast, in progeny testing, progenies of selected plants are not available in the generation of selection, thus the two components of crop yield are evaluated independently in successive generations. In progeny testing, the confounding effects of soil heterogeneity on plot yields are minimized through the utilization of specific types of field designs (e.g., RCB or lattice designs). In sibling testing, the confounding and favorable effects of soil heterogeneity on single-plant yields are minimized and exploited, respectively, by the honeycomb field designs.

Another difference is that traditional breeding considers plant density an essential component of crop yield potential (Yan and Wallace 1995). In prognostic breeding, plant density is not considered a component of crop yield. Cultivars developed by prognostic breeding tend to yield well over a wider range of plant densities, from very low to very high. Prognostic breeding provides an integrated methodology where inbreeders, outbreeders, and clonally propagated crops are treated similarly. The difference is that plants of sibling lines are selfed in inbreeders, outcrossed in outbreeders, and cloned in clonally propagated crops. Prognostic breeding, by ensuring precision whole-plant field phenotyping for crop yield potential, constitutes a valuable tool for molecular breeding and genomics. Accordingly, prognostic breeding offers great potential for the genetic betterment of crops by increasing the precision of whole-plant field phenotyping, maximizing selection efficiency, minimizing selection cost, and accelerating the time-interval needed to release superior cultivars.

LITERATURE CITED

Austin, R.B., J. Bingham, R.D. Blackwell, L.T. Evans, M.A. Ford, C.L. Morgan, and M. Taylor. 1980. Genetic improvements in winter wheat yields since 1900 and associated physiological changes. J. Agr. Sci. (Cambr.) 94:675–689.
Bänziger, M., G.O. Edmeades, and H.R. Lafitte. 1999. Selection for drought tolerance increases maize yields across a range of nitrogen levels. Crop Sci. 39:1035–1040.
Boerma, H.R., R.S. Hussey, D.V. Phillips, E.D. Wood, and S.L. Finnerty. 1994. Registration of Haskell soybean. Crop Sci. 34:541.
Borlaug, N.E. 1983. Contributions of conventional plant breeding to food production. Science 219:689–693.
Borlaug, N.E. 2007. Sixty-two years of fighting hunger: personal recollections. Euphytica 157:287–297.
Brisson, N., P. Gate, D. Gouache, G. Charmet, F.X. Oury, and F. Huard. 2010. Why are wheat yields stagnating in Europe? A comprehensive data analysis for France. Field Crops Res. 119:201–212.
Brunner, S., S. Tingley, A. Rafalski, K. Fengler, and M. Morgante. 2005. Evolution of DNA sequence nonhomologies among maize inbreds. Plant Cell 17:343–360.
Cox, T.S., J.P. Shroyer, L. Ben-Hui, R.G. Sears, and T.J. Martin. 1988. Genetic improvement in agronomic traits of hard red winter wheat cultivars from 1919 to 1987. Crop Sci. 28:756–760.
Cullis, C.A. 1990. DNA rearrangements in response to environmental stress. Adv. Genet. 28:73–97.
Douches, D.S., D. Maas, K. Jastrzebski, and R.W. Chase. 1996. Assessment of potato breeding progress in the USA over the last century. Crop Sci. 36:1544–1552.
Dudley, J.W., and R.J. Lambert. 2004. 100 generations of selection for oil and protein in corn. Plant Breed. Rev. 24:79–110.
Duvick, D.N. 1984. Genetic contributions to yield gains of U.S. hybrid maize, 1930 to 1980. p. 15–47. In: W.R. Fehr (Ed.), Genetic contributions to yield gains of five major crop plants. CSSA Special Pub. 7. CSSA, Madison, WI.
Duvick, D.N., J.S.C. Smith, and M. Cooper. 2004. Long-term selection in a commercial hybrid maize breeding program. Plant Breed. Rev. 24:109–151.
Edmeades, G.O., G.S. McMaster, J.W. White, and H. Campos. 2004. Genomics and the physiologist: bridging the gap between genes and crop response. Field Crops Res. 90:5–18.
Edmeades, G., T. Fischer, and D. Byerlee. 2010. Can we feed the world in 2050? Proceedings of 15th ASA Conference on Food Security from Sustainable Agriculture, Nov. 15–19, 2010, Lincoln, New Zealand. www.agronomy.org.au.
Falconer, D.S. 1989. Introduction to quantitative genetics. Wiley, New York.

Fan, M., J. Shen, L. Yuan, R. Jiang, X. Chen, W.J. Davies, and F. Zhang. 2011. Improving crop productivity and resource use efficiency to ensure food security and environmental quality in China. J. Exp. Bot. 63:13–24.

FAO. 2009. How to feed the World 2050: global agriculture towards 2050.

Fasoula, D.A. 1990. Correlations between auto-, allo- and nil-competition and their implications in plant breeding. Euphytica 50:57–62.

Fasoula, D.A. 2012. Nonstop selection for high and stable crop yield by two prognostic equations to reduce yield losses. Agriculture 2:211–227.

Fasoula, D.A., and V.A. Fasoula. 2005. Bridging the productivity gap between maize inbreds and hybrids by replacing gene and genome dichotomization with gene and genome integration. Maydica 50:49–61.

Fasoula, D.A., and A. Lithourgidis. 2008. Selection within cotton cultivars (*Gossypium hirsutum* L.) under the Hellenic climatic conditions maximizes crop yield. Proceedings of 12th National Hellenic Conference in Genetics and Plant Breeding, Naousa, Greece, Oct. 8–10, 2008. p. 51–55.

Fasoula, D.A., and V.A. Fasoula. 1997a. Competitive ability and plant breeding. Plant Breed. Rev. 14:89–138.

Fasoula, D.A., and V.A. Fasoula. 1997b. Gene action and plant breeding. Plant Breed. Rev. 15:315–374.

Fasoula, V.A. 2006. A novel equation paves the way for an everlasting revolution with cultivars characterized by high and stable crop yield and quality. Proceedings of 11th National Hellenic Conference in Genetics and Plant Breeding, Orestiada, Greece, 31 Oct. – 2 Nov. 2006. p. 7–14.

Fasoula, V.A. 2008. Two novel whole-plant field phenotyping equations maximize selection efficiency. p. 361–365. In: J. Prohens, and M.L. Badenes (eds.), Proceedings of 18th Eucarpia General Congress, Modern Variety Breeding for Present and Future Needs, Valencia, Spain, Sept. 9–12, 2008.

Fasoula, V.A. 2009. Selection of high yielding plants belonging to entries of high homeostasis maximizes efficiency in maize breeding. p. 28. XXI Eucarpia International Conference in Maize and Sorghum Breeding in the Genomics Era, Bergamo, Italy, June 21–24, 2009.

Fasoula, V.A., and H.R. Boerma. 2005a. Divergent selection at ultra-low plant density for seed protein and oil content within soybean cultivars. Field Crops Res. 91:217–229.

Fasoula, V.A., and H.R. Boerma. 2005b. Selection at ultra-low plant density for high yield per plant within elite soybean cultivars. ASA-CSSA-SSSA International Annual Meetings, Salt Lake City, Utah, USA., Nov. 6–10, 2005.

Fasoula, V.A., and H.R. Boerma. 2007. Intra-cultivar variation for seed weight and other agronomic traits within three elite soybean cultivars. Crop Sci. 47:367–373.

Fasoula, V.A., and D.A. Fasoula. 2000. Honeycomb breeding: principles and applications. Plant Breed. Rev. 18:177–250.

Fasoula, V.A., and D.A. Fasoula. 2002. Principles underlying genetic improvement for high and stable crop yield potential. Field Crops Res. 75:191–209.

Fasoula, V.A., and D.A. Fasoula. 2003. Partitioning crop yield into genetic components. p. 321–327. In: M.S. Kang (ed.), Handbook of formulas and software for plant geneticists and breeders. Food Products Press, Binghamton, NY.

Fasoula, V.A., and M. Tollenaar. 2005. The impact of plant population density on crop yield and response to selection in maize. Maydica 50:39–48.

Fasoula, V.A., H.R. Boerma, J.L. Yates, D.R. Walker, S.L. Finnerty, G.B. Rowan, and E.D. Wood. 2007a. Registration of five soybean germplasm lines selected within the cultivar 'Benning' differing in seed and agronomic traits. J. Plant Reg. 1:156–157.

Fasoula, V.A., H.R. Boerma, J.L. Yates, D.R. Walker, S.L. Finnerty, G.B. Rowan, and E.D. Wood. 2007b. Registration of six soybean germplasm lines selected within the cultivar 'Haskell' differing in seed and agronomic traits. J. Plant Reg. 1:160–161.

Fasoula, V.A., H.R. Boerma, J.L. Yates, D.R. Walker, S.L. Finnerty, G.B. Rowan, and E.D. Wood. 2007c. Registration of seven soybean germplasm lines selected within the cultivar 'Cook' differing in seed and agronomic traits. J. Plant Reg. 1:158–159.

Fasoulas, A.C. 2000. Building up resistance to *Verticillium* wilt in cotton through honeycomb breeding. p. 120–124. In: F.M. Gillham (ed.) New frontiers in cotton research. Proceedings of 2nd World Cotton Research Conference Sept. 6–12, 1998. Athens, Greece.

Fasoulas, A.C., and V.A. Fasoula. 1995. Honeycomb selection designs. Plant Breed. Rev. 13:87–139.

Feil, B. 1992. Breeding progress in small grain cereals—a comparison of old and modern cultivars. Plant Breed. 108:1–11.

Fischer, R.A., and G.O. Edmeades. 2010. Breeding and cereal yield progress. Crop Sci. 50: S85–S98.

Fok, A.C.M. 2000. Cotton yield stagnation: addressing a common effect of various causes. p. 38–45. In: F.M. Gillham (ed.), New frontiers in cotton research. Proceedings of 2nd World Cotton Research Conference. Sept. 6–12, 1998. Athens, Greece.

Graybosch, R.A., and C.J. Peterson. 2010. Genetic improvement in winter wheat yields in the great plains of North America, 1959–2008. Crop Sci. 50:1882–1890.

Gregory, P.J., and T.S. George. 2011. Feeding nine billion: the challenge to sustainable crop production. J. Exp. Bot. 62:5233–5239.

Greveniotis, V., I.S. Tokatlidis, V.A. Fasoula, and S. Zotis. 2010. Selection at ultra-low plant density for high yield and stability favours additive gene action in maize. p. 557–558. In: J. Wery et al. (eds.), Proc. Agro2010 the XIth ESA Congress, Montpellier, France, Aug. 29 – Sept. 3, 2010.

Greveniotis, V., V.A. Fasoula, I.I. Papadopoulos, P.J. Bebeli, and I.S. Tokatlidis. 2011. Bridging the productivity gap between maize inbreds and hybrids via selection of plants excelling in crop yield genetic potential. In: XXII Eucarpia Maize and Sorghum Conference on Resources in Maize and Sorghum Breeding, Opatija, Croatia, June 19–22, 2011.

Grierson, C.S., S.R. Barnes, M.W. Chase, M. Clarke, D. Grierson, K.J. Edwards, G.J. Jellis, J.D. Jones, S. Knapp, G. Oldroyd, G. Poppy, P. Temple, R. Williams, and R. Bastow. 2011. One hundred important questions facing plant science research. New Phytol. 192:6–12.

Haun, W.J., D.L. Hyten, W.W. Xu, D.J. Gerhardt, T.J. Albert, T. Richmond, J.A. Jeddeloh, G. Jia, N.M. Springer, C.P. Vance, and R.M. Stupar. 2011. The composition and origins of genomic variation among individuals of the soybean reference cultivar Williams 82. Plant Physiol. 155:645–655.

Higgs, R.L., and W.A. Russell. 1968. Genetic variation in quantitative characters in maize inbred lines, I. Variation among and within Corn Belt seed sources of six inbreds. Crop Sci. 8:345–348.

Hill Jr., R.R., J.S. Shenk, and R.F. Barnes. 1988. Breeding for yield and quality. p. 809–825. In: A.A. Hanson et al. (eds.), Alfalfa and alfalfa improvement. Agron. Monogr. 29, ASA-CSSA-SSSA, Madison, WI.

Holland, J.B., and E.T. Bingham. 1994. Genetic improvement for yield and fertility of alfalfa cultivars representing different eras of breeding. Crop Sci. 34:953–957.

Janick, J. 1999. Exploitation of heterosis: uniformity and stability. p. 319–333. In: The genetics and exploitation of heterosis in crops. ASA-CSSA-SSSA, Madison, WI.

Kyriakou, D.T., and A.C. Fasoulas. 1985. Effects of competition and selection pressure on yield response in winter rye (*Secale cereale* L.). Euphytica 34:883–895.

Li, Y., X. Ma., T. Wang, Y. Li, C. Liu, Z. Liu, B. Sun, Y. Shi, Y. Song, M. Carlone, D. Bubeck, H. Bhardwaj, D. Whitaker, W. Wilson., E. Jones, K. Wright, S. Sun, W. Niebur, and S. Smith. 2011. Increasing maize productivity in China by planting hybrids with germplasm that responds favorably to higher planting densities. Crop Sci. 51:2391–2400.

Licker, R., M. Johnston, J.A. Foley, C. Barford, C.J. Kucharik, C. Monfreda, and N. Ramankutty. 2010. Mind the gap: how do climate and agricultural management explain the "yield gap" of croplands around the world? Global Ecol. Biogeography 19:769–782.

Lobell, D.B., K.G. Cassman, and C.B. Field. 2009. Crop yield gaps: their importance, magnitudes, and causes. Annu. Rev. Environ. Resour. 34:179–204.

Lolle S., J.L. Victor, J.M. Young, and R.E. Pruitt. 2005. Genome-wide non-mendelian inheritance of extra-genomic information in *Arabidopsis*. Nature 434:505–509.

Lonnquist, J.H. 1964. A modification of the ear-to-row procedure for the improvement of maize populations. Crop Sci. 4:227–228.

Lonnquist, J.H. 1967. Mass selection for prolificacy in maize. Der Zuchter 37:185–187.

McClintock, B. 1984. The significance of the responses of the genome to challenge. Science 226:792–801.

Mauromoustakos, A., V.A. Fasoula, and K. Thompson. 2006. Honeycomb designs computing and analysis. International Biometric Society, Eastern North American Region. Tampa, FL, March 26–29, 2006.

Meredith Jr., W.R., and R.R. Bridge. 1984. Genetic contributions to yield changes in upland cotton. p. 75–87. In: W.R. Fehr (ed.), Genetic contributions to yield gains of five major crop plants. CSSA Special Pub. 7 CSSA, Madison, WI.

Miller, F.R., and Y. Kebede. 1984. Genetic contributions to yield gains in sorghum, 1950 to 1980. p. 1–14. In: W.R. Fehr (ed.), Genetic contributions to yield gains of five major crop plants. CSSA Special Pub. 7 CSSA, Madison, WI.

Mishra, G.P., R.K. Singh, S. Raghwendra, and S.B. Singh. 2010. Molecular and physiological approaches for drought management of crops. p. 367–408. In: Singh et al. (eds.). Molecular plant breeding: principle, method, and application. Studium Press LLC.

Ortiz, R., D. Mowbray, C. Dowswell, and S. Rajaram 2007. Dedication: Norman E. Borlaug. The humanitarian plant scientist who changed the world. Plant Breed. Rev. 28:1–37.

Ortiz, R., K.D. Sayre, B. Govaerts, R. Gupta, G.V. Subbarao, T. Ban, D. Hodson, J.M. Dixon, J. I. Ortiz-Monasterio, and M. Reynolds. 2008a. Climate change: can wheat beat the heat? Agr. Ecosystems Environ. 126:46–58.

Ortiz, R., H.J. Braun, J. Crossa, J.H. Crouch, G. Davenport, J. Dixon, S. Dreisigacker, E. Duveiller, Z. He, J. Huerta, A.K. Joshi, M. Kishii, P. Kosina, Y. Manes, M. Mezzalama, A. Morgounov, J. Murakami, J. Nicol, G. Ortiz Ferrara, JI Ortiz-Monasterio, T.S. Payne, R.J. Pena, M.P. Reynolds, K.D. Sayre, R.C. Sharma, R.P. Singh, J. Wang, M. Warburton, H. Wu, and M. Iwanaga. 2008b. Wheat genetic resources enhancement by the International Maize and Wheat Improvement Center (CIMMYT). Genet. Resour. Crop Evol. 55:1095–1140.

Peleman, J.D., and J.R.van derVoort. 2003. Breeding by design. Trends Plant Sci. 8:330–334.

Peltonen-Sainio, P., L. Jauhiainen, and I.P. Laurila. 2009. Cereal yield trends in northern European conditions: changes in yield potential and its realization. Field Crops Res. 110:85–90.

Peloquin, S. 1995. Genetic mechanisms of genome evolution and speciation in autopolyploid plants: the potato model. p. 16–26. In: W.J. Raupt and B.S. Gill (eds.), Classical and molecular cytogenetic analysis. Kansas State University, Manhattan, KS, USA.

Peng, S., K.G. Cassman, S.S. Virmani, J. Sheehy, and G.S. Khush. 1999. Yield potential trends of tropical rice since the release of IR8 and the challenge of increasing rice yield potential. Crop Sci. 39:1552–1559.

Peng, S., Q. Tang, and Y. Zou. 2009. Current status and challenges of rice production in China. Plant Prod. Sci. 12:3–8.

Rajaram, S. 1999. Approaches for breaching yield stagnation in wheat. Genome 42: 629–634.

Rasmusson, D.C., and R.L. Phillips. 1997. Plant breeding progress and genetic diversity from de novo variation and elevated epistasis. Crop Sci. 37:303–310.

Russell, W.A., G.F. Sprague, and L.H. Penny. 1963. Mutations affecting quantitative characters in long-time inbred lines of maize. Crop Sci. 3:175–178.

Riday, H., and E.C. Brummer. 2002. Forage yield heterosis in alfalfa. Crop Sci. 42:716–723.

Sato, S. and N. Uphoff. 2007. A review of on-farm evaluations of system of rice intensification (SRI) methods in eastern Indonesia. CAB Rev. 2:1–12.

Schmidt, J.W. 1984. Genetic contributions to yield gains in wheat. p. 89–101. In: W.R. Fehr (ed.), Genetic contributions to yield gains of five major crop plants. CSSA Special Pub. 7 CSSA, Madison, WI.

Specht, J.E., and J.H. Williams. 1984. Contribution of genetic technology to soybean productivity-retrospect and prospect. p. 49–74. In: W.R. Fehr (ed.), Genetic contribution to yield gains of five major crop plants. CSSA Special Pub. 7 CSSA, Madison, WI.

Sprague, G.F., W.A. Russell, and L.H. Penny. 1960. Mutations affecting quantitative traits in selfed progeny of doubled monoploid maize stocks. Genetics 45:855–865.

Supit, I., C.A.van Diepen, A.J.W.de Wit, P. Kabat, B. Baruth, and F. Ludwig. 2010. Recent changes in the climatic yield potential of various crops in Europe. Agr. Syst. 103: 683–694.

Swaminathan, M.S. 2007. Can science and technology feed the world in 2025? Field Crops Res. 104:3–9.

Tilman, D., C. Balzer, J. Hill, and B.L. Befort. 2011. Global food demand and the sustainable intensification of agriculture. Proc. Natl. Acad. Sci, USA 108:20260–20264.

Tokatlidis, I.S., M. Koutsika-Sotiriou, and A.C. Fasoulas. 2001. The development of density-independent hybrids in maize. Maydica 46:21–25.

Tokatlidis, I.S., I.N. Xynias, J.T. Tsialtas, and I.I. Papadopoulos. 2006. Single-plant selection at ultra-low density to improve stability of a bread wheat cultivar. Crop Sci. 46:90–97.

Tokatlidis, I.S., V. Has, V Melidis, I. Has, I. Mylonas, G. Evgenidis, A. Copandean, E. Ninou, and V.A. Fasoula. 2011. Maize hybrids less dependent on high plant densities improve resource-use efficiency in rainfed and irrigated conditions. Field Crops Res. 120:345–351.

Tollenaar, M., S.P. Nissanka, A. Aguilera, S.F. Weise, and C.J. Swanton. 1994a. Effect of weed interference and soil nitrogen on four maize hybrids. Agron. J. 86:596–601.

Tollenaar, M., L.M. Dwyer, and D.E. McCullough. 1994b. Physiological basis of genetic improvement in corn. p. 183–236. In: G.A. Slafer (ed.), Genetic improvement of field crops. Marcel Dekker, New York.

Tollenaar, M., and J. Wu. 1999. Yield improvement in temperate maize is attributable to greater stress tolerance. Crop Sci. 39:1597–1604.

Tollenaar, M., J. Ying, and D.N. Duvick. 2000. Genetic gain in corn hybrids from the Northern and Central Corn Belt. p. 53–62. 55th Annual Corn and Sorghum Research Conference, Vol. 55 American Seed Trade Association, Chicago, IL.

Tollenaar, M., and E.A. Lee. 2011. Strategies for enhancing grain yield in maize. Plant Breed. Rev. 34:37–52.

Trethowan, R.M., M.P. Reynolds, J.I. Ortiz-Monasterio, and R. Ortiz. 2007. The genetic basis of the green revolution in wheat production. Plant Breed. Rev. 28:39–59.

Vollmann, J., J. Winkler, C.N. Fritz, H. Grausgruber, and P. Ruckenbauer. 2000. Spatial field variations in soybean (*Glycine max* [L.] Merr.) performance trials affect agronomic characters and seed composition. Euro. J. Agron. 12:13–22.

Waddington, S.R., J.K. Ransom, M. Osmanzai, and D.A. Saunders. 1986. Improvement in the yield potential of bread wheat adapted to Northwest Mexico. Crop. Sci. 26:698–703.

Wych, R.D., and D.C. Rasmusson. 1983. Genetic improvement in malting barley cultivars since 1920. Crop Sci. 23:1037–1040.

Wych, R.D., and D.D. Stuthman. 1983. Genetic improvement in Minnesota-adapted oat cultivars released since 1923. Crop Sci. 23:879–881.

Yan, W., and D.H. Wallace. 1995. Breeding for negatively associated traits. Plant Breed. Rev. 13:141–177.

Yates, J.L., H.R. Boerma, and V.A. Fasoula. 2012. SSR-marker analysis of the intracultivar phenotypic variation discovered within 3 soybean cultivars. J. Hered. 103:570–578.

Subject Index

Almond breeding, 37:207–258

Biography, Bikram Gill, 1–34
Breeding:
  almond, 207–258
  cotton, 322–327
  loquat, 259–296
  maize, 123–205, 327–335
  prognosis, 297–347
  soybean, 315–322
  stress resistance, 123–205
  wheat. 1–34; 35–122

Cotton breeding, 322–327
Cytogenetics:
  wheat 1–34; 35–122

Disease & pest resistance, maize, 123–205

Fruit and nut breeding:
  almond, 207–238
  loquat,, 259–296

Gill, Bikram (biography), 1–34
Grain breeding:
  maize, 123–205., 327–335
  wheat, 1–34, 35–122

Honeycomb:
  breeding, 297–347
  selection, 297–347

Legume breeding, soybean, 315–322
Loquat breeding, 259–296

Maize, breeding, 123–205, 327–335

Selection, prognosis, 297–347
Stress resistance, maize, 123–205
Synthetic wheat, 1–34, 35–122

Wheat:
  breeding, 35–122
  cytogenetics, 1–34, 35–122

Plant Breeding Reviews, Volume 37, First Edition. Edited by Jules Janick.
© 2013 Wiley-Blackwell. Published 2013 by John Wiley & Sons, Inc.

Cumulative Subject Index
(Volumes 1–37)

A

Adaptation:
  blueberry, rabbiteye, 5:351–352
  durum wheat, 5:29–31
  genetics, 3:21–167
  raspberry, 32:53–54, 153–184
  testing, 12:271–297
Aglaonema breeding, 23:267–269
Allelopathy, 30:231–258
Alexander, Denton, E. (biography), 22:1–7
Alfalfa:
  honeycomb breeding, 18:230–232
  inbreeding, 13:209–233
  in vitro culture, 2:229–234
  somaclonal variation, 4:123–152
  unreduced gametes, 3:277
Allard, Robert W. (biography), 12:1–17
Allium cepa, see Onion
Alliums transgenics, 35:210–213
Almond:
  breeding, 37:207–258
  breeding self-compatible, 8:313–338
  domestication, 25:290–291
  transformation, 16:103
Alocasia breeding, 23:269
Alstroemaria, mutation breeding, 6:75
Amaranth:
  breeding, 19:227–285
  cytoplasm, 23:191
  genetic resources, 19:227–285
Animals, long term selection, 24(2):169–210, 211–234

Aneuploidy:
  alfalfa, 10:175–176
  alfalfa tissue culture, 4:128–130
  petunia, 1:19–21
  wheat, 10:5–9
Anther culture:
  cereals, 15:141–186
  maize, 11:199–224
Anthocyanin
  maize aleurone, 8:91–137
  pigmentation, 25:89–114
Anthurium breeding, 23:269–271
Antifungal proteins, 14:39–88
Antimetabolite resistance, cell selection, 4:139–141, 159–160
Apomixis:
  breeding, 18:13–86
  genetics, 18:13–86
  reproductive barriers, 11:92–96
  rice, 17:114–116
Apple:
  domestication, 25:286–289
  fire blight resistance, 29:315–358
  genetics, 9:333–366
  rootstocks, 1:294–394
  transformation, 16:101–102
Apricot:
  domestication, 25:291–292
  transformation, 16:102
Arabidopsis, 32:114–123
Arachis, see Peanut
Artichoke breeding, 12:253–269
Avena sativa, see Oat
Avocado domestication, 25:307

Plant Breeding Reviews, Volume 37, First Edition. Edited by Jules Janick.
© 2013 Wiley-Blackwell. Published 2013 by John Wiley & Sons, Inc.

Azalea, mutation breeding,
  6:75–76

B

Bacillus thuringensis, 12:19–45
Bacteria, long-term selection, 24
  (2):225–265
Bacterial diseases:
  apple rootstocks, 1:362–365
  cell selection, 4:163–164
  cowpea, 15:238–239
  fire blight, 29:315–358
  maize, 27:156–159
  potato, 19:113–122
  raspberry, 6:281–282; 32:219–221
  soybean, 1:209–212
  sweet potato, 4:333–336
  transformation fruit crops, 16:110
Banana:
  breeding, 2:135–155
  domestication, 25:298–299
  transformation, 16:105–106
Barley:
  anther culture, 15:141–186
  breeding methods, 5:95–138
  diversity, 21:234–235
  doubled haploid breeding, 15:141–186
  gametoclonal variation, 5:368–370
  haploids in breeding, 3:219–252
  molelcular markers, 21:181–220
  photoperiodic response, 3:74, 89–92, 99
  vernalization, 3:109
Bean (*Phaseolus*):
  breeding, 1:59–102; 10:199–269;
    23:21–72; 36:357–426
  breeding mixtures, 4:245–272
  breeding (tropics), 10:199–269
  heat tolerance, 10:149
  in vitro culture, 2:234–237
  long-term selection, 24(2):69–74
  photoperiodic response, 3:71–73,
    86–92; 16:102–109
  protein, 1:59–102
  rhizobia interaction, 23:21–72
  seed color genetics, 28:239–315
Beet (table) breeding, 22:357–388
Beta, see Beet
Biochemical markers, 9:37–61
Biography:
  Alexander, Denton E., 22:1–7

Allard, Robert W., 12:1–17
Bliss, Frederick A., 27:1–14
Borlaug, Norman E., 28:1–37
Bringhurst, Royce S., 9:1–8
Burton, Glenn W., 3:1–19
Coyne, Dermot E., 23:1–19
Daubeny, H. A., 32:21–37
Downey, Richard K., 18:1–12
Draper, Arlen D., 13:1–10
Dudley, J.W., 24(1):1–10
Duvick, Donald N., 14:1–11
Frey, Kenneth, J. 34:1–36
Gabelman, Warren H., 6:1–9
Gill, Bikram, 37:1–34
Goodman, Major M., 33:1–29
Hallauer, Arnel R., 15:1–17
Harlan, Jack R., 8:1–17
Hymowitz, Theodore, 29:1–18
Jahn, Margaret, M., 35:1–17.
Jennings, D., 32:2–21
Jones, Henry A., 1:1–10
Laughnan, John R. 19:1–14
Munger, Henry M., 4:1–8
Ortiz, Rodomiro, 36:1–84
Peloquin, Stanley J., 25:1–19
Rédei, George, P., 26:1–33
Ryder, Edward J., 16:1–14
Salamini, Francesco, 30:1–47
Sears, Ernest Robert, 10:1–22
Simmonds, Norman W., 20:1–13
Sprague, George F., 2:1–11
Vogel, Orville A., 5:1–10
Vuylsteke, Dirk R., 21:1–25
Weinberger, John H., 11:1–10
Yuan, Longping, 17:1–13
Biotechnology:
  Cucurbitaceae, 27:213–244
  Douglas-fir, 27:331–336
  politics, 25:21–55
  Rosaceae, 27:175–211
Birdsfoot trefoil, tissue culture,
  2:228–229
Blackberry, 8:249–312, 29:19–144
  mutation breeding, 6:79
Black walnut, 1:236–266
Bliss, Frederick A. (biography), 27:1–14
Blueberry:
  breeding, 5:307–357;13:1–10;
    30:353–414
  domestication, 25:304

highbush, 30:353–414
rabbiteye, 5:307–357
Borlaug, Norman, E.(biography), 28:1–37
Bramble (*see* Blackberry, Raspberry):
  domestication, 25:303–304
  transformation, 16:105
Brachiaria, apomixis, 18:36–39, 49–51
Brassica, *see* Cole crops
  carinata 35:57–65
  cytogenetics, 31:21–187
  domestication, 35:19–84
  evolution, 31: 21–87; 35:19–84
  history, 35:19–84
  juncea, 35:58–65
  napus, *35:65–67*, *see* Canola, Rutabaga
  nigra, 35:38–41
  oleracea, 35:41–45
  rapa, *35: 51–47*, See also Canola
  transgenics: 35: 199–205
Brassicaceae:
  incompatibility, 15:23–27
  molecular mapping, 14:19–23
Breeding:
  Aglaonema, 23:267–269
  alfalfa via tissue culture, 4:123–152
  allelopathy, 30:231–258
  alliums, 35:210–213
  almond, 8:313–338, 37:207–258
  Alocasia, 23:269
  amaranth, 19:227–285
  apomixis, 18:13–86
  apple, 9:333–366
  apple fire blight resistance, 29:315–358
  apple rootstocks, 1:294–394
  banana, 2:135–155
  barley, 3:219–252; 5:95–138; 26:125–169
  bean, 1:59–102; 4:245–272; 23:21–7;
    36:357–426
  beet (table), 22:357–388
  biochemical markers, 9:37–61
  blackberry, 8:249–312; 29:19–144
  black walnut, 1:236–266
  blueberry, 5:307–357; 30:353–414
  brassicas, 35:19–84, 199–205
  bromeliad, 23:275–276
  cactus, 20:135–166
  Calathea, 23:276
  carbon isotope
    discrimination, 12:81–113
  carrot, 19:157–190, 35:219–220
  cassava, 2:73–134; 31:247–275, 35:216
  cell selection, 4:153–173
  cereal stress resistance, 33:115–144
  chestnut, 4:347–397; 33:305–339;
    36:427–503
  chimeras, 15:43–84
  chrysanthemum, 14:321–361
  citrus, 8:339–374; 30:323–352
  coffee, 2:157–193; 30:415–447
  coleus, 3:343–360
  competitive ability, 14:89–138
  cotton, 37:322–327
  cowpea, 15:215–274, 35:215
  cucumber, 6:323–359
  Cucurbitaceae 27:213–244
  cucurbits, 27:213–244; 35:196–199
  currant, 29:145–175
  cytoplasmic DNA, 12:175–210
  diallel analysis, 9:9–36
  Dieffenbachia, 23:271–272
  doubled haploids, 15:141–186;
    25:57–88
  Dougas-fir, 27:245–253
  Dracaena, 23:277
  drought tolerance, maize, 25:173–253
  durum wheat, 5:11–40
  eggplant, 35:187–191
  Epepremnum, 23:272–273
  epigenetics, 30:49–177
  epistasis, 21:27–92
  exotic maize, 14:165–187
  fern, 23:276
  fescue, 3:313–342
  Ficus, 23:276
  fire blight resistance, 29:315–358
  flower color, 25:89–114
  foliage plant, 23:245–290
  forest tree, 8:139–188
  fruit crops, 25:255–320
  garlic, 6:81; 23:11–214
  gene action 15:315–374
  genotype x environment
    interaction, 16:135–178
  gooseberry, 29:145–175
  grain legumes, 33:157–304
  grapefruit, 13:345–363
  grasses, 11:251–274
  guayule, 6:93–165
  heat tolerance, 10:124–168
  Hedera, 23:279–280

Breeding: (*Continued*)
  herbicide-resistant crops, 11:155–198
  heritability, 22:9–111
  heterosis, 12:227–251
  homeotic floral mutants, 9:63–99
  honeycomb, 13:87–139; 18:177–249
  human nutrition, 31:325–392
  hybrid, 17:225–257
  hybrid wheat, 2:303–319; 3:169–191
  induced mutations, 2:13–72
  insect and mite resistance in cucurbits, 10:199–269
  isozymes, 6:11–54
  legumes, 26:171–357; 33:157–304
  lettuce, 16:1–14; 20:105–133; 35:205–210
  loquat, 37:259–296
  maize, 1:103–138, 139–161; 4:81–122; 9:181–216; 11:199–224; 14:139–163, 165–187, 189–236; 25:173–253; 27:119–173; 28:59–100; 31:223–245; 33:9–16; 34:37–182, 83–113, 131–160; 37:123–205, 327–335
  marker-assisted selection, 33:145–217, 219–256; 34:247–358
  meiotic mutants, 28:163–214
  melon, 35:85–150
  millets, 35:247–374
  mitochondrial genetics, 25:115–238
  molecular markers, 9:37–61, 10:184–190; 12:195–226; 13:11–86; 14:13–37, 17:113–114, 179, 212–215; 18:20–42; 19:31–68, 21:181–220, 23:73–174; 24(1):293–309; 26:292–299; 31:210–212, 33:145–217, 219–256; 34:247–348; 35:332–344
  mosaics, 15:43–84
  mushroom, 8:189–215
  negatively associated traits, 13:141–177
  nutrition enhancement, 36:169–211
  oat, 6:167–207
  oil palm, 4:175–201; 22:165–219
  onion, 20:67–103; 35:210–213
  ornamental transgenesis, 28:125–216
  palms, 23:280–281
  papaya, 26:35–78
  pasture legumes, 5:237–305
  pea, snap, 212:93–138
  peanut, 22:297–356; 30:295–322; 36:293–356

  pear fire blight resistance, 29:315–358
  pearl millet, 1:162–182
  perennial rye, 13:265–292
  persimmon, 19:191–225
  Philodendron, 23:2
  phosphate efficiency, 29:394–398
  plantain, 2:150–151; 14:267–320; 21:211–25
  potato, 3:274–277; 9:217–332; 16:15–86; 19:59–155, 25:1–19; 35:191–196
  prognosis, 37:297–347
  proteins in maize, 9:181–216
  quality protein maize (QPM), 9:181–216
  raspberry, 6:245–321; 32:1–37, 39–53
  recurrent restricted phenotypic selection, 9:101–113
  recurrent selection in maize, 9:115–179; 14:139–163
  rice, 17:15–156; 23:73–174
  rol genes, 26:79–103
  Rosaceae, 27:175–211
  rose, 17:159–189; 31:227–334
  rubber (*Hevea*), 29:177–283
  rutabaga, 8:217–248
  sesame, 16:179–228
  snap pea, 21:93–138
  somatic hybridization, 20:167–225
  sorghum drought tolerance, 31:189–222
  sorghum male sterility, 25:139–172
  soybean, 1:183–235; 3:289–311; 4:203–243; 21:212–307; 30:250–294; 37:315–322l
  soybean fatty acids, 30:259–294
  soybean hybrids, 21:212–307
  soybean nodulation, 11:275–318
  soybean recurrent selection, 15:275–313
  spelt, 15:187–213
  statistics, 17:296–300
  strawberry, 2:195–214
  stress resistance, 37:123–30
  sugarcane, 16:272–273; 27:15–158
  supersweet sweet corn, 14:189–236
  sweet cherry, 9:367–388
  sweet corn, 1:139–161; 14:189–236; 35:213–215
  sweet potato, 4:313–345; 35:217–218
  Syngonium, 23:274
  tomato, 4:273–311
  transgene technology, 25:105–108
  triticale, 5:41–93; 8:43–90

vegetable crop transgenics, 35:151–246
Vigna, 8:19–42
virus resistance, 12:47–79
wheat, 2:303–319; 3:169–191;
    5:11–40; 11:225–234; 13:293–343,
    28:1–37, 39–78; 36:85–165; 37:
    11–24, 35–122
wheat, rust resistance, 13:293–343
white clover, 17:191–223
wild relatives, 30:149–230
wild rice, 14:237–265
Bringhurst, Royce S. (biography), 9:1–8
Broadbean, in vitro culture, 2:244–245
Bromeliad breeding, 23:275–276
Brown, Anthony, H.D.
    (biography), 31:1–20
Burton, Glenn W. (biography), 3:1–19

C

Cactus:
    breeding, 20:135–166
    domestication, 20:135–166
Cajanus, in vitro culture, 2:224
Calathea breeding, 23:276
Canola, R.K. Downey, designer, 18:1–12
Carbohydrates, 1:144–148
Carbon isotope discrimination,
    12:81–113
Carica papaya, see Papaya
Carnation, mutation breeding, 6:73–74
Carrot:
    breeding, 19: 157–190
    transgenics, 35:219–220
Cassava:
    breeding, 2:73–134; 31:247–275
    long-term selection, 24(2):74–79
    transgenics: 35:216
Castanea, see Chestnut
Cell selection, 4:139–145, 153–173
Cereal breeding, see Grain breeding
Cereals:
    diversity, 21:221–261
    stress resistance, 33:31–114.
Cherry, see Sweet cherry
    domestication, 25:202–293
Chestnut breeding, 4:347–397; 33:305–339
Chickpea, in vitro culture, 2:224–225
Chimeras and mosaics, 15:43–84
Chinese cabbage, heat tolerance, 10:152
Chromosome, petunia, 1:13–21, 31–33

Chrysanthemum:
    breeding, 14:321–361
    mutation breeding, 6:74
Cicer, see Chickpea
Citrus:
    breeding (seedlessness), 30:323–352
    domestication, 25:296–298
    protoplast fusion, 8:339–374
Clonal repositories, see National Clonal
    Germplasm Repository
Clone identification (DNA), 34:221–295
Clover:
    in vitro culture, 2:240–244
    molecular genetics, 17:191–223
Coffea arabica, see Coffee
Coffee, 2:157–193; 30:415–437
Cold hardiness:
    breeding nectarines and
        peaches, 10:271–308
    wheat adaptation, 12:124–135
Cole crops:
    Chinese cabbage, heat tolerance, 10:152
    gametoclonal variation, 5:371–372
    rutabaga, 8:217–248
Coleus, 3:343–360
Competition, 13:158–165
Competitive ability breeding, 14:89–138
Controlling elements, see Transposable
    elements
Corn, see Maize; Sweet corn
Cotton:
    breeding, 37:322–327
    heat tolerance, 10:151
Cowpea:
    breeding, 15:215–274
    heat tolerance, 10:147–149
    in vitro culture, 2:245–246
    photoperiodic response, 3:99
    transgenics, 35:215
Coyne, Dermot E. (biography), 23:1–19
Cranberry domestication, 25:304–305
Crop domestication and selection, 24
    (2):1–44
Cryopreservation, 7:125–126, 148–151,
    167
    buds, 7:168–169
    genetic stability, 7:125–126
    meristems, 7:168–169
    pollen, 7:171–172
    seed, 7:148–151,168

Cucumber, breeding, 6:323–359
Cucumis sativus, see Cucumber
Cucumis melo, see Melon
Cucurbitaceae:
  insect and mite resistance, 10:309–360
  mapping, 27:213–244
Cucurbits:
  mapping, 27:213–244
  transgenics: 35:196–199
Currant breeding, 29:145–175
Cybrids. 3:205–210; 20: 206–209
Cytogenetics:
  alfalfa, 10:171–184
  blueberry, 5:325–326
  Brassica, 31:21–187; 35:25–36
  cassava, 2:94
  citrus, 8:366–370
  coleus, 3:347–348
  durum wheat, 5:12–14
  fescue, 3:316–319
  Glycine, 16:288–317
  guayule, 6:99–103
  maize mobile elements, 4:81–122
  maize-tripsacum hybrids, 20:15–66
  meiotic mutants, 28:163–214
  oat, 6:173–174
  polyploidy terminology, 26:105–124
  pearl millet, 1:167
  perennial rye, 13:265–292
  petunia, 1:13–21, 31–32
  potato, 25:1–19
  raspberry, 32: 135–137
  rose, 17:169–171
  rye, 13:265–292
  Saccharum complex, 16:273–275
  sesame, 16:185–189
  sugarcane, 27:74–78
  triticale, 5:41–93; 8:54
  wheat, 5:12–14; 10:5–15; 11:225–234; 37:1–24, 35–122
Cytoplasm:
  breeding, 23: 175–210; 25:115–138
  cybrids, 3:205–210; 20:206–209
  incompatibility, 25:115–138
  male sterility, 25:115–138,139–172
  molecular biology of male sterility, 10:23–51
  organelles, 2:283–302; 6:361–393
  pearl millet, 1:166
  petunia, 1:43–45
  sorghum male sterility, 25:139–172
  wheat, 2:308–319

D

Dahlia, mutation breeding, 6:75
Date palm domestication, 25:272–277
Daubeny, Hugh A. (biography), 32:21–37
Daucus, see Carrot
Diallel cross, 9:9–36
Dieffenbachia breeding, 23:271–272
Diospyros, see Persimmon
Disease and pest resistance:
  antifungal proteins, 14:39–88
  apple rootstocks, 1:358–373
  banana, 2:143–147
  barley, 26:135–169
  blackberry, 8:291–295
  black walnut, 1:251
  blueberry, rabbiteye, 5:348–350
  cassava, 2:105–114; 31:247–275
  cell selection, 4:143–145, 163–165
  chestnut blight, 4: 347–397; 33:305–339
  citrus, 8:347–349
  coffee, 2:176–181
  coleus, 3:353
  cowpea, 15:237–247
  durum wheat, 5:23–28
  fescue, 3:334–336
  herbicide-resistance, 11:155–198
  host-parasite genetics, 5:393–433
  induced mutants, 2:25–30
  lettuce, 1:286–287
  maize, 27:119–173; 31:223–245; 34:131–160; 37:123–205
  melon, 35: 86–150
  millets, 35:247–374
  ornamental transgenesis, 28:145–147
  peanut virus, 36:293–356
  papaya, 26:161–357
  potato, 9:264–285, 19:69–155
  raspberry, 6:245–321; 32:184–247
  rose, 31:277–324
  rutabaga, 8:236–240
  soybean, 1:183–235
  spelt, 15:195–198
  strawberry, 2:195–214
  verticillium wilt, 33:115–144
  virus resistance, 12:47–79
  wheat rust, 13:293–343

CUMULATIVE SUBJECT INDEX

Diversity:
  landraces, 21:221–261
  legumes, 26:171–357
  maize, 33:4–7
  melon, 35:85–150
  millets, 35:247–374
  raspberry, 32:54–58
DNA:
  clone identification, 34:221–295
  methylation, 18:87–176; 30:49–177
Doubled haploid breeding, 15:141–186; 25:57–88
Douglas-fir breeding, 27:245–353
Downey, Richard K. (biography), 18:1–12
Dracaena breeding, 23:277
Draper, Arlen D. (biography), 13:1–10
Drought resistance, *see also* Stress Resistance:
  cereals, 33:31–114
  durum wheat, 5:30–31
  maize, 25:173–253
  sorghum, 31:189–222
  soybean breeding, 4:203–243
  wheat adaptation, 12:135–146; 36:85–165
Dudley, J.W. (biography), 24(1):1–10
Durum wheat, 5:11–40
Duvick, Donald N. (biography), 14:1–11

E

Eggplant transgenics: 35:187–191
Elaeis, see Oil palm
Embryo culture:
  in crop improvement, 5:181–236
  oil palm, 4:186–187
  pasture legume hybrids, 5:249–275
Endosperm:
  balance number, 25:6–7
  maize, 1:139–161
  sweet corn, 1:139–161
Endothia parasitica, 4:355–357
Epepremnum breeding, 23:272–273
Epigenetics, 30:49–177
Epistasis, 21:27–92.
Escherichia coli, long-term selection, 24 (2):225–224
Evolution:
  Brassica, 31:21–187
  coffee, 2:157–193
  fruit, 25: 255–320
  grapefruit, 13:345–363
  maize, 20:15–66
  sesame, 16:189
Exploration, 7:9–11, 26–28, 67–94

F

Fabaceae, molecular mapping, 14:24–25
Fatty acid genetics and breeding, 30:259–294
Fern breeding, 23:276
Fescue, 3:313–342
Festuca, see Fescue
Fig domestication, 25:281–285
Fire blight resistance, 29:315–358
Flavonoid chemistry, 25:91–94
Floral biology:
  almond, 8:314–320
  blackberry, 8:267–269
  black walnut, 1:238–244
  cassava, 2:78–82
  chestnut, 4:352–353
  coffee, 2:163–164
  coleus, 3:348–349
  color, 25:89–114
  fescue, 3:315–316
  garlic: 23:211–244
  guayule, 6:103–105
  homeotic mutants, 9:63–99
  induced mutants, 2:46–50
  pearl millet, 1:165–166
  pistil in reproduction, 4:9–79
  pollen in reproduction, 4:9–79
  raspberry, 32:90–92
  reproductive barriers, 11:11–154
  rutabaga, 8:222–226
  sesame, 16:184–185
  sweet potato, 4:323–325
Flower:
  color genetics, 25:89–114
  color transgenesis, 28:28–139
Forage breeding:
  alfalfa inbreeding, 13:209–233
  diversity, 21:221–261
  fescue, 3:313–342
  perennials, 11:251–274
  white clover, 17:191–223
Foliage plant breeding, 23:245–290

Forest crop breeding:
    black walnut, 1:236–266
    chestnut, 4:347–397
    Douglas-fir, 27:245–353
    ideotype concept, 12:177–187
    molecular markers, 19:31–68
    quantitative genetics, 8:139–188
    rubber (*Hevea*), 29:177–283
Fragaria, see Strawberry
Frey, Kenneth J. (biography), 34:1–36.
Fruit, nut, and beverage crop breeding:
    almond, 8:313–338; 37:207–238
    apple, 9:333–366
    apple fire blight resistance,
        29:315–358
    apple rootstocks, 1:294–394
    banana, 2:135–155
    blackberry, 8:249–312; 29:19–144
    blueberry, 5:307–357; 13:1–10;
        30:323–414
    breeding, 25:255–320
    cactus, 20:135–166
    cherry, 9:367–388
    chestnut, 4:347–397; 33:305–339
    citrus, 8:339–374; 30:323–352
    coffee, 2:157–193; 30:415–437
    currant, 29:145–175
    domestication, 25:255–320
    fire blight resistance, 29:315–358
    genetic transformation, 16:87–134
    gooseberry, 29:145–175
    grapefruit, 13:345–363
    ideotype concept, 12:175–177
    incompatability, 28:215–237
    loquat, 37:259–296
    melon, 35:85–150
    mutation breeding, 6:78–79
    nectarine (cold hardy), 10:271–308
    origins, 25:255–320
    papaya, 26:35–78
    peach (cold hardy), 10:271–308
    pear fireblight resistance, 29:315–358
    persimmon, 19:191–225
    plantain, 2:135–155
    raspberry, 6:245–321; 32:1–353
    strawberry, 2:195–214
    sweet cherry, 9:367–388
Fungal diseases:
    apple rootstocks, 1:365–368
    banana and plantain, 2:143–145, 147

barley, *Fusarium* head
    blight, 26:125–169
cassava, 2:110–114
cell selection, 4:163–165
chestnut blight, 4:355–397; 33:305–339
coffee, 2:176–179
cowpea, 15:237–238
durum wheat, 5:23–27
Fusarium head blight
    (barley), 26:125–169
host-parasite genetics, 5:393–433
lettuce, 1:286–287
maize foliar, 27:119–173; 31:223–245
potato, 19:69–155
raspberry, 6:245–281; 32:184–221
rose, 31:277–324
soybean, 1:188–209
spelt, 15:196–198
strawberry, 2:195–214
sweet potato, 4:333–336
transformation, fruit crops, 16:111–112
verticillium wilt,
    Solanaceae, 33:115–144
wheat rust, 13:293–343
Fusarium head blight (barley),
    26:125–169

G
Gabelman, Warren H. (biography), 6:1–9
Gametes:
    almond, self compatibility, 7:322–330
    blackberry, 7:249–312
    competition, 11:42–46
    epigenetics, 30:49–177
    forest trees, 7:139–188
    maize aleurone, 7:91–137
    maize anthocynanin, 7:91–137
    mushroom, 7:189–216
    polyploid, 3:253–288
    rutabaga, 7:217–248
    transposable elements, 7:91–137
    unreduced, 3:253–288
Gametoclonal variation, 5:359–391
    barley, 5:368–370
    brassica, 5:371–372
    potato, 5:376–377
    rice, 5:362–364
    rye, 5:370–371
    tobacco, 5:372–376
    wheat, 5:364–368

CUMULATIVE SUBJECT INDEX 359

Garlic breeding, 6:81; 23:211–244
Genes:
  action, 15:315–374
  apple, 9:337–356
  Bacillus thuringensis, 12:19–45
  incompatibility, 15:19–42
  incompatibility in sweet
    cherry, 9:367–388
  induced mutants, 2:13–71
  lettuce, 1:267–293
  maize endosperm, 1:142–144
  maize protein, 1:110–120, 148–149
  petunia, 1:21–30
  quality protein in maize, 9:183–184
  Rhizobium, 23:39–47
  rol in breeding, 26:79–103
  rye perenniality, 13:261–288
  soybean, 1:183–235
  soybean nodulation, 11:275–318
  sweet corn, 1:142–144
  wheat rust resistance, 13:293–343
Genetic engineering (transgeneic breeding):
  bean, 1:89–91
  cereal stress resistance, 33:31–114
  DNA methylation, 18:87–176
  fire blight resistance, 29:315–358
  fruit crops, 16:87–134
  host-parasite genetics, 5:415–428
  legumes, 26:171–357
  maize mobile elements, 4:81–122
  ornamentals, 125–162
  papaya, 26:35–78.
  rol genes, 26:79–103
  salt resistance, 22:389–425
  sugarcane, 27:86–97
  transformation by particle
    bombardment, 13:231–260
  transgene technology, 25:105–108
  virus resistance, 12:47–79
Genetic load and lethal
    equivalents, 10:93–127
Genetics:
  adaptation, 3:21–167
  almond, self compatibility, 8:322–330
  amaranth, 19:243–248
  Amaranthus, see Amaranth
  apomixis, 18:13–86
  apple, 9:333–366
  Bacillus thuringensis, 12:19–45
  bean seed color: 28:219–315

  bean seed protein, 1:59–102
  beet, 22:357–376
  blackberry, 8:249–312; 29:19–144
  black walnut, 1:247–251
  blueberry, 13:1–10
  blueberry, rabbiteye, 5:323–325
  carrot, 19:164–171
  chestnut blight, 4:357–389
  chimeras, 15:43–84
  chrysanthemums, 14:321–361
  clover, white, 17:191–223
  coffee, 2:165–170
  coleus, 3:3–53
  cowpea, 15:215–274
  Cucurbitaceae, 27:213–344
  cytoplasm, 23:175–210
  DNA methylation, 18:87–176
  domestication, 25:255–320
  durum wheat, 5:11–40
  epigenetics, 30:49–177
  fatty acids in soybean, 30:259–294
  fire blight resistance, 29:315–358
  forest trees, 8:139–188
  flower color, 25:89–114
  fruit crop transformation, 16:87–134
  gene action, 15:315–374
  green revolution, 28:1–37, 39–78
  history, 24(1):11–40
  host-parasite, 5:393–433
  incompatibility:
    circumvention, 11:11–154
    molecular biology, 11:19–42;
      28:215–237
    sweet cherry, 9:367–388
  induced mutants, 2:51–54
  insect and mite resistance in
    Cucurbitaceae, 10:309–360
  isozymes, 6:11–54
  lettuce, 1:267–293
  maize adaptedness, 28:101–123
  maize aleurone, 8:91–137
  maize anther culture, 11:199–224
  maize anthocyanin, 8:91–137
  maize endosperm, 1:142–144\
  maize foliar diseases, 27:118–173
  maize male sterility, 10:23–51
  maize mobile elements, 4:81–122
  maize mutation, 5:139–180
  maize quality protein, 9:1183–184;
    34:83–113

Genetics: (*Continued*)
- maize seed protein, 1:110–120, 148–149
- maize soil acidity tolerance, 28:59–123
- mapping, 14:13–37
- markers to manage germplasm, 13:11–86
- maturity, 3:21–167
- meiotic mutants, 163–214
- metabolism and heterosis, 10:53–59
- millets, 247–374
- mitochondrial, 25:115–138.
- molecular mapping, 14:13–37
- mosaics, 15:43–84
- mushroom, 8:189–216
- oat, 6:168–174
- organelle transfer, 6:361–393
- overdominance, 17:225–257
- pea, 21:110–120
- pearl millet, 1:166, 172–180
- perennial rye, 13:261–288
- petunia, 1:1–58
- phosphate mechanisms, 29: 359–419
- photoperiod, 3:21–167
- plantain, 14:264–320
- polyploidy terminology, 26:105–124
- potato disease resistance, 19:69–165
- potato ploidy manipulation, 3:274–277; 16:15–86
- quality protein in maize, 9:183–184
- quantitative trait loci, 15:85–139
- quantitative trait loci in animals selection, 24(2):169–210, 211–224
- raspberry, 32 :9–353
- reproductive barriers, 11:11–154
- rhizobia, 23:21–72
- rice, hybrid, 17:15–156, 23:73–174
- Rosaceae, 27:175–211
- rose, 17:171–172
- rubber (*Hevea*), 29:177–283
- rutabaga, 8:217–248
- salt resistance, 22:389–425
- selection, 24(1):111–131, 143–151, 269–290
- snap pea, 21:110–120
- sesame, 16:189–195
- soybean, 1:183–235
- soybean nodulation, 11:275–318
- spelt, 15:187–213
- supersweet sweet corn, 14:189–236
- sweet corn, 1:139–161; 14:189–236
- sweet potato, 4:327–330
- temperature, 3:21–167
- tomato fruit quality, 4:273–311
- transposable elements, 8:91–137
- triticale, 5:41–93
- virus resistance, 12:47–79
- wheat gene manipulation, 11:225–234
- wheat male sterility, 2:307–308
- wheat molecular biology, 11:235–250
- wheat rust, 13:293–343
- white clover, 17:191–223
- yield, 3:21–167; 34:37–182

Genome:
- *Brassica*, 31:21–187; 35:25–36
- *Glycine*, 16:289–317
- Poaceae, 16:276–281

Genomics:
- coffee, 30:415–437
- grain legumes, 26:171–357

Genotype × environment, interaction, 16:135–178

Germplasm, *see also* National Clonal Germplasm Repositories; National Plant Germplasm System
- acquisition and collection, 7:160–161
- apple rootstocks, 1:296–299
- banana, 2:140–141
- blackberry, 8:265–267
- black walnut, 1:244–247
- *Brassica*, 31:21–187
- cactus, 20:141–145
- cassava, 2:83–94, 117–119; 31:247–275
- cereal stress resistance, 33:31–114
- chestnut, 4:351–352
- coffee, 2:165–172
- distribution, 7:161–164
- enhancement, 7:98–202
- evaluation, 7:183–198
- exploration and introduction, 7:9–18, 64–94
- genetic markers, 13:11–86
- guayule, 6:112–125
- isozyme, 6:18–21
- grain legumes, 26:171–357
- legumes, 26:171–357
- maintenance and storage, 7:95–110, 111–128, 129–158, 159–182; 13:11–86
- maize, 14:165–187; 33:9–16
- melon, 35:85–150
- management, 13:11–86
- millets, 35:247–374

oat, 6:174–176
peanut, 22:297–356
pearl millet, 1:167–170
plantain, 14:267–320
potato, 9:219–223
preservation, 2:265–282; 23:291–344
raspberry, 32:75–90
rights, 25:21–55
rutabaga, 8:226–227
sampling, 29:285–314
sesame, 16:201–204
spelt, 15:204–205
sweet potato, 4:320–323
triticale, 8:55–61
wheat, 2:307–313
wild relatives, 30:149–230
Gesneriaceae, mutation breeding, 6:73
Gill, Bikram (biography), 37:1–34
Gladiolus, mutation breeding, 6:77
Glycine, genomes, 16:289–317
Glycine max, see Soybean
Goodman, Major M. (biography), 33:1–29
Gooseberry breeding, 29:145–175
Grain breeding:
   amaranth, 19:227–285
   barley, 3:219–252, 5:95–138;
      26:125–169
   cereal stress resistance, 33:31–114
   diversity, 21:221–261
   doubled haploid breeding, 15:141–186
   ideotype concept, 12:173–175
   maize, 1:103–138, 139–161;
      5:139–180; 9:115–179, 181–216;
      11:199–224; 14:165–187; 22:3–4;
      24(1): 11–40, 41–59, 61–78;
      24(2): 53–64, 109–151; 25:173–253:
      27:119–173; 28:59–100, 101–123;
      31:223–245; 33:9–16. 34:37–82,
      83–113, 131–160; 37:123–205,
      327–335
   maize history, 24(2):31–59, 41–59,
      61–78
   millets, 35: 247–374
   oat, 6:167–207; 34:5–9
   pearl millet, 1:162–182
   rice, 17:15–156; 24(2):64–67
   sorghum, 25:139–172; 31:189–222
   spelt, 15:187–213
   transformation, 13:231–260
   triticale, 5:41–93; 8:43–90

wheat, 2:303–319; 5:11–40; 11:225–234,
   235–250; 13:293–343; 22:221–297; 24
   (2):67–69; 28:1–37, 39–78; 36:85–16;
   37:1–34, 35–122
wild rice, 14:237–265
Grape:
   domestication, 25:279–281
   transformation, 16:103–104
Grapefruit:
   breeding, 13:345–363
   evolution, 13:345–363
Grass breeding:
   breeding, 11:251–274
   mutation breeding, 6:82
   recurrent selection, 9:101–113
   transformation, 13:231–260
Growth habit, induced mutants,
   2:14–25
Guayule, 6:93–165

H

Hallauer, Arnel R. (biography),
   15:1–17
Haploidy, *see also* unreduced and
      polyploid gametes
   apple, 1:376
   barley, 3:219–252
   cereals, 15:141–186
   doubled, 15:141–186; 25:57–88
   maize, 11:199–224
   petunia, 1:16–18, 44–45
   potato, 3:274–277; 16:15–86
Harlan, Jack R. (biography), 8:1–17
Heat tolerance, *see also* Stress Resistance:
   breeding, 10:129–168
   wheat, 36: 85–165
Herbicide resistance:
   breeding needs, 11:155–198
   cell selection, 4:160–161
   decision trees, 18:251–303
   risk assessment, 18:251–303
   transforming fruit crops, 16:114
Heritability estimation, 22:9–111
Heterosis:
   gene action, 15:315–374
   overdominance, 17:225–257
   plant breeding, 12:227–251
   plant metabolism, 10:53–90
   rice, 17:24–33
   soybean, 21:263–320

Hevea, see Rubber
History:
  raspberry, 32:45–51
  raspberry improvement, 32:59–66, 309–314
Honeycomb:
  breeding, 18:177–249, 37:297–347
  selection, 13:87–139, 18:177–249, 37:297–347
Hordeum, see Barley
Host-parasite genetics, 5:393–433
Human nutrition:
  breeding 31:325–392
  enhanced food crops, 36:169–291
  quality protein maize, 34:97–101
Hyacinth, mutation breeding, 6:76–77
Hybrid and hybridization, *see also* Heterosis
  barley, 5:127–129
  blueberry, 5:329–341
  chemical, 3:169–191
  interspecific, 5:237–305
  maize high oil selection, 24(1):153–175
  maize history, 24(1): 31–59, 41–59, 61–78
  maize long-term selection, 24(2):43–64, 109–151
  raspberry, 32:92–94
  rice, 17:15–156
  soybean, 21:263–307
  verification, 34:193–205
  wheat, 2:303–319
Hymowitz, Theodore (biography), 29:1–18

I

Ideotype concept, 12:163–193
Inbreeding depression, 11:84–92
  alfalfa, 13:209–233
  cross pollinated crops, 13:209–233
Incompatibility:
  almond, 8:313–338
  molecular biology, 15:19–42, 28:215–237
  pollen, 4:39–48
  reproductive barrier, 11:47–70
  sweet cherry, 9:367–388
Incongruity, 11:71–83
Industrial crop breeding:
  guayule, 6:93–165
  rubber (*Hevea*), 29:177–283

  sugarcane, 27:5–118
Insect and mite resistance:
  apple rootstock, 1:370–372
  black walnut, 1:251
  cassava, 2:107–110
  clover, white, 17:209–210
  coffee, 2:179–180
  cowpea, 15:240–244
  Cucurbitaceae, 10:309–360
  durum wheat, 5:28
  maize, 6:209–243
  raspberry, 6:282–300; 32:221–242
  rutabaga, 8:240–241
  sweet potato, 4:336–337
  transformation fruit crops, 16:113
  wheat, 22:221–297
  white clover, 17:209–210
Intergeneric hybridization, papaya, 26:35–78
Interspecific hybridization:
  blackberry, 8:284–289
  blueberry, 5:333–341
  Brassica, 31:21–187
  cassava, 31:247–275
  citrus, 8:266–270
  issues, 34:161–220
  pasture legume, 5:237–305
  raspberry, 32:146–152
  rose, 17:176–177
  rutabaga, 8:228–229
  Vigna, 8:24–30
Intersubspecific hybridization, rice, 17:88–98
Introduction, 3:361–434; 7:9–11, 21–25
In vitro culture:
  alfalfa, 2:229–234; 4:123–152
  barley, 3:225–226
  bean, 2:234–237
  birdsfoot trefoil, 2:228–229
  blackberry, 8:274–275
  broadbean, 2:244–245
  cassava, 2:121–122
  cell selection, 4:153–173
  chickpea, 2:224–225
  citrus, 8:339–374
  clover, 2:240–244
  coffee, 2:185–187
  cowpea, 2:245–246
  embryo culture, 5:181–236, 249–275
  germplasm preservation, 7:125,162–167

introduction, quarantines, 3:411–414
legumes, 2:215–264
mungbean, 2:245–246
oil palm, 4:175–201
pea, 2:236–237
peanut, 2:218–224
petunia, 1:44–48
pigeon pea, 2:224
pollen, 4:59–61
potato, 9:286–288
raspberry, 32:120–122
sesame, 16:218
soybean, 2:225–228
Stylosanthes, 2:238–240
wheat, 12:115–162
wingbean, 2:237–238
zein, 1:110–111
Ipomoea, see Sweet potato
Isozymes, in plant breeding, 6:11–54

J

Jahn, Margaret M. (biography), 35:1–17
Jennings, Derek (biography), 32:2–21
Jones, Henry A. (biography), 1:1–10
Juglans nigra, see Black walnut

K

Karyogram, petunia, 1:13
Kiwifruit:
   domestication, 25:300–301
   transformation, 16:104

L

Lactuca sativa, see Lettuce
Landraces, diversity, 21:221–263
Laughnan, Jack R. (bibliography), 19:1–14
Legumes, *see also* Bean, Oilseed, Peanut, Soybean:
   breeding, 33:157–304; 37:315–322
   cowpea, 15:215–274
   genomics, 26:171–357; 33:157–304
   pasture legumes, 5:237–305
   Vigna, 8:19–42
Legume tissue culture, 2:215–264
Lethal equivalents and genetic load, 10:93–127
Lettuce:
   genes, 1:267–293
   breeding, 16:1–14; 20:105–133
   transgenics, 35:2–5-210

Lingonberry domestication, 25:300–301
Linkage:
   bean, 1:76–77
   isozymes, 6:37–38
   lettuce, 1:288–290
   maps, molecular markers, 9:37–61
   petunia, 1:31–34
Loquat breeding, 37:259–296
Lotus:
   hybrids, 5:284–285
   in vitro culture, 2:228–229
Lycopersicon, see Tomato

M

Maize:
   anther culture, 11:199–224; 15:141–186
   anthocyanin, 8:91–137
   apomixis, 18:56–64
   biotic resistance, 34:131–160
   breeding, 1:103–138, 139–161; 27:119–173; 33:9–16; 37:123–205, 327–335
   carbohydrates, 1:144–148
   cytoplasm, 23:189
   diversity, 33:4–7
   doubled haploid breeding, 15:141–186
   drought tolerance, 25:173–253
   exotic germplasm utilization, 14:165–187
   foliar diseases, 27:119–173
   germplasm, 33:9–16
   high oil, 22:3–4; 24(1):153–175
   history of hybrids, 23(1): 11–40, 41–59, 61–78
   honeycomb breeding, 18:226–227
   hybrid breeding, 17:249–251
   insect resistance, 6:209–243
   isozymes, 33:7–8
   long-term selection 24(2):53–64, 109–151
   male sterility, 10:23–51
   marker-assisted selection. 24(1):293–309
   mobile elements, 4:81–122
   mutations, 5:139–180
   origins, 20:15–66
   origins of hybrids, 24(1):31–50, 41–59, 61–78
   overdominance, 17:225–257
   physiological changes with selection, 24(1):143–151

Maize: (*Continued*)
   protein, storage, 1:103–138
   protein, quality, 9:181–216; 34:83–113
   recurrent selection, 9:115–179;
     14:139–163
   RFLF changes with selection, 24
     (1):111–131
   selection for oil and protein, 24
     (1):79–110, 153–175
   soil acidity tolerance, 28:59–100
   supersweet sweet corn, 14:189–236
   transformation, 13:235–264
   transposable elements, 8:91–137
   unreduced gametes, 3:277\
   yield, 27–182
   vegetative phase change, 131–160
Male sterility:
   chemical induction, 3:169–191
   coleus, 3:352–353
   genetics, 25:115–138, 139–172
   lettuce, 1:284–285
   molecular biology, 10:23–51
   pearl millet, 1:166
   petunia, 1:43–44
   rice, 17:33–72
   sesame, 16:191–192
   sorghum, 25:139–172
   soybean, 21:277–291
   wheat, 2:303–319
Malus spp, *see* Apple
Malus ×*domestica*, *see* Apple
Malvaceae, molecular mapping, 14:25–27
Mango:
   domestication, 25:277–279
   transformation, 16:107
Manihot esculenta, *see* Cassava
Mapping:
   Cucurbitaceae, 27:213–244
   Rosaceae, 27:175–211
Marker-assisted selection, *see* Selection
   conventional breeding, 33:145–217
   gene pyramiding, 33:210–256
   millets, 35:332–344
   strategies, 34:247–348
Medicago, *see also* Alfalfa
   in vitro culture, 2:229–234
Meiosis:
   mutants, 28:239–115
   petunia, 1:14–16
Melon, landraces of India, 35:85–150

Metabolism and heterosis, 10:53–90
Microprojectile bombardment,
   transformation, 13:231–260
Millets, genetic and genomic
   resources, 35:247–374
Mitochondrial genetics, 6:377–380;
   25:115–138
Mixed plantings, bean
   breeding, 4:245–272
Mobile elements, *see also* transposable
   elements:
   maize, 4:81–122; 5:146–147
Molecular biology:
   apomixis, 18:65–73
   comparative mapping, 14:13–37
   cytoplasmic male sterility, 10:23–51
   DNA methylation, 18:87–176
   herbicide-resistant crops, 11:155–198
   incompatibility, 15:19–42
   legumes, 26:171–357
   molecular markers, 9:37–61,
     10:184–190; 12:195–226; 13:11–86;
     14:13–37; 17:113–114, 179, 212–215;
     18:20–42; 19:31–68, 21:181–220,
     23:73–174, 24(1)203–309;
     26:292–299; 33:145–217, 219–256;
     34:247–358; 35:332–344
   papaya, 26:35–78
   raspberry, 32:126–134
   rol genes, 26:79–103
   salt resistance, 22:389–425
   somaclonal variation, 16:229–268
   somatic hybridization, 20:167–225
   soybean nodulation, 11:275–318
   strawberry, 21:139–180
   transposable (mobile)
     elements, 4:81–122; 8:91–137
   virus resistance, 12:47–79
   wheat improvement, 11:235–250
Molecular markers, 9:37–61, 10:184–190;
   12:195–226; 13:11–86; 14:13–37;
   17:113–114, 179, 212–215; 18:20–42;
   19:31–68, 21:181–220, 23:73–174;
   33:145–217, 219–256; 34:247–358
   alfalfa, 10:184–190
   apomixis, 18:40–42
   barley, 21:181–220
   clover, white, 17:212–215
   forest crops, 19:31–68
   fruit crops, 12:195–226

maize selection, 24(1):293–309
mapping, 14:13–37
millets, 35:332–344
plant genetic resource
  mangement, 13:11–86
rice, 17:113–114, 23:73–124
rose, 17:179
somaclonal variation, 16:238–243
strategies, 34:247–358
wheat, 21:181–220
white clover, 17:212–215
Monosomy, petunia, 1:19
Mosaics and chimeras, 15:43–84
Mungbean, 8:32–35
  in vitro culture, 2:245–246
  photoperiodic response, 3:74, 89–92
Munger, Henry M. (biography), 4:1–8
Musa, see Banana, Plantain
Mushroom, breeding and
  genetics, 8:189–215
Mutants and mutation:
  alfalfa tissue culture, 4:130–139
  apple rootstocks, 1:374–375
  banana, 2:148–149
  barley, 5:124–126
  blackberry, 8:283–284
  cassava, 2:120–121
  cell selection, 4:154–157
  chimeras, 15:43–84
  coleus, 3:355
  cytoplasmic, 2:293–295
  gametoclonal variation, 5:359–391
  homeotic floral, 9:63–99
  induced, 2:13–72
  long term selection variation, 24 (1):227–247
  maize, 1:139–161, 4:81–122; 5:139–180
  mobile elements, see Transposable elements
  mosaics, 15:43–84
  petunia, 1:34–40
  sesame, 16:213–217
  somaclonal variation, 4:123–152; 5:147–149
  sweet corn, 1:139–161
  sweet potato, 4:371
  transposable elements, 4:181–122; 8:91–137
  tree fruits, 6:78–79

vegetatively-propagated crops, 6:55–91
zein synthesis, 1:111–118
Mycoplasma diseases,
  raspberry, 6:253–254

N

National Clonal Germplasm Repository (NCGR), 7:40–43
  cryopreservation, 7:125–126
  genetic considerations, 7:126–127
  germplasm maintenance and storage, 7:111–128
  identification and label verification, 7:122–123
  in vitro culture and storage, 7:125
  operations guidelines, 7:113–125
  preservation techniques, 7:120–121
  virus indexing and plant health, 7:123–125
National Plant Germplasm System (NPGS),
  see also Germplasm
  history, 7:5–18
  information systems, 7:57–65
  operations, 7:19–56
  preservation of genetic resources, 23:291–34
National Seed Storage Laboratory (NSSL), 7:13–14, 37–38, 152–153
Nectarines, cold hardiness breeding, 10:271–308
Nematode resistance:
  apple rootstocks, 1:368
  banana and plantain, 2:145–146
  coffee, 2:180–181
  cowpea, 15:245–247
  raspberry, 32:235–237
  soybean, 1:217–221
  sweet potato, 4:336
  transformation fruit crops, 16:112–113
Nicotiana, see Tobacco
Nodulation, soybean, 11:275–318
Nutrition (human):
  enhanced crops, 36:169–291
  plant breeding, 31:325–392

O

Oat breeding, 6:167–207; 34:5–9
Oil palm:
  breeding, 4:175–201, 22:165–219
  in vitro culture, 4:175–201

Oilseed breeding:
   canola, 18:1–20
   oil palm, 4:175–201; 22:165–219
   peanut, 22:295–356; 30:295–322
   sesame, 16:179–228
   soybean, 1:183–235; 3:289–311; 4:203–245; 11:275–318; 15:275–313
Olive domestication, 25:277–279
Onion, breeding history, 20:57–103
Opuntia, see Cactus
Organelle transfer, 2:283–302; 3:205–210; 6:361–393
Ornamentals breeding:
   chrysanthemum, 14:321–361
   coleus, 3:343–360
   petunia, 1:1–58
   rose, 17:159–189; 31:277–324
   transgenesis, 28:125–162
Ornithopus, hybrids, 5:285–287
Ortiz, Rodomiro (bibliography): 36:1–84
Orzya, see Rice
Overdominance, 17:225–257
Ovule culture, 5:181–236

P

Palm (Arecaceae):
   foliage breeding, 23:280–281
   oil palm breeding, 4:175–201; 22:165–219.
Panicum maximum, apomixis, 18:34–36, 47–49
Patents, raspberry, 32: 108–115
Papaya:
   breeding, 26:35–78
   domestication, 25:307–308
   transformation, 16:105–106
Parthenium argentatum, see Guayule
Paspalum apomixis, 18:51–52
Paspalum notatum, see Pensacola bahiagrass
Passionfruit transformation, 16:105
Pasture legumes, interspecific hybridization, 5:237–305
Pea:
   breeding, 21:93–138
   flowering, 3:81–86, 89–92
   in vitro culture, 2:236–237
Peach:
   cold hardiness breeding, 10:271–308

   domestication, 25:294–296
   transformation, 16:102
Peanut:
   breeding, 22:297–356; 30:295–322; 36:293–356
   in vitro culture, 2:218–224
Pear:
   domestication, 25:289–290
   transformation, 16:102
Pearl millet:
   apomixis, 18:55–56
   breeding, 1:162–182
Pecan transformation, 16:103
Peloquin, Stanley, J. (biography), 25:1–19
Pennisetum americanum, see Pearl millet
Pensacola bahiagrass, 9:101–113
   apomixis, 18:51–52
   selection, 9:101–113
Pepino transformation, 16:107
Peppermint, mutation breeding, 6:81–82
Perennial grasses, breeding, 11:251–274
Perennial rye breeding, 13:261–288
Persimmon:
   breeding, 19:191–225
   domestication, 25:299–300
Petunia spp., genetics, 1:1–58
Phaseolin, 1:59–102
Phaseolus vulgaris, see Bean
Philodendrum breeding, 23:273
Phosphate molecular mechanisms, 29:359–419
Phytophthora fragariae, 2:195–214
Pigeon pea, in vitro culture, 2:224
Pineapple domestication, 25:305–307
Pistil, reproductive function, 4:9–79
Pisum, see Pea
Plantain:
   breeding, 2:135–155; 14:267–320; 21:1–25
   domestication, 25: 298
Plant breeders rights, 25:21–55
Plant breeding:
   epigenetics, 30:49–177
   politics, 25:21–55
   prediction, 19:15–40
Plant exploration, 7:9–11, 26–28, 67–94
Plant introduction, 3:361–434; 7:9–11, 21–25
Plastid genetics, 6:364–376, see also Organelle

Plum:
    domestication, 25:293–294
    transformation, 16:103–140
Poaceae:
    molecular mapping, 14:23–24
    Saccharum complex, 16:269–288
Pollen:
    reproductive function, 4:9–79
    storage, 13:179–207
Polyploidy, *see also* Haploidy
    alfalfa, 10:171–184
    alfalfa tissue culture, 4:125–128
    apple rootstocks, 1:375–376
    banana, 2:147–148
    barley, 5:126–127
    blueberry, 13:1–10
    Brassica, 35:34–36
    citrus, 30:322–352
    gametes, 3:253–288
    isozymes, 6:33–34
    petunia, 1:18–19
    potato, 16:15–86; 25:1–19
    reproductive barriers, 11:98–105
    sweet potato, 4:371
    terminology, 26:105–124
    triticale, 5:11–40
Pomegranate domestication, 25:285–286
Population genetics, *see* Quantitative Genetics
Potato:
    breeding, 9:217–332, 19:69–165
    cytoplasm, 23:187–189
    disease resistance breeding, 19:69–165
    gametoclonal variation, 5:376–377
    heat tolerance, 10:152
    honeycomb breeding, 18:227–230
    mutation breeding, 6:79–80
    photoperiodic response, 3:75–76, 89–92
    ploidy manipulation, 16:15–86
    transgenics, 35:191–196
    unreduced gametes, 3:274–277
Propagation, raspberry, 32:116–126
Protein:
    antifungal, 14:39–88
    bean, 1:59–102
    induced mutants, 2:38–46
    maize, 1:103–138, 148–149; 9:181–216
Protoplast fusion, 3:193–218; 20: 167–225
    citrus, 8:339–374
    mushroom, 8:206–208

Prunus:
    amygdalus, see Almond
    avium, see Sweet cherry
Pseudograin breeding, amaranth, 19:227–285
Psophocarpus, in vitro culture, 2:237–238

Q

Quality protein maize. 9:181–216; 34:83–113
Quantitative genetics:
    epistasis, 21:27–92
    forest trees, 8:139–188
    gene interaction, 24(1):269–290
    genotype x environment interaction, 16:135–178
    heritability, 22:9–111
    maize RFLP changes with selection, 24(1):111–131
    mutation variation, 24(1): 227–247
    overdominance, 17:225–257
    population size & selection, 24(1):249–268
    selection limits, 24(1):177–225
    statistics, 17:296–300
    trait loci (QTL), 15:85–139; 19:31–68
    variance, 22:113–163
Quantitative trait loci (QTL), 15:85–138; 19:31–68
    animal selection, 24(2):169–210, 211–224
    marker-assisted selection, 33:145–217, 219–256
    selection limits: 24(1):177–225
Quarantines, 3:361–434; 7:12, 35–37

R

Rabbiteye blueberry, 5:307–357
Raspberry, breeding and genetics, 6:245–321, 32:1–353
Recurrent restricted phenotypic selection, 9:101–113
Recurrent selection, 9:101–113, 115–179; 14:139–163
    soybean, 15:275–313
Red stele disease, 2:195–214
Rédei, George P. (biography), 26:1–33.
Regional trial testing, 12:271–297

Reproduction:
  barriers and circumvention, 11:11–154
  foliage plants, 23:255–259
  garlic, 23:211–244
Rhizobia, 23:21–72
Rhododendron, mutation
  breeding, 6:75–76
Ribes, see Currant, Gooseberry
Rice, *see also* Wild rice:
  anther culture, 15:141–186
  apomixis, 18:65
  cytoplasm, 23:189
  doubled haploid breeding, 15:141–186
  gametoclonal variation, 5:362–364
  heat tolerance, 10:151–152
  honeycomb breeding, 18:224–226
  hybrid breeding, 17:1–15, 15–156;
    23:73–174
  long-term selection 24(2): 64–67
  molecular markers, 17:113–114;
    23:73–174
  photoperiodic response, 3:74, 89–92
Rosa, see Rose
Rosaceae, synteny, 27:175–211
Rose breeding, 17:159–189; 31:277–324
Rubber (*Hevea*) breeding, 29:177–283
Rubus, see Blackberry, Raspberry
Rust, wheat, 13:293–343
Rutabaga, 8:217–248
Ryder, Edward J. (biography), 16:1–14
Rye:
  gametoclonal variation, 5:370–371
  perennial breeding, 13:261–288
  triticale, 5:41–93

S

Saccharum complex, 16:269–288
Salamini, Francisco (biography), 30:1–47
Salt resistance:
  cell selection, 4:141–143
  cereals, 33:31–114
  durum wheat, 5:31
  yeast systems, 22:389–425
Sears, Ernest R. (biography), 10:1–22
Secale, see Rye
Seed:
  apple rootstocks, 1:373–374
  banks, 7:13–14, 37–40, 152–153
  bean, 1:59–102; 28:239–315
  citrus, 30:322–350

garlic, 23:211–244
lettuce, 1:285–286
maintenance and storage, 7:95–110,
  129–158, 159–182
maize, 1:103–138, 139–161, 4:81–86
pearl millet, 1:162–182
protein, 1:59–138, 148–149
raspberry, 32:94–101
rice production, 17:98–111, 118–119,
  23:73–174
soybean, 1:183–235, 3:289–311
synthetic, 7:173–174
variegation, 4:81–86
wheat (hybrid), 2:313–317
Selection, *see also* Breeding
  bacteria, 24(2): 225–265
  bean, 24(2): 69–74
  cell, 4:139–145, 153–173
  crops of the developing world, 24
    (2):45–88
  divergent selection for maize ear
    length, 24(2):153-168
  domestication, 24(2):1–44
  Escherichia coli, 24(2): 225–265
  gene interaction, 24(1):269–290
  genetic models, 24(1):177–225
  honeycomb design, 13:87–139;
    18:177–249
  limits, 24(1):177–225
  maize high oil, 24(1):153–175
  maize history, 24(1):11–40, 41–59, 61–78
  maize inbreds, 28:101–123
  maize long term, 24(1):79–110, 111–131,
    133–151; 24(2):53- 64, 109–151
  maize oil & protein, 24(1):79–110,
    153–175
  maize physiological changes, 24
    (1):133–151
  maize RFLP changes, 24(1):111–131
  marker assisted, 9:37–61, 10:184–190;
    12:195–226; 13:11–86; 14:13–37;
    17:113–114, 179, 212–215; 18:20–42;
    19:31–68, 21:181–220, 23:73–174, 24
    (1):293–309; 26:292–299; 31:210–212,
    33:145–217, 219–256; 34:247–348,
    35:332–344
  millets, 35:332–344
  mutation variation, 24(1):227–268
  population size, 24(1):249–268
  prediction, 19: 15–40

productivity gains in US crops, 24(2):89–106
prognosis, 37:297–347
quantitative trait loci, 24(1):311–335
raspberry, 32:102–108, 143–146
recurrent restricted
  phenotypic, 9:101–113
recurrent selection in maize, 9:115–179; 14:139–163
rice, 24(2): 64–67
wheat, 24(2): 67–69
Sesame breeding, 16:179–228
Sesamum indicum, see Sesame
Simmonds, N.W. (biography), 21:1–13
Snap pea breeding, 21:93–138
Solanaceae:
  incompatibility, 15:27–34
  molecular mapping, 14:27–28
  verticillium wilt, 33:115–144
Solanum tuberosum, see Potato
Somaclonal variation, see also Gametoclonal variation
  alfalfa, 4:123–152
  isozymes, 6:30–31
  maize, 5:147–149
  molecular analysis, 16:229–268
  mutation breeding, 6:68–70
  rose, 17:178–179
  transformation interaction, 16:229–268
  utilization, 16:229–268
Somatic embryogenesis, 5:205–212; 7:173–174
  oil palm, 4:189–190
Somatic genetics, see also Gametoclonal variation; Somaclonal variation:
  alfalfa, 4:123–152
  legumes, 2:246–248
  maize, 5:147–149
  organelle transfer, 2:283–302
  pearl millet, 1:162–182
  petunia, 1:43–46
  protoplast fusion, 3:193–218
  wheat, 2:303–319
Somatic hybridization, see also Protoplast fusion 20:167–225
Sorghum:
  Drought tolerance, 31:189–222
  male sterility, 25:139–172
  photoperiodic response, 3:69–71, 97–99
  transformation, 13:235–264

Southern pea, see Cowpea
Soybean:
  cytogenetics, 16:289–317
  disease resistance, 1:183–235
  drought resistance, 4:203–243
  fatty acid manipulation, 30:259–294
  genetics and evolution, 29:1–18
  hybrid breeding, 21:263–307
  in vitro culture, 2:225–228
  nodulation, 11:275–318
  photoperiodic response, 3:73–74
  recurrent selection, 15:275–313
  semidwarf breeding, 3:289–311
Spelt, agronomy, genetics, breeding, 15:187–213
Sprague, George F. (biography), 2:1–11
Sterility, see also Male sterility, 11:30–41
Starch, maize, 1:114–118
Statistics:
  advanced methods, 22:113–163
  history, 17:259–316
Strawberry:
  biotechnology, 21: 139–180
  domestication, 25:302–303
  red stele resistance breeding, 2:195–214
  transformation, 16:104
Stenocarpella ear rot, 31:223–245
Stress resistance, see also Drought and Heat Resistance:
  cell selection, 4:141–143, 161–163
  cereals, 33:31–114
  maize, 37:1223–205
  transformation fruit crops, 16:115
Stylosanthes, in vitro culture, 2:238–240
Sugarcane:
  breeding, 27:15–118
  mutation breeding, 6:82–84
  Saccharum complex, 16:269–288
Synteny, Rosaceae, 27:175–211
Sweet cherry:
  Domestication, 25:202–293
  pollen-incompatibility and self-fertility, 9:367–388
  transformation, 16:102
Sweet corn, see also Maize:
  endosperm, 1:139–161
  supersweet (*shrunken2*), 14:189–236
  transgenics, 35:213–215

Sweet potato:
   breeding, 4:313–345; 6:80–81
   transgenics, 35: 217–218
Synthetic wheat, 1–134, 35–122

T

Tamarillo transformation, 16:107
Taxonomy:
   amaranth, 19:233–237
   apple, 1:296–299
   banana, 2:136–138
   blackberry, 8:249–253
   brassicas. 35:19–83
   cassava, 2:83–89
   chestnut, 4:351–352
   chrysanthemum, 14:321–361
   clover, white, 17:193–211
   coffee, 2:161–163
   coleus, 3:345–347
   fescue, 3:314
   garlic, 23:211–244
   Glycine, 16:289–317
   guayule, 6:112–115
   oat, 6:171–173
   pearl millet, 1:163–164
   petunia, 1:13
   plantain, 2:136; 14:271–272
   raspberry, 32:51–52
   rose, 17:162–169
   rutabaga, 8:221–222
   Saccharum complex, 16:270–272
   sweet potato, 4:320–323
   triticale, 8:49–54
   Vigna, 8:19–42
   white clover, 17:193–211
   wild rice, 14:240–241
Testing:
   adaptation, 12:271–297
   honeycomb design, 13:87–139
Tissue culture, *see* In vitro culture
Tobacco, gametoclonal
      variation, 5:372–376
Tomato:
   breeding for quality, 4:273–311
   heat tolerance, 10:150–151
Toxin resistance, cell selection, 4:163–165
Transformation and transgenesis
   alfalfa, 10:190–192
   alliums, 35:210–213
   allelopathy, 30:231–258

   barley, 26:155–157
   brassicas, 35:199–205
   carrot, 35:219–220
   cassava, 35:216
   cereals, 13:231–260; 33:31–114
   cowpea, 35;215
   cucurbits, 35:196–199
   eggplant, 35:187–191
   fire blight resistance, 29:315–358
   fruit crops, 16:87–134
   lettuce, 35:205–210
   mushroom, 8:206
   ornamentals, 28:125–162
   papaya, 26:35–78
   potato, 35:191–196
   raspberry, 16:105; 32:133–134
   rice, 17:179–180
   somaclonal variation, 16:229–268
   sugarcane, 27:86–97
   sweet corn, 35:213–215
   tomato, 35:164–187
   sweet potato, 35:217–218
   vegetable crops, 35:1511–246
   white clover, 17:193–211
Transpiration efficiency, 12:81–113
Trilobium, long-term selection, 24
      (2):211–224
Transposable elements, 4:81–122;
      5:146–147; 8:91–137
Tree crops, ideotype concept, 12:163–193
Tree fruits, *see* Fruit, nut, and beverage crop
      breeding
Trifolium, see Clover, White Clover
Trifolium hybrids, 5:275–284
   in vitro culture, 2:240–244
Tripsacum:
   apomixis, 18:51
   maize ancestry, 20:15–66
Trisomy, petunia, 1:19–20
Triticale, 5:41–93; 8:43–90
Triticosecale, see Triticale
Triticum:
   Aestivum, see Wheat
   Turgidum, see Durum wheat
Tulip, mutation breeding, 6:76

U

United States National Plant Germplasm
      System, *see* National Plant
      Germplasm System

Unreduced and polyploid
  gametes, 3:253–288; 16:15–86
Urd bean, 8:32–35

V

Vaccinium, see Blueberry,
Variance estimation, 22:113–163
Vegetable, rootstock, and tuber breeding:
  alliums transgenics, 35:210–213
  artichoke, 12:253–269
  bean, 1:59–102; 4:245–272, 24(2):69–74;
    28:239–315; 36:357–426
  bean (tropics), 10:199–269
  beet (table), 22:257–388
  brassica transgenics, 35:19–84, 199–205
  carrot 19: 157–190, 35; 219–220
  cassava, 2:73–134; 24(2):74–79;
    31:247–275; 35:216; 36:427–503
  cowpea, 35:215
  cucumber, 6:323–359
  cucurbit, 10:309–360; 35:196–199
  eggplant transgenics, 35:187–191
  lettuce, 1:267–293; 16:1–14; 20:105:-
    133; 35:205–210
  melon, 35:85–150
  mushroom, 8:189–215
  onion, 20:67–103
  pea, 21:93–138
  peanut, 22:297–356; 36:293–356
  potato, 9:217–232; 16:15–86l;
    19:69–165; 35:191–196
  rutabaga, 8:217–248
  snap pea, 21:93–138
  Solanaceae, verticillium
    wilt, 33:115–144
  tomato, 4:273–311, 35:164–187
  sweet corn, 1:139–161; 14:189–236;
    35:213–215
  sweet potato, 4:313–345; 6:80–8135:
    213–215
  vegetable crop transgenics:
    151–246
  verticillium wilt,
    Solanaceae, 22:115–144
Verticillium wilt, Solanaceae,
  33:115–144
Vicia, in vitro culture, 2:244–245
Vigna, see Cowpea, Mungbean
  in vitro culture, 2:245–246;
    8:19–42

Virus diseases:
  apple rootstocks, 1:358–359
  clover, white, 17:201–209
  coleus, 3:353
  cowpea, 15:239–240
  indexing, 3:386–408, 410–411,
    423–425
  in vitro elimination, 2:265–282
  lettuce, 1:286
  maize, 27:142–156
  papaya, 26:35–78
  peanut, 36:293–356
  potato, 19:122–134
  raspberry, 6:247–254; 32:242–247
  resistance, 12:47–79
  soybean, 1:212–217
  sweet potato, 4:336
  transformation fruit crops, 16:108–110
  white clover, 17:201–209
Vogel, Orville A. (biography), 5:1–10
Vuylsteke, Dirk R. (biography), 21:1–25

W

Walnut (black), 1:236–266
Walnut transformation, 16:103
Weinberger, John A. (biography), 11:1–10
Wheat:
  anther culture, 15:141–186
  apomixis, 18:64–65
  breeding, 37:35–122
  chemical hybridization, 3:169–191
  cold hardiness adaptation, 12:124–135
  cytogenetics, 10:5–15; 37:1–34, 35–122
  cytoplasm, 23:189–190
  diversity, 21:236–237
  doubled haploid breeding, 15:141–186
  drought tolerance, 12:135–146;
    36:85–165
  durum, 5:11–40
  gametoclonal variation, 5:364–368
  gene manipulation, 11:225–234
  green revolution, 28; 1–37, 39–58
  heat tolerance, 10:152; 36:85–165
  hybrid, 2:303–319; 3:185–186
  insect resistance, 22:221–297
  in vitro adaptation, 12:115–162
  long-term selection, 24(2):67–69
  molecular biology, 11:235–250
  molecular markers, 21:191–220
  photoperiodic response, 3:74

rust interaction, 13:293–343
triticale, 5:41–93
vernalization, 3:109
White clover, molecular
    genetics, 17:191–223
Wild rice, breeding, 14:237–265
Winged bean, in vitro culture,
    2:237–238

Y

Yeast, salt resistance, 22:389–425
Yuan, Longping (biography), 17:1–13

Z

Zea mays, see Maize, Sweet corn
Zein, 1:103–138
Zizania palustris, see Wild rice

Cumulative Contributor Index
(Volumes 1–37)

Abbott, A.G., 27:175
Abdalla, O.S., 8:43; 37;35
Acquaah, G., 9:63
Aldwinckle, H.S., 1:294; 29:315
Alexander, D.E., 24(1):53
Anderson, N.O., 10:93; 11:11
Andersson, M.S., 36:169
Aronson, A.I., 12:19
Aruna, R., 30:295
Arús, P., 27:175
Ascher, P.D., 10:9
Ashok Kumar, A., 31:189
Ashri, A., 16:179
Atlin, G.N., 34:83

Babu, R., 34:83
Baddu-Apraku, B., 37:123
Badenes, M.L., 37:259
Baggett, J.R., 21:93
Bajic, V., 33:31
Balaji, J., 26:171
Balyan, H.S., 36:85
Baltensperger, D.D., 19:227; 35:247
Barker, T., 25:173
Bartels, D., 30:1
Basnizki, J., 12:253
Bassett, M.J., 28:239
Becerra-López-Lavalle, L.A., 36:427
Beck, D.L., 17:191
Beebe, S., .23:21-72; 36:357
Beineke, W.F., 1:236
Bell, A.E., 24(2):211
Bhatnagar-Mathur, P., 36:293
Below, F.E., 24(1):133

Bertin, C. 30:231
Bertioli, D.J., 30:179
Berzonsky, W.A., 22:221
Bhat, S.R., 31:21; 35:19
Bingham, E.T., 4:123; 13:209
Binns, M.R., 12:271
Bird, R. McK., 5:139
Bjarnason, M., 9:181
Blair, M.W., 26; 30:179; 36:169
Bliss, F.A., 1:59; 6:1
Boase, M.R., 14:321
Bonnett, D., 37:35
Borlaug, N.E., 5:1
Boyer, C.D., 1:139
Bravo, J.E., 3:193
Brennan, R., 32:1
Brenner, D.M., 19:227
Bressan, R.A., 13:235; 14:39; 22:389
Bretting, P.K., 13:11
Broertjes, C., 6:55
Brown, A.H.D., 21:221
Brown, J.W.S., 1:59
Brown, S.K., 9:333, 367
Buhariwalla, H.K., 26:171
Bünger, L., 24(2):169
Burnham, C.R., 4:347
Burton, G.W., 1:162; 9:101
Burton, J.W., 21:263
Byrne, D., 2:73

Camadro, E.L., 26:105
Campbell, K.G., 15:187
Campos, H., 25:173
Cantrell, R.G., 5:11

Plant Breeding Reviews, Volume 37, First Edition. Edited by Jules Janick.
© 2013 Wiley-Blackwell. Published 2013 by John Wiley & Sons, Inc.

Cardinal, A.J., 30:259
Carputo, D., 25:1; 26:105; 28:163
Carvalho, A., 2:157
Casas, A.M., 13:235
Ceballos, H., 36:427
Cervantes-Martinez, C.T., 22:9
Chandler, M.A., 34:131
Chen, J., 23:245
Cherry, M., 27:245.
Chew, P.S., 22:165
Choo, T.M., 3:219; 26:125
Chopra, V.L., 31:21
Christenson, G.M., 7:67
Christie, B.R., 9:9
Clark, J.R., 29:19
Clark, R.L., 7:95
Clarke, A.E., 15:19
Clegg, M.T., 12:1
Clément-Demange, A., 29:177
Clevidence, B.A., 31:325
Comstock, J.G., 27:15
Condon, A.G., 12:81
Conicella, C., 28:163
Conner, A.J., 34:161
Consiglio, F., 28:163
Cooper, M, 24(2):109; 25:173
Cooper, R.L., 3:289
Cornu, A., 1:11
Costa, W.M., 2:157
Cregan, P., 12:195
Crouch, J.H., 14:267; 26:171; 36:1
Crow, J.F., 17:225
Cummins, J.N., 1:294

Dambier, D. 30:323
Dana, S., 8:19
Das, B., 34:83
Dean, R.A., 27:213
De Groote, H., 34:83
De Jong, H., 9:217
Dekkers, J.C.M., 24(1):311
Deroles, S.C., 14:321
Dhillon, B.S., 14:139
Dhillon, N.P.S., 35:85
Diao, X., 35:247
Dias, J.S., 35:151
D'Hont, A., 27:15
Dickmann, D.I., 12:163
Ding, H., 22:221

Dirlewanger, E., 27:175
Dodds, P.N., 15:19
Dolan, D., 25:175
Donini, P., 21:181
Dowswell, C., 28:1
Doyle, J.J., 31:1
Draper, A.D., 2:195
Drew, R., 26:35
Dudley, J.W. 24(1):79
Dumas, C., 4:9
Duncan, D.R., 4:153
Duvick, D.N., 24(2):109
Dwivedi, S.L., 26:171; 30:179; 33:311; 35:247; 36:169

Ebert, A.W., 30:415
Echt, C.S., 10:169
Edmeades, G., 25:173
Ehlers, J.D., 15:215
England, F., 20:1
Eubanks, M.W., 20:15
Evans, D.A., 3:193; 5:359
Everett, L.A., 14:237
Ewart, L.C., 9:63

Fakorede, M.A.B., 37:123
Farquhar, G.D., 12:81
Fasoula, D.A., 14:89; 15:315; 18:177
Fasoula, V.A., 13:87; 14:89; 15:315; 18:177; 37:297
Fasoulas, A.C., 13:87
Fazuoli, L.C., 2:157
Fear, C.D., 11:1
Ferris, R.S.B., 14:267
Finn, C.E., 29:19
Flore, J.A., 12:163
Forsberg, R.A., 6:167
Forster, B.P., 25:57
Forster, R.L.S., 17:191
Fowler, C., 25:21
Frei, U., 23:175
French, D.W., 4:347
Friesen, D.K., 28:59; 34:83
Froelicher, Y. 30:323
Frusciante, L., 25:1; 28:163
Fukunaga, K., 35:247

Gai, J., 21:263
Gahlaut, V., 36:85
Galiba, G., 12:115

CUMULATIVE CONTRIBUTOR INDEX

Galletta, G.J., 2:195
Garcia-Mas, J., 35:85
Gao, Y., 33:115
Gehring, C., 33:31
Gepts, P., 24(2):1
Glaszmann, J.G., 27:15
Gmitter, F.G., Jr., 8:339; 13:345
Gold, M.A., 12:163
Goldman, I.L. 19:15; 20:67; 22:357; 24(1):61; 24(2):89; 35:1
Goldway, M., 28:215
Gonsalves, D., 26:35
Goodnight, C.J., 24(1):269
Gordon, S.G., 27:119
Gosman, N., 37:35
Gradziel, T.M., 15:43; 37:207
Gressel, J., 11:155; 18:251
Gresshof, P.M., 11:275
Griesbach, R.J., 25:89
Griffin, W.B., 34:161
Grombacher, A.W., 14:237
Grosser, J.W., 8:339
Grumet, R., 12:47
Gudin, S., 17:159
Guimarães, C.T., 16:269
Gupta, P.K., 33:145; 36:1
Gustafson, J.P., 5:41; 11:225
Guthrie, W.D., 6:209

Habben, J., 25:173
Haley, S.D., 22:221
Hall, A.E., 10:129; 12:81; 15:215
Hall, H.K., 8:249; 29:19; 32:1, 39
Hallauer, A.R., 9:115; 14:1,165; 24(2):153
Hamblin, J., 4:245
Hancock, J.F., 13:1
Hancock, J.R., 9:1
Hanna, W.W., 13:179
Harlan, J.R., 3:1
Harris, M.O., 22:221
Hasegawa, P.M. 13:235; 14:39; 22:389
Hash, C., 35:247
Havey, M.J., 20:67
Haytowitz, D.B., 31:325
Henny, R.J., 23:245
Hershey, C., 36:427
Hill, W.G., 24(2):169
Hillel, J., 12:195
Hjalmarsson, I., 29:145

Hoa, T.T.T., 29:177
Hodgkin, T., 21:221
Hokanson, S.C., 21:139; 31:277
Holbrook, C.C., 22:297
Holden, J.M., 31:325
Holland, J.B., 21:27; 22:9; 33:1
Hor, T.Y., 22:165
Howe, G.T., 27:245
Hummer, K., 32:1, 39
Hunt, L.A., 16:135
Hutchinson, J.R., 5:181
Hymowitz, T., 8:1; 16:289

Iván Ortiz-Monasterio, J., 28:39

Jackson, S.A., 33:257
Jain, A., 29:359
Jamieson, A.R., 32:39
Janick, J., 1:xi; 23:1; 25:255; 37:259
Jansky, S., 19:77
Jayaram, Ch., 8:91
Jayawickrama, K., 27:245
Jenderek, M.M., 23:211
Jifon, J., 27:15
Johnson, A.A.T., 16:229; 20:167
Johnson, G.R., 27:245
Johnson, R., 24(1):293
Jones, A., 4:313
Jones, J.S., 13:209
Joobeur, T., 27:213
Ju, G.C., 10:53

Kang, H., 8:139
Kann, R.P., 4:175
Kapazoglou, A., 30:49
Karmakar, P.G., 8:19
Kartha, K.K., 2:215, 265
Kasha, K.J., 3:219
Kaur, H., 30:231
Kazi, A.G., 37:35
Keep, E., 6:245
Keightley, P.D., 24(1):227
Kirti, P.B., 31:21
Kleinhofs, A., 2:13
Knox, R.B., 4:9
Koebner, R.M.D., 21:181
Kollipara, K.P., 16:289
Koncz, C., 26:1
Kononowicz, A.K., 13:235
Konzak, C.F., 2:13

Kovačević, N.M., 30:49
Krikorian, A.D., 4:175
Krishnamani, M.R.S., 4:203
Kronstad, W.E., 5:1
Kuehnle, A.R., 28:125
Kulakow, P.A., 19:227
Kulwal, P.L., 36:85
Kumar, A., 33:145
Kumar, J., 33:145

Lagudah, E.S., 37:35
Lamb, R.J., 22:221
Lambert, R.J., 22:1; 24(1):79:153
Lamborn, C., 21:93
Lamkey, K.R., 15:1; 24(1):xi; 24(2):xi; 31:223
Lavi, U., 12:195
Layne, R.E.C., 10:271
Lebowitz, R.J., 3:343
Lee, E.A., 34:37
Lee, M., 24(2):153
Lehmann, J.W., 19:227
Lenski, R.E., 24(2):225
Levings, III, C.S., 10:23
Lewers, K.R., 15:275
Li, J., 17:1,15
Liang, G.L, 37:259
Liedl, B.E., 11:11
Lin, C.S., 12:271
Lin, S., 37:259
Lockwood, D.R., 29:285
Lovell, G.R., 7:5
Lower, R.L., 25:21
Lukaszewski, A.J., 5:41
Luro, F., 30:323
Lyrene, P.M., 5:307; 30:353

Maas, J. L., 21:139
Mackenzie, S.A., 25:115
Maheswaran, G., 5:181
Maizonnier, D., 1:11
Malnoy, M., 29:285
Marcotrigiano, M., 15:43
Martin, F.W., 4:313
Martinez-Gómez, P., 37:207
Matsumoto, T.K. 22:389
May, G.D., 33:257
McCoy, T.J., 4:123; 10:169
McCreight, J.D., 1:267; 16:1; 35:85
McDaniel, R.G., 2:283

McKeand, S.E., 19:41
McKenzie, R.I.H., 22:221
McRae, D.H., 3:169
Medina-Filho, H.P., 2:157
Mejaya, I.J., 24(1):53
Michler, C.H., 33:305
Mikkilineni, V., 24(1):111
Miles, D., 24(2):211
Miles, J.W., 24(2):45
Miller, R., 14:321
Ming, R., 27:15; 30:415
Mir, R.R., 33:145
Mirkov, T.E., 27:15
Mobray, D., 28:1
Mondragon Jacobo, C., 20:135
Monti, L.M., 28:163
Monforte, A.J., 35: 85
Moose, S.P., 24(1):133
Morgan, E.R., 34:161
Morrison, R.A., 5:359
Mowder, J.D., 7:57
Mroginski, L.A., 2:215
Mudalige-Jayawickrama, 28:125
Mujeeb-Kazi, A., 37:35
Muir, W.M., 24(2):211
Mumm, R.H., 24(1):1
Murphy, A.M., 9:217
Mutschler, M.A., 4:1
Myers, J.R., 21:93
Myers, O., Jr., 4:203
Myers, R.L., 19:227.

Namkoong, G., 8:1
Narro León, L.A., 28:59
Nassar, N.M.A., 31:248
Navazio, J., 22:357
Nelson, P.T., 33:1
Neuffer, M.G., 5:139
Newbigin, E., 15:19
Nielen, S., 30:179
Nigam, S.N., 30:295; 36:293
Nikki Jennings, S. 32:1, 39
Nybom, H., 34:221
Nyquist, W.E., 22:9

Ogbonnaya, F.C., 37:35
Ohm, H.W., 22:221
Ollitrault, P., 30:323
O'Malley, D.M., 19:41

CUMULATIVE CONTRIBUTOR INDEX

Ortiz, R., 14:267; 16:15; 21:1; 25:1, 139; 26:171; 28:1, 39; 30:179; 31:248; 33:31; 35:151
Osborn, T.C., 27:1

Palacios, N., 34:83
Palmer, R.G., 15:275, 21:263; 29:1; 31:1
Pandey, S., 14:139; 24(2):45; 28:59; 35:85
Pardo, J.M., 22:389
Parliman, B.J., 3:361
Paterson, A.H., 14:13; 26:15
Patterson, F.L., 22:221
Peairs, F.B., 22:221
Pedersen, J.F., 11:251
Peiretti, E.G., 23:175
Peloquin, S.J., 26:105
Perdue, R.E., Jr., 7:67
Peterson, P.A., 4:81; 8:91
Pfeiffer, W., 36:169
Pickering, R., 34:161
Pitrat, M., 35:85
Pixley, K.V., 34:83
Polidoros, A.N., 18:87; 30:49
Pollak, L.M. 31:325
Porter, D.A., 22:221
Porter, R.A., 14:237
Powell, W., 21:181
Prakash, S., 31:21; 35:19
Prasad, M., 35:247
Prasada Rao, J.D.V.J., 36:293
Prasartsee, V., 26:35
Pratt, R.C., 27:119
Pretorius, Z.A., 31:223
Priyadarshan, P.M., 29:177

Quiros, C.F., 31:21

Raghothama, K.G. 29:359
Rai, K.N., 36:169
Rai, M. 27:15
Raina, S.K. 15:141
Rajaram, S. 28:1
Rakow, G., 18:1
Ramage, R.T. 5:95
Ramash, S., 31:189
Ramesh, S. 25:139
Ramming, D.W. 11:1
Ratcliffe, R.H., 22:221
Raupp, W.J., 37:1

Ray, D.T., 6:93
Reddy, B.V.S., 25:139; 31:189
Redei, G.P., 10:1; 24(1):11
Reimann-Phillipp, R., 13:265
Reinbergs, E., 3:219
Reitsma, K.R., 35:85
Reynolds, M.P., 28:39
Rhodes, D., 10:53
Richards, C.M., 29:285
Richards, R.A., 12:81
Riedeman, E.S., 34:131
Roath, W.W., 7:183
Robinson, R.W., 1:267; 10:309
Robertson, L., 34:1
Rochefored, T.R., 24(1):111
Ron Parra, J., 14:165
Roos, E.E., 7:129
Ross, A.J., 24(2):153
Rossouw, J.D., 31:223
Rotteveel, T., 18:251
Rowe, P., 2:135
Russell, W.A., 2:1
Rutter, P.A., 4:347
Ryder, E.J., 1:267; 20:105

Sahi, S.V., 2:359
Sahrawat, K.L., 36:169
Samaras, Y., 10:53
Sanjana Reddy, P., 31 :189
Sansavini, S., 16:87
Santra, D., 35:247
Sapir, G., 28:215
Saunders, J.W., 9:63
Savidan, Y., 18:13
Sawhney, R.N., 13:293
Schaap, T., 12:195
Schaber, M.A., 24(2):89
Schneerman, M.C., 24(1):133
Schnell, R.J., 27:15
Schroeck, G., 20:67
Schussler, J., 25:173
Scott, D.H., 2:195
Seabrook, J.E.A., 9:217
Sears, E.R., 11:225
Seebauer, J.R., 24(1):133
Senthilvel, S., 36:247
Serraj, R., 26:171
Shands, Hazel L., 6:167
Shands, Henry L., 7:1, 5
Shannon, J.C., 1:139

Shanower, T.G., 22:221
Sharma, A., 35:85
Sharma, K.K., 36:293
Shattuck, V.I., 8:217; 9:9
Shaun, R., 14:267
Sidhu, G.S., 5:393
Silva, da, J., 27:15
Silva, H.D., 31:223
Simmonds, N.W., 17:259
Simon, P.W., 19:157; 23:211; 31:325
Singh, B.B., 15:215
Singh, P.K., 35:85
Singh, R.J., 16:289
Singh, S.P., 10:199
Singh, Z., 16:87
Slabbert, M.M., 19:227
Sleper, D.A., 3:313
Sleugh, B.B., 19
Smith, J.S.C., 24(2):109
Smith, K.F., 33:219
Smith, S.E., 6:361
Snoeck, C., 23:21
Sobral, B.W.S., 16:269
Socias i Company, R., 8:313
Soh, A.C., 22:165
Sondahl, M.R., 2:157
Sorrells, M.E., 37:35
Spoor, W., 20: 1
Stafne, E.T., 29:19
Stalker, H.T., 22:297; 30:179
Steadman, J.R., 23:1
Steffensen, D. M., 19:1
Stern, R.A., 28:215
Stevens, M.A., 4:273
Stoner, A.K., 7:57
Stuber, C.W., 9:37; 12:227
Subudhi, P., 33:31
Sugiura, A., 19:191
Sun, H., 21:263
Suzaki, J.Y., 26:35

Tai, G.C.C., 9:217
Talbert, L.E., 11:235
Tan, C.C., 22:165
Tani, E., 30:49
Tarn, T.R., 9:217
Tehrani, G., 9:367
Teshome, A., 21:221
Tew, T.L., 27:15
Thomas, W.T.B., 25:57

Thompson, A.E., 6:93
Thro, A.M., 34:1
Thudi, M., 33:257
Tiefenthaler, A.E., 24(2):89
Timmerman-Vaughan, G.M., 34:161
Tollenaar, M., 34:37
Towill, L.E., 7:159, 13:179
Tracy, W.F., 14:189; 24(2):89; 34:131
Trethowan, R.M., 28:39
Tripathi, S., 26:35
Troyer, A.F., 24(1):41; 28:101
Tsaftaris, A.S., 18:87; 30:49
Tsujimoto, H., 37:35
Tsai, C.Y., 1:103
Twumasi-Afriyie, S., 83

Ullrich, S.E., 2:13
Upadhyaya, H.D., 26:171; 39:179; 33:31; 35:247
Uribelarrea, M., 24(1):133

Vanderleyden, J., 23:21
Van Ginkel, M. 34:297
Van Harten, A.M., 6:55
Varshney, R.K., 33:257
Varughese, G., 8:43
Vasal, S.K., 9:181; 14:139
Vasconcelos, M.J., 29:359
Vega, F.E., 30:415
Vegas, A., 26:35
Veilleux, R., 3:253; 16:229; 20:167; 33:115
Venkatachalam, P., 29:177
Villareal, R.L., 8:43
Vivak, B., 34:83
Vogel, K.P., 11:251
Volk, G.M., 23:291; 29:285
Vuylsteke, D., 14:267

Wallace, B., 29:145
Wallace, D.H., 3:21; 13:141
Walsh, B. 24(1):177
Wan, Y., 11:199
Wang, W., 37:259
Wang, Y.-H., 27:213
Waters, C., 23:291
Weber, C.A., 32:39
Weber, K., 24(1):249
Weeden, N.F., 6:11
Wehner, T.C., 6:323

Weising, K., 34:221
Welander, M., 26:79
Wenzel, G. 23:175
Weston, L.A. 30:231
Westwood, M.N., 7:111
Wheeler, N.C., 27:245
Whitaker, T.W., 1:1
Whitaker, V.M., 31:277
White, D.W.R., 17:191
White, G.A., 3:361; 7:5
Widholm, J.M., 4:153; 11:199
Widmer, R.E., 10:93
Widrlechner, M.P., 13:11
Wilcox, J.R., 1:183
Williams, E.G., 4:9; 5:181, 237
Williams, M.E., 10:23
Williamson, B., 32:1
Wilson, J.A., 2:303
Woeste, K.E., 33:305
Wong, G., 22:165
Woodfield, D.R., 17:191
Worthen, L.M., 33:305
Wright, D., 25:173
Wright, G.C., 12:81
Wu, K.-K., 27:15

Wu, L., 8:189
Wu, R., 19:41
Wu, X.-M. 35:19

Xin, Y., 17:1
Xu, S., 22:113
Xu, S.S., 37:35
Xu, Y., 15:85; 23:73

Yamada, M., 19:191
Yamamoto, T., 27:175
Yan, W., 13:141
Ye, G., 33:219; 34:297
Yang, W.-J., 10:53
Yonemori, K., 19:191
Yopp, J.H., 4:203
Yun, D.-J., 14:39

Zhang, Z., 37:259
Zeng, Z.-B., 19:41
Zhu, L.-H., 26:79
Zimmerman, M.J.O., 4:245
Zinselmeier, C., 25:173
Zitter, T.A., 33:115
Zohary, D., 12:253